T0235676

Python for Probability, Statistics, and Machine
Learning

José Unpingco

Python for Probability, Statistics, and Machine Learning

Third Edition

 Springer

José Unpingco
San Diego, CA, USA

ISBN 978-3-031-04650-6   ISBN 978-3-031-04648-3   (eBook)
https://doi.org/10.1007/978-3-031-04648-3

This Springer imprint is published by the registered company Springer Nature Switzerland AG
The registered company address is: Gewerbestrasse 11, 6330 Cham, Switzerland

To Irene, Nicholas, and Daniella, for all their patient support.

Preface to the Third Edition

This third edition is updated for Python version 3.8+ but does not use any new syntax and should be compatible with Python 3.6+ also. More importantly, many existing sections have been revised based on feedback from the first and second versions. The book has been adopted into university-level curricula in data science and machine learning worldwide, including the University of California, San Diego. It has also been translated into multiple languages. With this in mind, I reedited significant portions for clarity to hopefully ease the translation burden of this edition and make it easier to understand overall. Almost all the figures have been updated for clarity.

The statistics chapter has doubled in size and now covers important but hard-to-find material, such as categorical data analysis and missing data imputation. The machine learning chapter has been updated, and new sections covering gradient tree boosting have been added, along with a section on interpreting machine learning models. The introduction now includes a discussion of the Xarray module for multidimensional dataframes. Overall, the book is now about one-third larger than the second edition.

As before, there are more *Programming Tips* that the illustrate effective Python modules and methods for scientific programming and machine learning. There are over 650 run-able code blocks that have been tested for accuracy, so you can try these out for yourself in your own codes. This edition features over 200 graphical visualizations generated using Python that illustrate the concepts that are developed both in code and in mathematics. We also discuss and use key Python modules, such as Numpy, Scikit-learn, Sympy, Scipy, Lifelines, CvxPy, Theano, Matplotlib, Pandas, Tensorflow, Statsmodels, Xarray, Seaborn, and Keras.

As with the first and second editions, all of the key concepts are developed mathematically and are reproducible in the given Python, to provide the reader multiple perspectives on the material. This book is not designed to be exhaustive and reflects the author's eclectic industrial background. The focus remains on concepts and fundamentals for day-to-day work using Python in the most expressive way possible. You can reach the author with comments at github.com/unpingco by opening an issue on the project.

Acknowledgements I would like to acknowledge the Python community as a whole, for all their contributions that made this book possible. Hans Petter Langtangen was the author of the Doconce [22] document preparation system that was used to write this text. Thanks to Geoffrey Poore [36] for his work with PythonTeX and , both key technologies were used to produce this book.

San Diego, CA, USA   José Unpingco
March, 2022

Preface to the Second Edition

This second edition is updated for Python version 3.6+. Furthermore, many existing sections have been revised for clarity based on feedback from the first version. The book is now over 30 percent larger than the original with new material about important probability distributions, including key derivations and illustrative code samples. Additional important statistical tests are included in the statistics chapter including the Fisher exact test and the Mann-Whitney-Wilcoxon Test. A new section on survival analysis has been included. The most significant addition is the section on deep learning for image processing with a detailed discussion of gradient descent methods that underpin all deep learning work. There is also substantial discussion regarding generalized linear models. As before, there are more *Programming Tips* that the illustrate effective Python modules and methods for scientific programming and machine learning. There are 445 run-able code blocks that have been tested for accuracy, so you can try these out for yourself in your own codes. Over 158 graphical visualizations (almost all generated using Python) illustrate the concepts that are developed both in code and in mathematics. We also discuss and use key Python modules, such as Numpy, Scikit-learn, Sympy, Scipy, Lifelines, CvxPy, Theano, Matplotlib, Pandas, Tensorflow, Statsmodels, and Keras.

As with the first edition, all of the key concepts are developed mathematically and are reproducible in Python, to provide the reader multiple perspectives on the material. There are multiple As before, this book is not designed to be exhaustive and reflects the author's eclectic industrial background. The focus remains on concepts and fundamentals for day-to-day work using Python in the most expressive way possible.

Preface to the First Edition

This book will teach you the fundamentals concepts that underpin probability and statistics and illustrate how they relate to machine learning via the Python language and its powerful extensions. This is not a good *first* book in any of these topics, because we assume that you already had a decent undergraduate-level introduction to probability and statistics. Furthermore, we also assume that you have a good grasp of the basic mechanics of the Python language itself. Having said that, this book is appropriate if you have this basic background and want to learn how to use the scientific Python toolchain to investigate these topics. On the other hand, if you are comfortable with Python, perhaps through working in another scientific field, then this book will teach you the fundamentals of probability and statistics and how to use these ideas to interpret machine learning methods. Likewise, if you are a practicing engineer using a commercial package (e.g., Matlab, IDL), then you will learn how to effectively use the scientific Python toolchain by reviewing concepts you are already familiar with.

The most important feature of this book is that everything in it is reproducible using Python. Specifically, all of the code, all of the figures, and (most of) the text are available in the downloadable supplementary materials that correspond to this book as IPython Notebooks. IPython Notebooks are *live* interactive documents that allow you to change parameters, recompute plots, and generally tinker with all of the ideas and code in this book. I *urge* you to download these IPython Notebooks and follow along with the text to experiment with the topics covered. I guarantee doing this will boost your understanding, because the IPython Notebooks allow for interactive widgets, animations, and other intuition-building features that help make many of these abstract ideas concrete. As an open-source project, the entire scientific Python toolchain, including the IPython Notebook, is freely available. Having taught this material for many years, I am convinced that the only way to learn is to experiment as you go. The text provides instructions on how to get started installing and configuring your scientific Python environment.

This book is not designed to be exhaustive and reflects the author's eclectic background in industry. The focus is on fundamentals and intuitions for day-to-day work, especially when you must explain the results of your methods to a nontechnical audience. We have tried to use the Python language in the most expressive way possible while encouraging good Python coding practices.

Contents

Chapter 1
Getting Started with Scientific Python

Python is fundamental to data science and machine learning, as well as an ever-expanding list of areas including cybersecurity, and web programming. The fundamental reason for Python's widespread use is that it provides the software *glue* that permits easy exchange of methods and data across core routines typically written in Fortran or C.

Python is a language geared toward scientists and engineers who may not have formal software development training. It is used to prototype, design, simulate, and test without *getting in the way*, because Python provides an inherently easy and incremental development cycle, interoperability with existing codes, access to a large base of reliable open-source codes, and a hierarchical compartmentalized design philosophy. Python is known for enhancing user productivity, because it reduces the development time (i.e., time spent programming).

Python is an *interpreted* language. This means that Python codes run on a Python *virtual machine* that provides a layer of abstraction between the code and the platform it runs on, thus making codes portable across different platforms. For example, the same script that runs on a Windows laptop can also run on a Linux-based supercomputer or on a mobile phone. This makes programming easier, because the virtual machine handles the low-level details of implementing the business logic of the script on the underlying platform.

Python is a dynamically typed language, which means that the interpreter itself figures out the representative types (e.g., floats, integers) interactively or at run-time. This is in contrast to a language like Fortran that has compilers that study the code from beginning to end, perform many compiler-level optimizations, link intimately with the existing libraries on a specific platform, and then create an executable that is henceforth liberated from the compiler. As you may guess, the compiler's access to the details of the underlying platform means that it can utilize optimizations that exploit chip-specific features and cache memory. Because the virtual machine abstracts away these details, it means that the Python language does not have programmable access to these kinds of optimizations. So, where is

© The Author(s), under exclusive license to Springer Nature Switzerland AG 2022
J. Unpingco, *Python for Probability, Statistics, and Machine Learning*,
https://doi.org/10.1007/978-3-031-04648-3_1

the balance between the ease of programming the virtual machine and these key numerical optimizations that are crucial for scientific work?

The balance comes from Python's native ability to bind to compiled Fortran and C libraries. This means that you can send intensive computations to compiled libraries directly from the interpreter. This approach has two primary advantages. First, it gives you the fun of programming in Python, with its expressive syntax and lack of visual clutter. This is a particular boon to scientists who typically want to *use* software as a tool as opposed to developing software as a product. The second advantage is that you can mix and match different compiled libraries from diverse research areas that were not otherwise designed to work together. This works because Python makes it easy to allocate and fill memory in the interpreter, pass it as input to compiled libraries, and then recover the output back in the interpreter.

Moreover, Python provides a multiplatform solution for scientific codes. As an open-source project, Python itself is available anywhere you can build it, even though it typically comes standard nowadays, as part of many operating systems. This means that once you have written your code in Python, you can just transfer the script to another platform and run it, as long as the third-party compiled libraries are also available there. What if the compiled libraries are absent? Building and configuring compiled libraries across multiple systems used to be a painstaking job, but as scientific Python has matured, a wide range of libraries have now become available across all of the major platforms (i.e., Windows, MacOS, Linux, Unix) as prepackaged distributions.

Finally, Scientific Python facilitates maintainability of scientific codes, because Python syntax is clean and free of semicolon litter and other visual distractions that make code hard to read and easy to obfuscate. Python has many built-in testing, documentation, and development tools that ease maintenance. Scientific codes are usually written by scientists unschooled in software development, so having solid software development tools built into the language itself is a particular boon.

1.1   Installation and Setup

The easiest way to get started is to download the freely available Anaconda distribution provided by Anaconda (anaconda.com), which is available for all of the major platforms. On Linux, even though most of the toolchain is available via the built-in Linux package manager, it is still better to install the Anaconda distribution, because it provides its own powerful package manager (i.e., conda) that can keep track of changes in the software dependencies of the packages that it supports. If you do not have administrator privileges, there is also a corresponding Miniconda distribution that does not require these privileges. Regardless of your platform, we recommend Python version 3.6 or better.

You may have encountered other Python variants on the web, such as IronPython (Python implemented in C#) and Jython (Python implemented in Java). In this text, we focus on the C-implementation of Python (i.e., known as *CPython*), which is, by far, the most popular implementation. These other Python variants permit specialized, native interaction with libraries in C# or Java (respectively), which is still possible (but clunky) using CPython. Even more Python variants exist that implement the low-level machinery of Python differently for various reasons, beyond interacting with native libraries in other languages. Most notable of these is Pypy that implements a just-in-time compiler (JIT) and other powerful optimizations that can substantially speed up *pure* Python codes. The downside of Pypy is that its coverage of some popular scientific modules (e.g., Matplotlib, Scipy) is limited or nonexistent, which means that you cannot use those modules in code meant for Pypy.

If you want to install a Python module that is not available via the conda manager, the pip installer is available. This installer is the main one used outside of the scientific computing community. The key difference between the two installers is that conda implements a satisfiability solver that checks for conflicts in versions among and between installed packages. This can result in conda decreasing versions of certain packages to accommodate proposed package installation. The pip installer does not check for such conflicts and only checks if the proposed package already has its dependencies installed and will install them if not or remove conflicting incompatible modules. The following commandline uses pip to install the given Python module:

```
Terminal> pip install package_name
```

The pip installer will download the package you want and its dependencies and install them in the existing directory tree. This works beautifully in the case where the package in question is pure-Python, without any system-specific dependencies. Otherwise, this can be a real nightmare, especially on Windows, which lacks freely available Fortran compilers. If the module in question is a C-library, one way to cope is to install the freely available Visual Studio Community Edition, which usually has enough to compile many C-codes. This platform dependency is the problem that conda was designed to solve by making the binary dependencies of the various platforms available instead of attempting to compile them. On a Windows system, if you installed Anaconda and registered it as the default Python installation (it asks during the install process), then you can use the high-quality Python *wheel* files on Christoph Gohlke's laboratory site at the University of California, Irvine, where he kindly makes a long list of scientific modules available.[1] Failing this, you can try the conda-forge site, which is a community-powered repository of modules that conda is capable of installing but which are not formally supported by Anaconda. Notably, conda-forge allows you to share Scientific Python configurations with your remote colleagues using authentication so that you can be sure that you are downloading and running code from users you trust.

---

[1] Wheel files are a Python distribution format that you download and install using pip as in pip install file.whl. Christoph names files according to Python version (e.g., cp27 means Python 2.7) and chipset (e.g., amd32 vs. Intel win32).

Again, if you are on Windows and none of the above works, then you may want to consider installing a full virtual machine solution, as provided by VMWare's `Player` or Oracle's `VirtualBox` (both freely available under liberal terms) or with the Windows Subsystem for Linux (WSL) that is built into Windows 10. Using either of these, you can set up a Linux machine running on top of Windows, which should cure these problems entirely! The great part of this approach is that you can share directories between the virtual machine and the Windows system so that you don't have to maintain duplicate data files. Anaconda Linux images are also available on the cloud by Platform as a Service (PaaS) providers like Amazon Web Services and Microsoft Azure. For the vast majority of users, especially newcomers to Python, the Anaconda distribution should be more than enough on any platform. It is just worth highlighting the Windows-specific issues and associated workarounds early on. There are other well-maintained Scientific Python Windows installers like `WinPython` and `PythonXY`. These provide the `spyder` integrated development environment, which is very Matlab-like environment for transitioning Matlab users.

1.2 Numpy

As we touched upon earlier, to use a compiled scientific library, the memory allocated in the Python interpreter must somehow reach this library as input. Furthermore, the output from these libraries must likewise return to the Python interpreter. This two-way exchange of memory is essentially the core function of the Numpy (numerical arrays in Python) module. Numpy is the de facto standard for numerical arrays in Python. It arose as an effort by Travis Oliphant and others to unify the preexisting numerical arrays in Python. In this section, we provide an overview and some tips for using Numpy effectively, but for much more detail, Travis' freely available book [34] is a great place to start.

Numpy provides specification of byte-sized arrays in Python. For example, below we create an array of three numbers, each of four-bytes long (32 bits at 8 bits per byte) as shown by the `itemsize` property. The first line imports Numpy as np, which is the recommended convention. The next line creates an array of 32 bit floating-point numbers. The `itemize` property shows the number of bytes per item.

```
>>> import numpy as np # recommended convention
>>> x = np.array([1,2,3],dtype=np.float32)
>>> x
array([1., 2., 3.], dtype=float32)
>>> x.itemsize
4
```

In addition to providing uniform containers for numbers, Numpy provides a comprehensive set of universal functions (i.e., *ufuncs*) that process arrays element-wise without additional looping semantics. Below, we show how to compute the element-wise sine using Numpy:

```
>>> np.sin(x)
array([0.841471 , 0.9092974, 0.14112  ], dtype=float32)
```

This computes the sine of the input array `[1,2,3]`, using Numpy's unary function, `np.sin`. There is another sine function in the built-in `math` module, but the Numpy version is faster, because it does not require explicit looping (i.e., using a `for` loop) over each of the elements in the array. That looping happens in the compiled `np.sin` function itself. Otherwise, we would have to do looping explicitly as in the following:

```
>>> from math import sin
>>> [sin(i) for i in [1,2,3]] # list comprehension
[0.8414709848078965, 0.9092974268256817, 0.1411200080598672]
```

Numpy uses common-sense casting rules to resolve the output types. For example, if the inputs had been an integer type, the output would still have been a floating-point type. In this example, we provided a Numpy array as input to the sine function. We could have also used a plain Python list instead, and Numpy would have built the intermediate Numpy array (e.g., `np.sin([1,1,1])`). The Numpy documentation provides a comprehensive (and very long) list of available *ufuncs*.

Numpy arrays come in many dimensions. For example, the following shows a two-dimensional 2x3 array constructed from two conforming Python lists.

```
>>> x = np.array([ [1,2,3],[4,5,6] ])
>>> x.shape
(2, 3)
```

Numpy is limited to 32 dimensions unless you build it for more.[2] Numpy arrays follow the usual Python slicing rules in multiple dimensions as shown below where the `:` colon character selects all elements along a particular axis.

```
>>> x = np.array([ [1,2,3],[4,5,6] ])
>>> x[:,0] # 0th column
array([1, 4])
>>> x[:,1] # 1st column
array([2, 5])
>>> x[0,:] # 0th row
array([1, 2, 3])
>>> x[1,:] # 1st row
array([4, 5, 6])
```

You can also select subsections of arrays by using slicing as shown below.

```
>>> x = np.array([ [1,2,3],[4,5,6] ])
>>> x
array([[1, 2, 3],
       [4, 5, 6]])
>>> x[:,1:] # all rows, 1st thru last column
array([[2, 3],
       [5, 6]])
>>> x[:,::2] # all rows, every other column
```

---

[2] See `arrayobject.h` in the Numpy source code.

```
array([[1,  3],
       [4,  6]])
>>> x[:,::-1] # reverse order of columns
array([[3,  2,  1],
       [6,  5,  4]])
```

1.2.1 Numpy Arrays and Memory

Some interpreted languages implicitly allocate memory. For example, in Matlab, you can extend a matrix by simply tacking on another dimension as in the following Matlab session:

```
>> x=ones(3,3)
x =
      1        1        1
      1        1        1
      1        1        1
>> x(:,4)=ones(3,1)  % tack on extra dimension
x =
      1        1        1        1
      1        1        1        1
      1        1        1        1
>> size(x)
ans =
      3        4
```

This works because Matlab arrays use pass-by-value semantics so that slice operations actually copy parts of the array as needed. By contrast, Numpy uses pass-by-reference semantics so that slice operations are *views* into the array without implicit copying. This is particularly helpful with large arrays that already strain available memory. In Numpy terminology, *slicing* creates views (no copying) and advanced indexing creates copies. Let us start with advanced indexing.

If the indexing object (i.e., the item between the brackets) is a non-tuple sequence object, another Numpy array (of type integer or boolean), or a tuple with at least one sequence object or Numpy array, then indexing creates copies. For the above example, to accomplish the same array extension in Numpy, you have to do something like the following:

```
>>> x = np.ones((3,3))
>>> x
array([[1., 1., 1.],
       [1., 1., 1.],
       [1., 1., 1.]])
>>> x[:,[0,1,2,2]] # notice duplicated last dimension
array([[1., 1., 1., 1.],
       [1., 1., 1., 1.],
       [1., 1., 1., 1.]])
>>> y = x[:,[0,1,2,2]] # same as above, but do assign it to y
```

Because of advanced indexing, the variable y has its own memory, because the
relevant parts of x were copied. To prove it, we assign a new element to x and see
that y is not updated.

```
>>> x[0,0]=999 # change element in x
>>> x          # changed
array([[999.,    1.,    1.],
       [  1.,    1.,    1.],
       [  1.,    1.,    1.]])
>>> y            # not changed!
Array([[1., 1., 1., 1.],
       [1., 1., 1., 1.],
       [1., 1., 1., 1.]])
```

However, if we start over and construct y by slicing (which makes It a view) as
shown below, then the change we made *does* affect y, because a view is just a
window into the same memory.

```
>>> x = np.ones((3,3))
>>> y = x[:2,:2] # view of upper left piece
>>> x[0,0] = 999 # change value
>>> x
array([[999.,    1.,    1.],
       [  1.,    1.,    1.],
       [  1.,    1.,    1.]])
>>> y
array([[999.,    1.],
       [  1.,    1.]])
```

If you want to explicitly force a copy without any indexing tricks, you can do
y=x.copy(). The code below works through another example of advanced
indexing versus slicing.

```
>>> x = np.arange(5) # create array
>>> x
array([0, 1, 2, 3, 4])
>>> y = x[[0,1,2]] # index by integer list to force copy
>>> y
array([0, 1, 2])
>>> z = x[:3]      # slice creates view
>>> z              # note y and z have same entries
array([0, 1, 2])
>>> x[0] = 999     # change element of x
>>> x
array([999,    1,    2,    3,    4])
>>> y              # note y is unaffected,
array([0, 1, 2])
>>> z              # but z is (it's a view).
array([999,    1,    2])
```

In this example, y is a copy, not a view, because it was created using advanced
indexing whereas z was created using slicing. Thus, even though y and z have the

same entries, only z is affected by changes to x. The `flags` property of Numpy arrays can help sort this out until you get used to it.

Manipulating memory using views is particularly powerful for signal and image processing algorithms that require overlapping fragments of memory. The following is an example of how to use advanced Numpy to create overlapping blocks that do not actually consume additional memory:

```
>>> from numpy.lib.stride_tricks import as_strided
>>> x = np.arange(16,dtype=np.int64)
>>> y = as_strided(x,(7,4),(16,8)) # overlapped entries
>>> y
array([[ 0,   1,   2,   3],
       [ 2,   3,   4,   5],
       [ 4,   5,   6,   7],
       [ 6,   7,   8,   9],
       [ 8,   9,  10,  11],
       [10,  11,  12,  13],
       [12,  13,  14,  15]])
```

The above code creates a range of integers and then overlaps the entries to create a 7x4 Numpy array. The final argument in the `as_strided` function are the strides, which are the steps in bytes to move in the row and column dimensions, respectively. Thus, the resulting array steps 8 bytes in the column dimension and 16 bytes in the row dimension. Because the integer elements in the Numpy array are 8 bytes, this is equivalent to moving by one element in the column dimension and by two elements in the row dimension. The second row in the Numpy array starts at 16 bytes (two elements) from the first entry (i.e., 2) and then proceeds by 8 bytes (by one element) in the column dimension (i.e, 2, 3, 4, 5). The important part is that memory is reused in the resulting 7x4 Numpy array. The code below demonstrates this by reassigning elements in the original x array. The changes show up in the y array because they point at the same allocated memory.

```
>>> x[::2]=99 # assign every other value
>>> x
array([99,   1, 99,   3, 99,   5, 99,   7, 99,   9, 99, 11, 99, 13,
       99, 15])
>>> y # the changes appear because y is a view
array([[99,   1, 99,   3],
       [99,   3, 99,   5],
       [99,   5, 99,   7],
       [99,   7, 99,   9],
       [99,   9, 99,  11],
       [99,  11, 99,  13],
       [99,  13, 99,  15]])
```

Bear in mind that `as_strided` does not check that you stay within memory block bounds. So, if the size of the target matrix is not filled by the available data, the remaining elements will come from whatever bytes are at that memory location. In other words, there is no default filling by zeros or other strategy that defends memory block bounds. One defense is to explicitly control the dimensions as in the following code:

```
>>> n = 8 # number of elements
>>> x = np.arange(n) # create array
>>> k = 5 # desired number of rows
>>> y = as_strided(x,(k,n-k+1),(x.itemsize,)*2)
>>> y
array([[0, 1, 2, 3],
       [1, 2, 3, 4],
       [2, 3, 4, 5],
       [3, 4, 5, 6],
       [4, 5, 6, 7]])
```

1.2.2 Numpy Matrices

Matrices in Numpy are similar to Numpy arrays, but they can only have two dimensions. They implement row-column matrix multiplication as opposed to element-wise multiplication. If you have two matrices you want to multiply, you can either create them directly or convert them from Numpy arrays. For example, the following shows how to create two matrices and multiply them.

```
>>> import numpy as np
>>> A = np.matrix([[1,2,3],[4,5,6],[7,8,9]])
>>> x = np.matrix([[1],[0],[0]])
>>> A*x
matrix([[1],
        [4],
        [7]])
```

This can also be done using arrays as shown below:

```
>>> A = np.array([[1,2,3],[4,5,6],[7,8,9]])
>>> x = np.array([[1],[0],[0]])
>>> A.dot(x)
array([[1],
       [4],
       [7]])
```

Numpy arrays support element-wise multiplication, not row-column multiplication. You must use Numpy matrices for this kind of multiplication, unless use the inner product np.dot, which also works in multiple dimensions (see np.tensordot for more general dot products). Python 3.x has a new @ notation for matrix multiplication, so we can redo the last calculation as follows:

```
>>> A @ x
array([[1],
       [4],
       [7]])
```

It is unnecessary to cast all multiplicands to matrices for multiplication. In
the next example, everything until the last line is a Numpy array, and thereafter,
we cast the array as a matrix with np.matrix, which then uses row-column
multiplication. It is unnecessary to cast the x variable as a matrix, because the left-
to-right order of the evaluation takes care of that automatically. If we need to use A
as a matrix elsewhere in the code, then we should bind it to another variable instead
of recasting it every time. If you find yourself casting back and forth for large arrays,
passing the copy=False flag to matrix avoids the expense of making a copy.

```
>>> A = np.ones((3,3))
>>> type(A) # array not matrix
<class 'numpy.ndarray'>
>>> x = np.ones((3,1)) # array not matrix
>>> A*x
array([[1., 1., 1.],
       [1., 1., 1.],
       [1., 1., 1.]])
>>> np.matrix(A)*x # row-column multiplication
matrix([[3.],
        [3.],
        [3.]])
```

1.2.3 Numpy Broadcasting

Numpy broadcasting is a powerful way to make implicit multidimensional grids for
expressions. It is probably the single most powerful feature of Numpy and the most
difficult to grasp. Proceeding by example, consider the vertices of a two-dimensional
unit square as shown below:

```
>>> X,Y = np.meshgrid(np.arange(2),np.arange(2))
>>> X
array([[0, 1],
       [0, 1]])
>>> Y
array([[0, 0],
       [1, 1]])
```

Numpy's meshgrid creates two-dimensional grids. The X and Y arrays have
corresponding entries that match the coordinates of the vertices of the unit square
(e.g., (0, 0), (0, 1), (1, 0), (1, 1)). To add the x- and y-coordinates, we could use X
and Y as in X+Y shown below. The output is the sum of the vertex coordinates of
the unit square.

```
>>> X+Y
array([[0, 1],
       [1, 2]])
```

Because the two arrays have compatible shapes, they can be added together element-wise. It turns out we can skip a step here and not bother with meshgrid to implicitly obtain the vertex coordinates by using broadcasting as shown below:

```
>>> x = np.array([0,1])
>>> y = np.array([0,1])
>>> x
array([0, 1])
>>> y
array([0, 1])
>>> x + y[:,None] # add broadcast dimension
array([[0, 1],
       [1, 2]])
>>> X+Y
array([[0, 1],
       [1, 2]])
```

On line 7 the None Python singleton tells Numpy to make copies of y along this dimension to create a conformable calculation. Note that np.newaxis can be used instead of None to be more explicit. The following lines show that we obtain the same output as when we used the X+Y Numpy arrays. Note that without broadcasting, x+y=array([0, 2]), which is not what we are trying to compute. Let us continue with a more complicated example where we have differing array shapes.

```
>>> x = np.array([0,1])
>>> y = np.array([0,1,2])
>>> X,Y = np.meshgrid(x,y)
>>> X
array([[0, 1],
       [0, 1],
       [0, 1]])
>>> Y
array([[0, 0],
       [1, 1],
       [2, 2]])
>>> X+Y
array([[0, 1],
       [1, 2],
       [2, 3]])
>>> x+y[:,None] # same as with meshgrid
array([[0, 1],
       [1, 2],
       [2, 3]])
```

In this example, the array shapes are different, so the addition of x and y is not possible without Numpy broadcasting. The last line shows that broadcasting generates the same output as using the compatible array generated by meshgrid. This shows that broadcasting works with different array shapes. For the sake of comparison, on line 3, meshgrid creates two conformable arrays, X and Y. On the last line, x+y[:,None] produces the same output as X+Y without the meshgrid.

We can also put the None dimension on the x array as x[:,None]+y, which would give the transpose of the result.

Broadcasting works in multiple dimensions also. The output shown has shape (4,3,2). On the last line, the x+y[:,None] produces a two-dimensional array which is then broadcast against z[:,None,None], which duplicates itself along the *two* added dimensions to accommodate the two-dimensional result on its left (i.e., x + y[:,None]). The caveat about broadcasting is that it can potentially create large, memory-consuming, intermediate arrays. There are methods for controlling this by reusing previously allocated memory, but that is beyond our scope here. Formulas in physics that evaluate functions on the vertices of high dimensional grids are great use cases for broadcasting.

```
>>> x = np.array([0,1])
>>> y = np.array([0,1,2])
>>> z = np.array([0,1,2,3])
>>> x+y[:,None]+z[:,None,None]
array([[[0, 1],
        [1, 2],
        [2, 3]],

       [[1, 2],
        [2, 3],
        [3, 4]],

       [[2, 3],
        [3, 4],
        [4, 5]],

       [[3, 4],
        [4, 5],
        [5, 6]]])
```

1.2.4 Numpy Masked Arrays

Numpy provides a powerful method to temporarily hide array elements without changing the shape of the array itself:

```
>>> from numpy import ma # import masked arrays
>>> x = np.arange(10)
>>> y = ma.masked_array(x, x<5)
>>> print (y)
[-- -- -- -- -- 5 6 7 8 9]
>>> print (y.shape)
(10,)
```

The elements in the array for which the logical condition (x<5) is true are masked, but the size of the array remains the same. This is particularly useful in plotting categorical data, where you may only want those values that correspond to a given category for part of the plot. Another common use is for image processing, wherein parts of the image may need to be excluded from subsequent processing. Creating a masked array does not force an implicit copy operation unless `copy=True` argument is used. For example, changing an element in x *does* change the corresponding element in y, even though y is a masked array:

```
>>> x[-1] = 99 # change this
>>> print(x)
[ 0  1  2  3  4  5  6  7  8 99]
>>> print(y) # masked array changed!
[-- -- -- -- -- 5 6 7 8 99]
```

1.2.5 Floating-Point Numbers

There are precision limitations when representing floating-point numbers on a computer with finite memory. For example, the following shows these limitations when adding two simple numbers:

```
>>> 0.1 + 0.2
0.30000000000000004
```

So, then, why is the output not 0.3? The issue is the floating point representation of the two numbers and the algorithm that adds them. To represent an integer in binary, we just write it out in powers of 2. For example, $230 = (11100110)_2$. Python can do this conversion using string formatting:

```
>>> print('{0:b}'.format(230))
11100110
```

To add integers, we just add up the corresponding bits and fit them into the allowable number of bits. Unless there is an overflow (the results cannot be represented with that number of bits), then there is no problem. Representing floating-point is trickier, because we have to represent these numbers as binary fractions. The IEEE 754 standard requires that floating-point numbers be represented as $\pm C \times 2^E$, where C is the significand (*mantissa*) and E is the exponent.

To represent a regular decimal fraction as binary fraction, we need to compute the expansion of the fraction in the following form $a_1/2 + a_2/2^2 + a_3/2^3 ...$ In other words, we need to find the a_i coefficients. We can do this using the same process we would use for a decimal fraction: just keep dividing by the powers of the fraction 1/2 and keep track of the whole and fractional parts. Python's `divmod` function can do most of the work for this. For example, to represent 0.125 as a binary fraction:

```
>>> a = 0.125
>>> divmod(a*2,1)
(0.0, 0.25)
```

The first item in the tuple is the quotient and the other is the remainder. If the quotient was greater than 1, then the corresponding a_i term is one and is zero otherwise. For this example, we have $a_1 = 0$. To get the next term in the expansion, we just keep multiplying by 2, which moves us rightward along the expansion to a_{i+1} and so on. Then,

```
>>> a = 0.125
>>> q,a = divmod(a*2,1)
>>> print (q,a)
0.0 0.25
>>> q,a = divmod(a*2,1)
>>> print (q,a)
0.0 0.5
>>> q,a = divmod(a*2,1)
>>> print (q,a)
1.0 0.0
```

The algorithm stops when the remainder term is zero. Thus, we have that $0.125 = (0.001)_2$. The specification requires that the leading term in the expansion be one. Thus, we have $0.125 = (1.000) \times 2^{-3}$. This means the significand is 1 and the exponent is -3.

Now, let us get back to our main problem 0.1+0.2 by developing the representation 0.1 by coding up the individual steps above.

```
>>> a = 0.1
>>> bits = []
>>> while a>0:
...     q,a = divmod(a*2,1)
...     bits.append(q)
...
>>> print (''.join(['%d'%i for i in bits]))
0001100110011001100110011001100110011001100110011001101
```

Note that the representation has an infinitely repeating pattern. This means that we have $(1.\overline{1001})_2 \times 2^{-4}$. The IEEE standard does not have a way to represent infinitely repeating sequences. Nonetheless, we can compute this:

$$\sum_{n=1}^{\infty} \frac{1}{2^{4n-3}} + \frac{1}{2^{4n}} = \frac{3}{5}$$

Thus, $0.1 = 1.6 \times 2^{-4}$. Per the IEEE 754 standard, for float type, we have 24 bits for the significand and 23 bits for the fractional part. Because we cannot represent the infinitely repeating sequence, we have to round off at 23 bits, 10011001100110011001101. Thus, whereas the significand's representation used to be 1.6, with this rounding, it is now

```
>>> b = '10011001100110011001101'
>>> 1+sum([int(i)/(2**n) for n,i in enumerate(b,1)])
1.600000023841858
```

Thus, we now have $0.1 \approx 1.600000023841858 \times 2^{-4} = 0.10000000149011612$. For the 0.2 expansion, we have the same repeating sequence with a different exponent, so that we have $0.2 \approx 1.600000023841858 \times 2^{-3} = 0.20000000298023224$. To add $0.1+0.2$ in binary, we must adjust the exponents until they match the higher of the two. Thus,

```
 0.11001100110011001100110
+1.10011001100110011001101
-------------------------
10.01100110011001100110011
```

Now, the sum has to be scaled back to fit into the significand's available bits, so the result is `1.0011001100110011001101` with exponent `-2`. Computing this in the usual way as shown below gives the result.

```
>>> k='0011001100110011001101010'
>>> print('%0.12f'%((1+sum([int(i)/(2**n)
...                         for n,i in enumerate(k,1)]))
  /2**2))
0.300000011921
```

which matches what we get with numpy

```
>>> import numpy as np
>>> print('%0.12f'%(np.float32(0.1) + np.float32(0.2)))
0.300000011921
```

The entire process proceeds the same for 64-bit floats. Python has a `fractions` and `decimal` modules that allow more exact number representations. The `decimal` module is particularly important for certain financial computations.

Roundoff Error Let us consider the example of adding $100,000,000$ and 10 in 32-bit floating point.

```
>>> print('{0:b}'.format(100000000))
101111101011110000100000000
```

This means that $100,000,000 = (1.0111110101111000010000000000)_2 \times 2^{26}$. Likewise, $10 = (1.010)_2 \times 2^3$. To add these, we have to make the exponents match as in the following:

```
 1.0111110101111000010000000000
+0.0000000000000000000000001010
--------------------------------
 1.0111110101111000010000001010
```

Now, we have to round off because we only have 23 bits to the right of the decimal point and obtain `1.0111110101111000010000`, thus losing the trailing `10` bits. This effectively makes the decimal $10 = (1010)_2$ that we started out with instead become $8 = (1000)_2$. Thus, using Numpy again,

```
>>> print(format(np.float32(100000000) +
        np.float32(10),'10.3f'))
100000008.000
```

The problem here is that the order of magnitude between the two numbers was so great that it resulted in loss in the significand's bits as the smaller number was right-shifted. When summing numbers like these, the Kahan summation algorithm (see `math.fsum()`) can effectively manage these roundoff errors.

```
>>> import math
>>> math.fsum([np.float32(100000000),np.float32(10)])
100000010.0
```

Cancellation Error Cancellation error (loss of significance) results when two nearly equal floating-point numbers are subtracted. Let us consider subtracting 0.1111112 and 0.1111111. As binary fractions, we have the following:

```
    1.11000111000111001000101 E-4
   -1.11000111000111000110111 E-4
   ---------------------------
    0.00000000000000000011100
```

As a binary fraction, this is 1.11 with exponent -23 or $(1.75)_{10} \times 2^{-23} \approx$ 0.00000010430812836. In Numpy, this loss of precision is shown in the following:

```
>>> print(format(np.float32(0.1111112)-np.float32(0.1111111),
    '1.17f'))
0.00000010430812836
```

To sum up, when using floating point, you must check for approximate equality using something like Numpy `allclose` instead of the usual Python equality (i.e., `==`) sign. This enforces error bounds instead of strict equality. Whenever practicable, use fixed scaling to employ integer values instead of decimal fractions. Double precision 64-bit floating-point numbers are much better than single precision and, reduces these problems for all but the strictest precision requirements. The Kahan algorithm is effective for summing floating point numbers across very large data without accruing roundoff errors. To minimize cancellation errors, re-factor the calculation to avoid subtracting two nearly equal numbers.

1.2.6   Numpy Optimizations and Prospectus

The Scientific Python community continues to push the frontier of scientific computing. Several important extensions to Numpy are under active development. First, Numba is a compiler that generates optimized machine code from pure-Python code using the LLVM compiler infrastructure. LLVM started as a research project at the University of Illinois to provide a target-independent compilation strategy for arbitrary programming languages and is now a well-established technology. The combination of LLVM and Python via Numba means that accelerating a block of Python code can be as easy as putting a `@numba.jit` decorator above the function definition, but this doesn't work for all situations. Numba can target general graphics processing units (GPGPUs) also.

The Dask project contains `dask.array` extensions for manipulating very large datasets that are too big to fit in a single computer's RAM (i.e., out-of-core) using Numpy semantics. Furthermore, `dask` includes extensions for Pandas dataframes (see Sect. 1.7). Roughly speaking, this means that `dask` understands how to unpack Python expressions and translate them for a variety of distributed back-end data services upon which the computing takes place. This means that `dask` separates the expression of the computation from the particular implementation on a given back end.

1.3   Matplotlib

Matplotlib is the primary visualization tool for scientific graphics in Python. Like all great open-source projects, it originated to satisfy a personal need. At the time of its inception, John Hunter primarily used Matlab for scientific visualization, but as he began to integrate data from disparate sources using Python, he realized he needed a Python solution for visualization, so he single-handedly wrote Matplotlib. Since those early years, Matplotlib has displaced the other competing methods for two-dimensional scientific visualization and today is a very actively maintained project, even without John Hunter, who sadly passed away in 2012.

John had a few basic requirements for Matplotlib:

- Plots should look publication quality with beautiful text.
- Plots should output PostScript for inclusion within documents and publication quality printing.
- Plots should be embeddable in a graphical user interface (GUI) for application development.
- The code should be mostly Python to allow for users to become developers.
- Plots should be easy to make with just a few lines of code for simple graphs.

Each of these requirements has been completely satisfied, and Matplotlib's capabilities have grown far beyond these requirements. In the beginning, to ease the transition from Matlab to Python, many of the Matplotlib functions were closely named after the corresponding Matlab commands.

The following shows the quickest way to draw a plot using Matplotlib and the plain Python interpreter. Later, we'll see how to do this even faster using IPython. The first line imports the requisite module as `plt`, which is the recommended

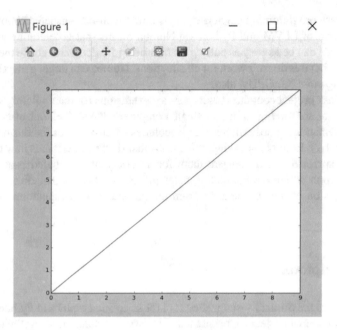

Fig. 1.1 The Matplotlib figure window. The icons on the bottom allow some limited plot-editing tools

convention. The next line plots a sequence of numbers generated using Python's range object. Note the output list contains a Line2D object. This is an *artist* in Matplotlib parlance. Finally, the plt.show() function draws the plot in a GUI figure window.

```
import matplotlib.pyplot as plt
plt.plot(range(10))
plt.show() # unnecessary in IPython (discussed later)
```

If you try this in your own plain Python interpreter (and you should!), you will see that you cannot type in anything further in the interpreter until the figure window (i.e., something like Fig. 1.1) is closed. This is because the plt.show() function preoccupies the interpreter with the controls in the GUI and *blocks* further interaction. As we discuss below, IPython provides ways to get around this blocking so you can simultaneously interact with the interpreter and the figure window.[3]

As shown in Fig. 1.1, the plot function returns a list containing the Line2D object. More complicated plots yield larger lists filled with *artists*. The terminology is that artists draw on the *canvas* contained in the Matplotlib figure. The final line is the plt.show function that provokes the embedded artists to render on

---

[3] You can also do this in the plain Python interpreter by doing import matplotlib;matplotlib.interactive(True).

the Matplotlib canvas. The reason this is a separate function is that plots may have dozens of complicated artists, and rendering may be a time-consuming task undertaken at the end, when all the artists have been mustered. Matplotlib supports plotting images, contours, and many others that we cover in detail in the following chapters.

Even though this is the quickest way to draw a plot in Matplotlib, it is not Recommended, because there are no handles to the intermediate products of the plot such as the plot's axis. While this is okay for a simple plot like this, later on we will see how to construct complicated plots using the recommended method.

One of the best ways to get started with Matplotlib is to browse the extensive online gallery of plots on the main Matplotlib site. Each plot comes with corresponding source code that you can use as a starting point for your own plots. In Sect. 1.4, we discuss special *magic* commands that make this particularly easy. The annual *John Hunter: Excellence in Plotting Contest* provides fantastic, compelling examples of scientific visualizations that are possible using Matplotlib.

1.3.1 Alternatives to Matplotlib

Even though Matplotlib is the most complete option for script-based plotting, there are some alternatives for specialized scientific graphics that may be of interest.

If you require real-time data display and tools for volumetric data rendering and complicated 3D meshes with isosurfaces, then PyQtGraph is an option. PyQtGraph is a pure-Python graphics and GUI library that depends on Python bindings for the Qt GUI library (i.e., PySide or PyQt4) and Numpy. This means that the PyQtGraph relies on these other libraries (especially Qt's GraphicsView framework) for the heavy-duty number crunching and rendering. This package is actively maintained, with solid documentation. You also need to grasp a few Qt-GUI development concepts to use this effectively.

An alternative that comes from the R community is ggplot, which is a Python port of the ggplot2 package that is fundamental to statistical graphics in R. From the Python standpoint, the main advantage of ggplot is the tight integration with the Pandas dataframe, which makes it easy to draw beautifully formatted statistical graphs. The downside of this package is that it applies un-Pythonic semantics based on the *Grammar of Graphics* [49], which is nonetheless a well-thought-out method for articulating complicated graphs. Of course, because there are two-way bridges between Python and R via the R2Py module (among others), it is workable to send Numpy arrays to R for native ggplot2 rendering and then retrieve the so-computed graphic back into Python. This is a workflow that is lubricated by the Jupyter Notebook (see below) via the rmagic extension. Thus, it is quite possible to get the best of both worlds via the Jupyter Notebook, and this kind of multi-language workflow is quite common in data analysis communities.

1.3.2   Extensions to Matplotlib

Initially, to encourage adoption of Matplotlib from Matlab, many of the graphical
sensibilities were adopted from Matlab to preserve the look and feel for transitioning
users. Modern sensibilities and prettier default plots are possible, because Matplotlib
provides the ability to drill down and tweak every element on the canvas. However,
this can be tedious to do and several alternatives offer relief. For statistical plots,
the first place to look is the `seaborn` module that includes a vast array of
beautifully formatted plots, including violin plots, kernel density plots, and bivariate
histograms. The `seaborn` gallery includes samples of available plots and the
corresponding code that generates them. Importing `seaborn` hijacks the default
settings for all plots, so you have to coordinate this if you only want to use `seaborn`
for some (not all) of your visualizations in a given session. You can find the defaults
for Matplotlib in the `matplotlib.rcParams` dictionary.

1.4   IPython

IPython [35] originated as a way to enhance Python's basic interpreter for smooth
interactive scientific development. In the early days, the most important enhance-
ment was tab completion for dynamic introspection of workspace variables. For
example, you can start IPython at the commandline by typing `ipython`, and then
you should see something like the following in your terminal:

```
Python 3.10.0 | packaged by conda-forge | (default, Nov 20
  2021, 02:25:18) [GCC 9.4.0]
Type 'copyright', 'credits' or 'license' for more information
IPython 8.0.1 -- An enhanced Interactive Python. Type '?' for
  help.

In [1]:
```

Next, creating a string as shown and hitting the TAB key after the dot character
initiates the introspection, showing all the functions and attributes of the string
object in x.

```
In [1]: x = 'this is a string'

In [2]: x.<TAB>
x.capitalize    x.encode      x.format      x.isalpha     x.isidentifier
x.casefold      x.endswith    x.format_map  x.isascii     x.islower
x.center        x.expandtabs  x.index       x.isdecimal   x.isnumeric
x.count         x.find        x.isalnum     x.isdigit     x.isprintable
```

To get help about any of these, you simply add the ? character at the end as shown
below:

```
In [2]: x.center?
Signature: x.center(width, fillchar=' ', /)
```

```
Docstring:
Return a centered string of length width.

Padding is done using the specified fill character (default is
  a space).
Type:        builtin_function_or_method
```

and IPython provides the built-in help documentation. Note that you can also get this documentation with `help(x.center)`, which works in the plain Python interpreter as well.

The combination of dynamic tab-based introspection and quick interactive help accelerates development, because you can keep your eyes and fingers in one place as you work. This was the original IPython experience, but IPython has since grown into a complete framework for delivering a rich scientific computing workflow that retains and enhances these fundamental features.

1.5   Jupyter Notebook

As you may have noticed investigating Python on the web, most Python users are Web developers, not scientific programmers, meaning that the Python stack is *very* well-developed for web technologies. The genius of the IPython development team was to leverage these technologies for scientific computing by embedding IPython in modern web browsers. In fact, this strategy has been so successful that IPython has moved into other languages beyond Python such as Julia and R as the *Jupyter* project. You can start the Jupyter Notebook with the following commandline:

```
Terminal> jupyter notebook
```

After starting the notebook, you should see something like the following in the terminal:

```
[I 16:08:21.213 NotebookApp] Serving notebooks from local
  directory: /home/user
[I 16:08:21.214 NotebookApp] The Jupyter Notebook is running
  at:
[I 16:08:21.214 NotebookApp] http://localhost:8888/?token=
  80281f0c324924d34a4e
[I 16:08:21.214 NotebookApp] Use Control-C to stop this
  server and shut down
```

The first line reveals where Jupyter looks for default settings. The next line shows where it looks for documents in the Jupyter Notebook format. The third line shows that the Jupyter Notebook started a web server on the local machine (i.e., 127.0.0.1) on port number 8888. This is the address your browser needs to connect to the Jupyter session, although your default browser should have opened automatically to this address. The port number and other configuration options are available either on the commandline or in the profile file shown in the first line. If you are on a Windows platform and you do not get this far, then the Window's

Fig. 1.2 The Jupyter Notebook dashboard

Fig. 1.3 A new Jupyter Notebook

firewall is probably blocking the port. For additional configuration help, see the main Jupyter site (`www.jupyter.org`).

When Jupyter starts, it initiates several Python processes that use the blazing-fast `ZeroMQ` message passing framework for interprocess communication, along with the WebSocket protocol for back-and-forth communication with the browser. To start Jupyter and get around your default browser, you can use the additional `--no-browser` flag and then manually type in the local host address `http://127.0.0.1:8888` into your favorite browser to get started. Once all that is settled, you should see something like Fig. 1.2:

You can create a new document by clicking the `New Notebook` button shown in Fig. 1.2. Then, you should see something like Fig. 1.3. To start using the Jupyter Notebook, you just start typing code in the shaded textbox and then hit `SHIFT+ENTER` to execute the code in that Jupyter *cell*. Figure 1.4 shows the

Fig. 1.4 Jupyter Notebook pulldown completion menu

dynamic introspection in the pulldown* menu when you type the TAB key after the x.. Context-based help is also available as before by using the ? suffix, which opens a help panel at the bottom of the browser window. There are many amazing features including the ability to share notebooks between different users and to run Jupyter Notebooks in the Amazon cloud, but these features go beyond our scope here. Check the jupyter.org website or peek at the mailing list for the latest work on these fronts.

The Jupyter Notebook supports high-quality mathematical typesetting using MathJaX, which is a JavaScript implementation of most of, as well as video and other rich content. The concept of consolidating mathematical algorithm descriptions and the code that implements those algorithms into a shareable document is more important than all of these amazing features. There is no understating the importance of this in practice, because the algorithm documentation (if it exists) is usually in one format and completely separate from the code that implements it. This common practice leads to un-synchronized documentation and code that renders one or the other useless. The Jupyter Notebook solves this problem by putting everything into a living shareable document based upon open standards and freely available software. Jupyter Notebooks can even be saved as static HTML documents for those without Python!

Finally, Jupyter provides a large set of magic commands for creating macros, profiling, debugging, and viewing codes. A full list of these can be found by typing in %lsmagic in Jupyter. Help on any of these is available using the ? character suffix. Some frequently used commands include the %cd command that changes the current working directory, the %ls command that lists the files in the current directory, and the %hist command that shows the history of previous commands (including optional searching). The most important of these for new users is probably the %loadpy command that can load scripts from the local disk or from the web. Using this to explore the Matplotlib gallery is a great way to experiment with and reuse the plots there.

1.6   Scipy

Scipy was the first consolidated module for a wide range of compiled libraries, all based on Numpy arrays. Scipy includes numerous special functions (e.g., Airy, Bessel, elliptical) as well as powerful numerical quadrature routines via the QUADPACK Fortran library (see scipy.integrate), where you will also find other quadrature methods. Some of the same functions appear in multiple places within Scipy itself as well as in Numpy. Additionally, Scipy provides access to the ODEPACK library for solving differential equations. Lots of statistical functions, including random number generators, and a wide variety of probability distributions are included in the scipy.stats module. Interfaces to the Fortran MINPACK optimization library are provided via scipy.optimize. These include methods for root-finding, minimization and maximization problems, with and without higher-order derivatives. Methods for interpolation are provided in the scipy.interpolate module via the FITPACK Fortran package. Some of the modules are so big that you do not get all of them with import scipy, because that would take too long to load. You may have to load some of these packages individually as import scipy.interpolate, for example.

As we discussed, the Scipy module is already packed with an extensive list of scientific codes. For that reason, the *Scikits* modules were originally established as a way to stage candidates that could eventually make it into the already stuffed Scipy module, but it turns out that many of these modules became so successful on their own that they will never be integrated into Scipy proper. Some examples include scikit-learn for machine learning and scikit-image for image processing.

1.7   Pandas

Pandas [30] is a powerful module that is optimized on top of Numpy and provides a set of data structures particularly suited to time series and spreadsheet-style data analysis (think of pivot tables in Excel). If you are familiar with the R statistical package, then you can think of Pandas as providing a Numpy-powered dataframe for Python.

1.7.1 Series

There are two primary data structures in Pandas. The first is the `Series` Object, which combines an index and corresponding data values.

```
>>> import pandas as pd # recommended convention
>>> x = pd.Series(index = range(5),data=[1,3,9,11,12])
>>> x
0     1
1     3
2     9
3     11
4     12
dtype: int64
```

The main thing to keep in mind with Pandas is that these data structures were originally designed to work with time series data. In that case, the `index` in the data structures corresponds to a sequence of ordered time stamps. In the general case, the index must be a sortable array-like entity. For example:

```
>>> x = pd.Series(index = ['a','b','d','z','z'],
   data=[1,3,9,11,12])
>>> x
a     1
b     3
d     9
z     11
z     12
dtype: int64
```

Note the duplicated z entries in the `index`. We can get at the entries in the `Series` in a number of ways. First, we can used the *dot* notation to select as in the following:

```
>>> x.a
1
>>> x.z
z     11
z     12
dtype: int64
```

We can also use the indexed position of the entries with `iloc` as in the following:

```
>>> x.iloc[:3]
a     1
b     3
d     9
dtype: int64
```

which uses the same slicing syntax as Numpy arrays. You can also slice across the `index`, even if it is not numeric with `loc` as in the following:

```
>>> x.loc['a':'d']
a    1
b    3
d    9
dtype: int64
```

which you can get directly from the usual slicing notation:

```
>>> x['a':'d']
a    1
b    3
d    9
dtype: int64
```

Unlike with regular Python, slicing this way *includes* the endpoints. While that is very interesting, the main power of Pandas comes from its power to aggregate and group data. In the following, we build a more interesting `Series` object:

```
>>> x = pd.Series(range(5),[1,2,11,9,10])
```

and then group it in the following:

```
>>> grp = x.groupby(lambda i:i%2) # odd or even
>>> grp.get_group(0) # even group
2     1
10    4
dtype: int64
>>> grp.get_group(1) # odd group
1     0
11    2
9     3
dtype: int64
```

The first line groups the elements of the `Series` object by whether or not the index is even or odd. The `lambda` function returns 0 or 1, depending on whether or not the corresponding index is even or odd, respectively. The next line shows the 0 (even) group, and then the one after shows the 1 (odd) group. Now that we have separate groups, we can perform a wide variety of summarizations on the group. You can think of these as reducing each group into a single value. For example, in the following, we get the maximum value of each group:

```
>>> grp.max() # max in each group
0    4
1    3
dtype: int64
```

Conspicuously, the operation above returns another `Series` object with an `index` corresponding to the [0,1] elements.

1.7.2 Dataframe

The Pandas `DataFrame` is an encapsulation of the `Series` that extends to two dimensions. One way to create a `DataFrame` is with dictionaries as in the following:

```
>>> df = pd.DataFrame({'col1': [1,3,11,2], 'col2': [9,23,0,2]})
```

The keys in the input dictionary are now the column headings (labels) of the `DataFrame`, with each corresponding column matching the list of corresponding values from the dictionary. Like the `Series` object, the `DataFrame` also has in index, which is the `[0,1,2,3]` column on the far left. We can extract elements from each column using the `iloc` method as discussed earlier as shown below:

```
>>> df.iloc[:2,:2] # get section
   col1   col2
0    1      9
1    3     23
```

or by directly slicing or by using the *dot* notation as shown below:

```
>>> df['col1'] # indexing
0     1
1     3
2    11
3     2
Name: col1, dtype: int64
>>> df.col1 # use dot notation
0     1
1     3
2    11
3     2
Name: col1, dtype: int64
```

Subsequent operations on the `DataFrame` preserve its column-wise structure as in the following:

```
>>> df.sum()
col1    17
col2    34
dtype: int64
```

where each column was totaled. Grouping and aggregating with the dataframe is even more powerful than with series. Let us construct the following dataframe:

```
>>> df = pd.DataFrame({'col1': [1,1,0,0], 'col2': [1,2,3,4]})
```

In the above dataframe, note that the `col1` column has only two entries. We can group the data using this column as in the following:

```
>>> grp = df.groupby('col1')
>>> grp.get_group(0)
   col1   col2
2     0      3
3     0      4
>>> grp.get_group(1)
   col1   col2
0     1      1
1     1      2
```

Each group corresponds to entries for which col1 was either of its two values. Now that we have grouped on col1, as with the series object, we can also functionally summarize each of the groups as in the following:

```
>>> grp.sum()
        col2
col1
0          7
1          3
```

where the sum is applied across each of the dataframes present in each group. The index of the output above is each of the values in the original col1.

The dataframe can compute new columns based on existing columns using the eval method as shown below:

```
>>> df['sum_col']=df.eval('col1+col2')
>>> df
   col1   col2   sum_col
0     1      1         2
1     1      2         3
2     0      3         3
3     0      4         4
```

You can assign the output to a new column to the dataframe as shown.[4] We can group by multiple columns as shown below:

```
>>> grp = df.groupby(['sum_col','col1'])
```

Doing the sum operation on each group gives the following:

```
>>> res = grp.sum()
>>> res
                   col2
sum_col col1
2          1          1
3          0          3
           1          2
4          0          4
```

---

[4] Note this kind of on-the-fly memory extension is not possible in regular Numpy. For example, x = np.array([1,2]); x[3]=3 generates an error.

This output is much more complicated than anything we have seen so far, so let us carefully walk through it. Below the headers, the first row 2 1 1 indicates that for `sum_col=2` and for all values of `col1` (namely, just the value 1), the value of `col2` is 1. For the next row, the same pattern applies except that for `sum_col=3`, there are now two values for `col1`, namely 0 and 1, which each have their corresponding two values for the sum operation in `col2`. This layered display is one way to look at the result. The layers above are not uniform. Alternatively, we can `unstack` this result to obtain the following tabular view of the previous result:

```
>>> res.unstack()
         col2
col1        0    1
sum_col
2         NaN  1.0
3         3.0  2.0
4         4.0  NaN
```

The NaN values indicate positions in the table where there is no entry. For example, for the pair (`sum_col=2,col2=0`), there is no corresponding value in the dataframe, as you may verify by looking at the penultimate code block. There is also no entry corresponding to the (`sum_col=4,col2=1`) pair. Thus, this shows that the original presentation in the penultimate code block is the same as this one, just without the abovementioned missing entries indicated by NaN.

We have barely scratched the surface of what Pandas is capable of, and we have completely ignored its powerful features for managing dates and times. The text by Mckinney [30] is a very complete and happily readable introduction to Pandas. The online documentation and tutorials at the main Pandas site are also great for diving deeper into Pandas.

1.8 Sympy

Sympy [8] is the main computer algebra module in Python. It is a pure-Python package with no platform dependencies. With the help of multiple *Google Summer of Code* sponsorships, it has grown into a powerful computer algebra system with many collateral projects that make it faster and integrate it tighter with Numpy and Jupyter. Sympy's online tutorial is excellent and allows interacting with its embedded code samples in the browser by running the code on the Google App Engine behind the scenes. This provides an excellent way to interact and experiment with Sympy.

If you find Sympy too slow or need algorithms that it does not implement, then SAGE is your next stop. The SAGE project is a consolidation of over 70 of the best open-source packages for computer algebra and related computation. Although Sympy and SAGE share code freely between them, SAGE is a specialized build of the Python kernel to facilitate deep integration with the underlying libraries. Thus, it is not a pure-Python solution for computer algebra (i.e., not as portable), and it is a proper superset of Python with its own extended syntax. The choice between SAGE and Sympy really depends on whether or not you intend *primarily* work in SAGE or just need *occasional* computer algebra support in your existing Python code.

An important new development regarding SAGE is the freely available SAGE Cloud (https://cloud.sagemath.com/), sponsored by the University of Washington that allows you to use SAGE entirely in the browser with no additional setup. Both SAGE and Sympy offer tight integration with the Jupyter Notebook for mathematical typesetting in the browser using MathJaX.

To get started with Sympy, you must import the module as usual:

```
>>> import sympy as S # might take awhile
```

which may take a bit because it is a big package. The next step is to create a Sympy variable as in the following:

```
>>> x = S.symbols('x')
```

Now we can manipulate this using Sympy functions and Python logic as shown below:

```
>>> p=sum(x**i for i in range(3)) # 2nd order polynomial
>>> p
x**2 + x + 1
```

Now, we can find the roots of this polynomial using Sympy functions:

```
>>> S.solve(p) # solves p == 0
[-1/2 - sqrt(3)*I/2, -1/2 + sqrt(3)*I/2]
```

There is also a `sympy.roots` function that provides the same output but as a dictionary.

```
>>> S.roots(p)
{-1/2 - sqrt(3)*I/2: 1, -1/2 + sqrt(3)*I/2: 1}
```

We can also have more than one symbolic element in any expression as in the following:

```
>>> from sympy.abc import a,b,c  # quick way to get common
 symbols
>>> p = a* x**2 + b*x + c
>>> S.solve(p,x) # specific solving for x-variable
[(-b + sqrt(-4*a*c + b**2))/(2*a), -(b + sqrt(-4*a*c + b**2))
 /(2*a)]
```

which is the usual quadratic formula for roots. Sympy also provides many mathematical functions designed to work with Sympy variables. For example:

```
>>> S.exp(S.I*a) #using Sympy exponential
exp(I*a)
```

We can expand this using `expand_complex` to obtain the following:

```
>>> S.expand_complex(S.exp(S.I*a))
I*exp(-im(a))*sin(re(a)) + exp(-im(a))*cos(re(a))
```

which gives us Euler's formula for the complex exponential. Note that Sympy does
not know whether or not a is itself a complex number. We can fix this by making
that fact part of the construction of a as in the following:

```
>>> a = S.symbols('a',real=True)
>>> S.expand_complex(S.exp(S.I*a))
I*sin(a) + cos(a)
```

Note the much simpler output this time, because we have forced the additional
condition on a.

A powerful way to use Sympy is to construct complicated expressions that you
can later evaluate using Numpy via the lambdify method. For example:

```
>>> y = S.tan(x) * x + x**2
>>> yf= S.lambdify(x,y,'numpy')
>>> y.subs(x,.1) # evaluated using Sympy
0.0200334672085451
>>> yf(.1) # evaluated using Numpy
0.020033467208545055
```

After creating the Numpy function with lambdify, you can use Numpy arrays as
input as shown:

```
>>> yf(np.arange(3)) # input is Numpy array
array([ 0.        ,  2.55740772, -0.37007973])
>>> [ y.subs(x,i).evalf() for i in range(3) ] # need extra
 work for Sympy
[0, 2.55740772465490, -0.370079726523038]
```

We can get the same output using Sympy, but that requires the extra programming
logic shown to do the vectorizing that Numpy performs natively.

Once again, we have merely scratched the surface of what Sympy is capable
of, and the online interactive tutorial is the best place to learn more. Sympy also
allows automatic mathematical typesetting within the Jupyter Notebook using , so
the so-constructed notebooks look almost publication-ready (see sympy.latex)
and can be made so with the jupyter nbconvert command. This makes it
easier to jump the cognitive gap between the Python code and the symbology of
traditional mathematics.

1.9 Xarray for High Dimensional Dataframes

Xarray implements multidimensional dataframes. Pandas used to do some of this
with the Panel object but gave up on it when Xarray came along. Xarray has
primarily been developed by the Earth sciences and astronomy communities, so

much of their documentation is geared toward those audiences that typically have large high dimensional cubes (i.e., not ragged/irregular) datasets.

There are two fundamental data structures in Xarray the `DataArray` and the `Dataset`. The `DataArray` contains a single embedded Numpy data cube and its corresponding coordinates. The `Dataset` holds multiple `DataArray` objects that share the same coordinates. In a rough analogy to Pandas, `DataArray` is to `Dataset` as Pandas `Series` is to `DataFrame`.

A `DataArray` has four attributes:

- `values`—multidimensional Numpy array
- `dims`—dimension names for each axis (e.g., x, y, z)
- `coords`—dictionary of arrays that label each point
- `attrs`—Python `OrderedDict` for metadata

Let's construct one to see how this works.

```
>>> import xarray as xr
>>> d = xr.DataArray(np.arange(1,10))
>>> d
<xarray.DataArray (dim_0: 9)>
array([1, 2, 3, 4, 5, 6, 7, 8, 9])
Dimensions without coordinates: dim_0
```

Xarray `dim` gives a dimension a descriptive name. Note that unlike Pandas, you cannot set these after the fact.

```
>>> d = xr.DataArray(np.arange(1,10), dims=['x'])
>>> d.dims
('x',)
```

It's more descriptive to define *coordinates*, which is similar to a Pandas index.

```
>>> d = xr.DataArray(np.arange(1,10),
...              coords= {'x': ['a', 'b', 'c', 'd', 'e', 'f',
             'g', 'h', 'i']})
>>> d
<xarray.DataArray (x: 9)>
array([1, 2, 3, 4, 5, 6, 7, 8, 9])
Coordinates:
  * x           (x) <U1 'a' 'b' 'c' 'd' 'e' 'f' 'g' 'h' 'i'
```

You can slice the array using these `coords` as in Pandas `loc` accessor:

```
>>> d.loc['b':'c']
<xarray.DataArray (x: 2)>
array([2, 3])
Coordinates:
  * x           (x) <U1 'b' 'c'
```

Basic plotting is built-in:

```
d.plot(marker='o')
```

You can slice using Numpy indexing as in the following:

```
>>> d[:3] # like .iloc[] in Pandas
<xarray.DataArray (x: 3)>
array([1, 2, 3])
Coordinates:
  * x           (x) <U1 'a' 'b' 'c'
```

The .sel() method is much more powerful for indexing using labels:

```
>>> d.sel(x='c')
<xarray.DataArray ()>
array(3)
Coordinates:
    x           <U1 'c'
```

There is a corresponding isel() also, if you still want to use the Numpy indexing. Slicing is not built into the syntax, so you have to use slice explicitly.

```
>>> d.sel(x= slice('d','f'))
<xarray.DataArray (x: 3)>
array([4, 5, 6])
Coordinates:
  * x           (x) <U1 'd' 'e' 'f'
```

Basic arithmetic operations are supported:

```
>>> d*2 + d**2/3
<xarray.DataArray (x: 9)>
array([ 2.33333333,  5.33333333,  9.        , 13.33333333,
   18.33333333,
       24.        , 30.33333333, 37.33333333, 45.        ])
Coordinates:
  * x           (x) <U1 'a' 'b' 'c' 'd' 'e' 'f' 'g' 'h' 'i'
```

Slicing using dates and times is supported:

```
>>> ts = pd.date_range(start='1/1/2020', end='1/08/2021')
>>> d = xr.DataArray(np.arange(len(ts)), coords={'t': ts})
>>> d.sel(t = '2020-12-13')
<xarray.DataArray ()>
array(347)
Coordinates:
    t           datetime64[ns] 2020-12-13
```

You still need the slice to get a time interval:

```
>>> d.sel(t = slice('3/10/2020','2020-12-13')) # time formats
different
<xarray.DataArray (t: 279)>array([
    69,  70,  71,  72,  73,  74,  75,  76,  77,  78,  79,  80,  81,
    82,  83,  84,  85,  86,  87,  88,  89,  90,  91,  92,  93,  94,
    95,  96,  97,  98,  99, 100, 101, 102, 103, 104, 105, 106, 107,
   108, 109, 110, 111, 112, 113, 114, 115, 116, 117, 118, 119, 120,
```

```
       121,  122,  123,  124,  125,  126,  127,  128,  129,  130,  131,  132,  133,
       134,  135,  136,  137,  138,  139,  140,  141,  142,  143,  144,  145,  146,
       147,  148,  149,  150,  151,  152,  153,  154,  155,  156,  157,  158,  159,
       160,  161,  162,  163,  164,  165,  166,  167,  168,  169,  170,  171,  172,
       173,  174,  175,  176,  177,  178,  179,  180,  181,  182,  183,  184,  185,
       186,  187,  188,  189,  190,  191,  192,  193,  194,  195,  196,  197,  198,
       199,  200,  201,  202,  203,  204,  205,  206,  207,  208,  209,  210,  211,
       212,  213,  214,  215,  216,  217,  218,  219,  220,  221,  222,  223,  224,
       225,  226,  227,  228,  229,  230,  231,  232,  233,  234,  235,  236,  237,
       238,  239,  240,  241,  242,  243,  244,  245,  246,  247,  248,  249,  250,
       251,  252,  253,  254,  255,  256,  257,  258,  259,  260,  261,  262,  263,
       264,  265,  266,  267,  268,  269,  270,  271,  272,  273,  274,  275,  276,
       277,  278,  279,  280,  281,  282,  283,  284,  285,  286,  287,  288,  289,
       290,  291,  292,  293,  294,  295,  296,  297,  298,  299,  300,  301,  302,
       303,  304,  305,  306,  307,  308,  309,  310,  311,  312,  313,  314,  315,
       316,  317,  318,  319,  320,  321,  322,  323,  324,  325,  326,  327,  328,
       329,  330,  331,  332,  333,  334,  335,  336,  337,  338,  339,  340,  341,
       342,  343,  344,  345,  346,  347])
Coordinates:
  * t            (t) datetime64[ns] 2020-03-10 2020-03-11 ...
    2020-12-12 2020-12-13
```

Interpolation is possible while indexing as long as `scipy` is installed. Interpolation also works in multiple dimensions. The `interp_like` method is also helpful. The `scipy` documentation has more details regarding the interpolation methods that are supported in the `interp` method (see Fig. 1.5).

```
>>> d = xr.DataArray(np.arange(10), coords = {'x':
        np.arange(10)})
>>> d.interp({'x': np.linspace(0,8,10)})
<xarray.DataArray (x: 10)>
array([0.      , 0.88888889, 1.77777778, 2.66666667,
    3.55555556,
        4.44444444, 5.33333333, 6.22222222, 7.11111111,
        8.      ])
Coordinates:
  * x            (x) float64 0.0 0.8889 1.778 2.667 3.556 ...
    5.333 6.222 7.111 8.0
```

Pandas Vs Xarray for Multilayered Dataframes `DataArray` handles multidimensional data better than Pandas. Consider the following dataframe with a `MultiIndex` along the rows:

```
>>> midx = pd.MultiIndex.from_product([['A', 'B' ],
...                                    ['normal','high']])
>>> df = pd.DataFrame(index = midx,
...             data = [[786,89],
...                     [207,25],
...                     [819,67],
...                     [176,22]],
...             columns = ['BP','CP'])
>>> # add names in separately
>>> df.index.set_names(['type','measure'],inplace=True)
```

Fig. 1.5 Xarray interpolation

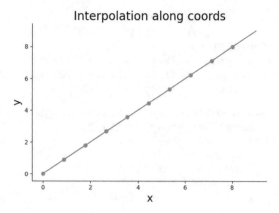

```
>>> df.columns.set_names(['cost'],inplace=True)
>>> df
cost            BP   CP
type measure
A     normal    786  89
      high      207  25
B     normal    819  67
      high      176  22
```

Notice that we had to add in the names of the levels separately. This is where the Xarray coords dictionary makes explicit and very useful. What if we want to sum across the type index using Pandas:

```
>>> df.groupby(level=1,axis=0).sum() # need the groupby
cost          BP    CP
measure
high          383   47
normal        1605  156
```

Even though you can cast this Pandas dataframe into DataArray using the constructor, it does not pick up the multiiindex correctly. Let's consider the same data using Xarray to build the DataArray object from scratch:

```
>>> data = np.array([[786,89],
...                  [207,25],
...                  [819,67],
...                  [176,22]])

>>> d = xr.DataArray(data.reshape((2,2,2)),
        coords=dict(type=['A','B'],measure=['normal','high'],
        cost=['BP','CP']))
>>> d
<xarray.DataArray (type: 2, measure: 2, cost: 2)>
array([[[786,    89],
        [207,    25]],
```

```
        [[819,    67],
         [176,    22]]])
Coordinates:
  * type        (type)  <U1 'A' 'B'
  * measure     (measure)  <U6 'normal' 'high'
  * cost        (cost)  <U2 'BP' 'CP'

>>> d.sel(type='A', measure='high')
<xarray.DataArray (cost: 2)>
array([207,   25])
Coordinates:
    type        <U1 'A'
    measure     <U6 'high'
  * cost        (cost)  <U2 'BP' 'CP'

>>> d.loc[:,'high','CP'] # using coords for slicing
<xarray.DataArray (type: 2)>
array([25, 22])
Coordinates:
  * type        (type)  <U1 'A' 'B'
    measure     <U6 'high'
    cost        <U2 'CP'
```

This the same summation as before with Pandas but no groupby needed:

```
>>> d.sum('type').values # no groupby needed as with Pandas
array([[1605,   156],
       [ 383,    47]])
```

Similarly, summation over multiple dimensions follows the same pattern:

```
>>> d.sum(['type','measure'])
<xarray.DataArray (cost: 2)>
array([1988,  203])
Coordinates:
  * cost        (cost)  <U2 'BP' 'CP'
```

However, doing this in Pandas requires complex and unreadable grouping and summing.

```
>>> df.groupby(level=1).sum().sum(axis=0) # far less clear
 with Pandas
cost
BP      1988
CP       203
dtype: int64
```

To normalized by the partial sum, broadcasting and alignment is handled automatically with Xarray:

```
>>> d/d.sum(['type','measure'])
<xarray.DataArray (type: 2, measure: 2, cost: 2)>
array([[[0.39537223, 0.43842365],
```

```
            [0.10412475, 0.12315271]],

       [[0.41197183, 0.33004926],
        [0.08853119, 0.10837438]]]])
Coordinates:
  * type       (type) <U1 'A' 'B'
  * measure    (measure) <U6 'normal' 'high'
  * cost       (cost) <U2 'BP' 'CP'
```

Computing the same result using Pandas is much less intelligible:

```
>>> df/df.groupby(level=1).sum().sum(axis=0)
cost                  BP        CP
type measure
A      normal    0.395372  0.438424
       high      0.104125  0.123153
B      normal    0.411972  0.330049
       high      0.088531  0.108374
```

Basic Calculations   Built-in methods facilitate common calculations such as *mean*:

```
>>> d.mean('type')
<xarray.DataArray (measure: 2, cost: 2)>
array([[802.5,  78. ],
       [191.5,  23.5]])
Coordinates:
  * measure    (measure) <U6 'normal' 'high'
  * cost       (cost) <U2 'BP' 'CP'
```

and *variance*:

```
>>> d.var('type')
<xarray.DataArray (measure: 2, cost: 2)>
array([[272.25, 121. ],
       [240.25,   2.25]])
Coordinates:
  * measure    (measure) <U6 'normal' 'high'
  * cost       (cost) <U2 'BP' 'CP'
```

```
>>> d.mean(['type','cost'])
<xarray.DataArray (measure: 2)>
array([440.25, 107.5 ])
Coordinates:
  * measure    (measure) <U6 'normal' 'high'
```

The same calculation is much harder to read in Pandas:

```
>>> df.groupby(level=1).mean().mean(axis=1)
measure
high      107.50
normal    440.25
dtype: float64
```

Groupby The Xarray `groupby` keeps the dimensions well-organized and largely works the same as Pandas `groupby`:

```
>>> d.groupby('type').sum('cost')
<xarray.DataArray (type: 2, measure: 2)>
array([[875, 232],
       [886, 198]])
Coordinates:
  * type       (type) <U1 'A' 'B'
  * measure    (measure) <U6 'normal' 'high'
```

Note that Xarray does do multiple groups that produce MultiIndex, like Pandas does. Xarray does support indexing via Pandas `MultiIndex`, however, but this is a separate issue from aggregations on groups.

Dataset The dataset holds multiple DataArray objects that share the same coordinates. In a rough analogy to Pandas, DataArray is to Dataset as Pandas Series is to DataFrame. Let's start with some random data:

```
>>> price = np.random.rand(2,3,4)
>>> cost = np.random.rand(2,3,4)*10
```

The `Dataset` constructor has more options, but these are the basics. Notice that the `price` and `cost` arrays have the same size and that the names of the dimensions are specified in the tuples. Because these two arrays have the same dimension, they can have the same `coords` as shown below:

```
>>> ds = xr.Dataset({'price': (['region', 'dates', 'store'],
        price),
...                   'cost' : (['region', 'dates', 'store'],
        cost),
...                  },
...             coords = {'region': ['usa','can'],
...                       'dates' :
        pd.date_range("2020-01-01", periods=3),
...                       'store' : ['a','b','c','d'],
...                  })
```

```
>>> ds
<xarray.Dataset>
Dimensions:   (region: 2, dates: 3, store: 4)
Coordinates:
  * region     (region) <U3 'usa' 'can'
  * dates      (dates) datetime64[ns] 2020-01-01 2020-01-02
    2020-01-03
  * store      (store) <U1 'a' 'b' 'c' 'd'
Data variables:
    price      (region, dates, store) float64 0.1461 0.8069 ...
    0.07674 0.6562
    cost       (region, dates, store) float64 1.753 4.63 0.9176
    ... 9.136 6.856
```

The individual "DataArray"'s that are embedded in the `Dataset` can be reference using attributes or keys:

```
>>> ds.price - ds['cost']
<xarray.DataArray (region: 2, dates: 3, store: 4)>
array([[[-1.6072773 , -3.82336401, -0.18647072, -0.21047595],
        [-0.93422365, -1.33608059, -4.23803009, -1.06452868],
        [-0.26095129, -0.95536878, -0.51543561, -4.54364682]],

       [[-8.51404155, -5.9075424 , -1.61412181, -3.40844964],
        [-8.41904718, -6.93693035, -0.87115019, -3.29709936],
        [-3.59088224, -4.84586913, -9.05939489, -6.20019101]]])
Coordinates:
  * region    (region) <U3 'usa' 'can'
  * dates     (dates) datetime64[ns] 2020-01-01 2020-01-02
    2020-01-03
  * store     (store) <U1 'a' 'b' 'c' 'd'
```

As with Pandas dataframes, variables can be dropped using `drop_vars`, which creates a new `Dataset` object.

```
>>> ds.drop_vars('price')
<xarray.Dataset>
Dimensions:   (region: 2, dates: 3, store: 4)
Coordinates:
  * region    (region) <U3 'usa' 'can'
  * dates     (dates) datetime64[ns] 2020-01-01 2020-01-02
    2020-01-03
  * store     (store) <U1 'a' 'b' 'c' 'd'
Data variables:
    cost      (region, dates, store) float64 1.753 4.63 0.9176
    ... 9.136 6.856
```

There is an `assign` function to create new variables in a new `Dataset`:

```
>>> ds.assign(profit = ds.price - ds.cost)
<xarray.Dataset>
Dimensions:   (region: 2, dates: 3, store: 4)
Coordinates:
  * region    (region) <U3 'usa' 'can'
  * dates     (dates) datetime64[ns] 2020-01-01 2020-01-02
    2020-01-03
  * store     (store) <U1 'a' 'b' 'c' 'd'
Data variables:
    price     (region, dates, store) float64 0.1461 0.8069 ...
    0.07674 0.6562
    cost      (region, dates, store) float64 1.753 4.63 0.9176
    ... 9.136 6.856
    profit    (region, dates, store) float64 -1.607 -3.823 ...
    -9.059 -6.2
```

As with Pandas, functions can also be used for `assign`:

```
>>> ds.assign(profit = lambda i: (i.price-i.cost))
<xarray.Dataset>
Dimensions:    (region: 2, dates: 3, store: 4)
Coordinates:
  * region     (region) <U3 'usa' 'can'
  * dates      (dates) datetime64[ns] 2020-01-01 2020-01-02
    2020-01-03
  * store      (store) <U1 'a' 'b' 'c' 'd'
Data variables:
    price      (region, dates, store) float64 0.1461 0.8069 ...
      0.07674 0.6562
    cost       (region, dates, store) float64 1.753 4.63 0.9176
      ... 9.136 6.856
    profit     (region, dates, store) float64 -1.607 -3.823 ...
      -9.059 -6.2
```

```
>>> ds.drop_dims('region') # dimensions can be droppedwake up
<xarray.Dataset>
Dimensions:    (dates: 3, store: 4)
Coordinates:
  * dates      (dates) datetime64[ns] 2020-01-01 2020-01-02
    2020-01-03
  * store      (store) <U1 'a' 'b' 'c' 'd'
Data variables:
    *empty*
```

For `Dataset`, `groupby` may not work as you expect unless you supply the `...` term, which indicates that you want to sum over all of the remaining dimensions.

```
>>> ds.groupby('region').sum(...)
<xarray.Dataset>
Dimensions:    (region: 2)
Coordinates:
  * region     (region) object 'can' 'usa'
Data variables:
    price      (region) float64 6.266 4.682
    cost       (region) float64 68.93 24.36
```

If you want to sum over the `dates` variable, then you can specify it in the aggregation `sum`:

```
>>> ds.groupby('region').sum('dates')
<xarray.Dataset>
Dimensions:    (region: 2, store: 4)
Coordinates:
  * region     (region) <U3 'usa' 'can'
  * store      (store) <U1 'a' 'b' 'c' 'd'
Data variables:
    price      (region, store) float64 1.015 1.345 0.9973 ...
      2.032 1.459 1.335
    cost       (region, store) float64 3.818 7.46 5.937 7.143
      ... 19.72 13.0 14.24
```

Xarray is extremely powerful and the right solution for dealing with high dimensional data cubes. We have just scratched the surface here, and Xarray can do much more with data alignment and database like operations (e.g., Pandas `merge`) via the `align` method. Much more excellent documentation is available on the main Xarray website.

1.10 Interfacing with Compiled Libraries

As we have discussed, Python for scientific computing really consists of gluing together different scientific libraries written in a compiled language like C or Fortran. Ultimately, you may want to use libraries not available with existing Python bindings. There are many, many options for doing this. The most direct way is to use the built-in `ctypes` module, which provides tools for providing input/output pointers to the library's functions just as if you were calling them from a compiled language. This means that you have to know the function signatures in the library *exactly*—how many bytes for each input and how many bytes for the output. You are responsible for building the inputs exactly the way the library expects and collecting the resulting outputs. Even though this seems tedious, Python bindings for vast libraries have been built this way.

If you want an easier way, then `SWIG` is an automatic wrapper generating tool that can provide bindings to a long list of languages, not just Python; therefore, if you need bindings for multiple languages, then this is your best and only option. Using `SWIG` consists of writing an interface file so that the compiled Python dynamically linked library (`.pyd` files) can be readily imported into the Python interpreter. Huge and complex libraries like Trilinos (Sandia National Labs) have been interfaced to Python using `SWIG`, so it is a well-tested option. `SWIG` also supports Numpy arrays.

However, the `SWIG` model assumes that you want to continue developing primarily in C/Fortran and you are hooking into Python for usability or other reasons. On the other hand, if you start developing algorithms in Python and then want to speed them up, then `Cython` is an excellent option, because it provides a mixed language that allows you to have both C-language and Python code intermixed. Like `SWIG`, you have to write additional files in this hybrid Python/C dialect to have `Cython` generate the C-code that you will ultimately compile. The best part of `Cython` is the profiler that can generate an HTML report showing where the code is slow and could benefit from translation to `Cython`. The Jupyter Notebook integrates nicely with `Cython` via its `%cython` magic command. This means you can write `Cython` code in a cell in Jupyter Notebook, and the notebook will handle all of the tedious details, like setting up the intermediate files to actually compile the `Cython` extension. `Cython` also supports Numpy arrays.

`Cython` and `SWIG` are just two of the ways to create Python bindings for your favorite compiled libraries. Other notable (but less popular) options include `FWrap`, `f2py`, `CFFI`, and `weave`. It is also possible to use Python's own API directly, but this is a tedious undertaking that is hard to justify, given the existence of so many well-developed alternatives.

1.11   Integrated Development Environments

For those who prefer integrated development environments (IDEs), there is a lot
to choose from. The most comprehensive is Enthought Canopy, which includes
a rich, syntax-highlighted editor, integrated help, debugger, and even integrated
training. If you are already familiar with Eclipse from other projects or do mixed-
language programming, then there is a Python plug-in called `PyDev` that contains
all usual features from Eclipse with a Python debugger. Wingware provides an
affordable professional-level IDE with multi-project management support and
unusually clairvoyant code completion that works even in debug mode. Another
favorite is PyCharm, which also supports multiple languages and is particularly
popular among Python web developers, because it provides powerful templates for
popular web frameworks like Django. Visual Studio Code has quickly developed a
strong following among Python newcomers because of its beautiful interface and
plug-in ecosystem. If you are a VIM user, then the Jedi plug-in provides excellent
code completion that works well with `pylint`, which provides static code analysis
(i.e., identifies missing modules and typos). Naturally, `emacs` has many related
plugins for developing in Python. Note that are many other options, but I have tried
to emphasize those most suitable for Python beginners.

1.12   Quick Guide to Performance and Parallel Programming

There are many options available to improve the performance of your Python codes.
The first thing to determine is what is limiting your computation. It could be
CPU speed (unlikely), memory limitations (out-of-core computing), or data transfer
speed (waiting on data to arrive for processing). If your code is pure-Python, then
you can try running it with `Pypy`, which is an alternative Python implementation
that employs a just-in-time compiler. If your code does not experience a massive
speedup with `Pypy`, then there is probably something external to the code that is
slowing it down (e.g., disk access or network access). If `Pypy` doesn't make any
sense because you are using many compiled modules that `Pypy` does not support,
then there are many diagnostic tools available.

Python has its own built-in profiler `cProfile` you can invoke from the
command line as in the following:

```
Terminal> python -m cProfile -o program.prof my_program.py
```

The output of the profiler is saved to the `program.prof` file. This file can be
visualized in `runsnakerun` to get a nice graphical picture of where the code
is spending the most time. The task manager on your operating system can also
provide clues as your program runs to see how it is consuming resources. The
`line_profiler` by Robert Kern provides an excellent way to see how the
code is spending its time by annotating each line of the code by its timings. In

combination with `runsnakerun`, this narrows down problems to the line level from the function level.

The most common situation is that your program is waiting on data from disk or from some busy network resource. This is a common situation in web programming, and there are lots of well-established tools to deal with this. Python has a `multiprocessing` module that is part of the standard library. This makes it easy to spawn child worker processes that can break off and individually process small parts of a big job. However, it is still your responsibility as the programmer to figure out how to distribute the data for your algorithm. Using this module means that the individual processes are to be managed by the operating system, which will be in charge of balancing the load.

The basic template for using `multiprocessing` is the following:

```python
# filename multiprocessing_demo.py
import multiprocessing
import time
def worker(k):
    'worker function'
    print('am starting process %d' % (k))
    time.sleep(10) # wait ten seconds
    print('am done waiting!')
    return


if __name__ == '__main__':
    for i in range(10):
        p = multiprocessing.Process(target=worker, args=(i,))
        p.start()
```

Then, you run this program at the terminal as in the following:

```
Terminal> python multiprocessing_demo.py
```

It is crucially important that you run the program from the terminal this way. It is not possible to do this interactively from within Jupyter, say. If you look at the process manager on the operating system, you should see a number of new Python processes loitering for 10 seconds. You should also see the output of the `print` statements above. Naturally, in a real application, you would be assigning some meaningful work for each of the workers and figuring out how to send partially finished pieces between individual workers. Doing this is complex and easy to get wrong, so Python 3 has the helpful `concurrent.futures`.

```python
# filename: concurrent_demo.py
from concurrent import futures
import time

def worker(k):
    'worker function'
    print ('am starting process %d' % (k))
    time.sleep(10) # wait ten seconds
    print ('am done waiting!')
    return
```

```
def main():
    with futures.ProcessPoolExecutor(max_workers=3) as executor:
        list(executor.map(worker,range(10)))

if __name__ == '__main__':
    main()
```

```
Terminal> python concurrent_demo.py
```

You should see something like the following in the terminal. Note that we explicitly restricted the number of processes to three.

```
am starting process 0
am starting process 1
am starting process 2
am done waiting!
am done waiting!
...
```

The `futures` module is built on top of `multiprocessing` and makes it easier to use for this kind of simple task. There are also versions of both that use threads instead of processes while maintaining the same usage pattern. The main difference between threads and processes is that processes have their own compartmentalized resources. The C-language Python (i.e., CPython) implementation uses a Global Interpreter Lock (GIL) that prevents threads from fighting over internal data structures. This is a course-grained locking mechanism, where one thread may individually run faster, because it does not have to keep track of all the bookkeeping involved in running multiple threads simultaneously. The downside is that you cannot run multiple threads simultaneously to speed up certain tasks.

There is no corresponding contention problem with processes, but these are somewhat slower to start up, because each process has to create its own private workspace for data structures that may be transferred between them. However, each process can certainly run independently and simultaneously once all that is set up. Certain alternative implementations of Python like IronPython use a finer-grain threading design rather than a GIL approach. As a final comment, on modern systems with multiple cores, it could be that multiple threads actually slow things down, because the operating system may have to switch threads between different cores. This creates additional overheads in the thread switching mechanism that ultimately slow things down.

Jupyter itself has a parallel programming framework built (`ipyparallel`) that is both powerful and easy to use. The first step is to fire up separate Jupyter engines at the terminal as in the following:

```
Terminal> ipcluster start --n=4
```

Then, in an Jupyter window, you can get the client:

```
In [1]: from ipyparallel import Client
   ...: rc = Client()
```

The client has a connection to each of the processes we started before using `ipcluster`. To use all of the engines, we assign the `DirectView` object from the client as in the following:

```
In [2]: dview = rc[:]
```

Now, we can apply functions for each of the engines. For example, we can get the process identifiers using the `os.getpid` function:

```
In [3]: import os
In [4]: dview.apply_sync(os.getpid)
Out[4]: [6824, 4752, 8836, 3124]
```

Once the engines are up and running, data can be distributed to them using `scatter`:

```
In [5]: dview.scatter('a',range(10))
Out[5]: <AsyncResult: finished>
In [6]: dview.execute('print(a)').display_outputs()
[stdout:0] [0, 1, 2]
[stdout:1] [3, 4, 5]
[stdout:2] [6, 7]
[stdout:3] [8, 9]
```

The `execute` method evaluates the given string in each engine. Now that the data have been sprinkled among the active engines, we can do further computing on them:

```
In [7]: dview.execute('b=sum(a)')
Out[7]: <AsyncResult: finished>
In [8]: dview.execute('print(b)').display_outputs()
[stdout:0] 3
[stdout:1] 12
[stdout:2] 13
[stdout:3] 17
```

In this example, we added up the individual a sub-lists available on each of the engines. We can gather up the individual results into a single list as in the following:

```
In [9]: dview.gather('b').result
Out[9]: [3, 12, 13, 17]
```

This is one of the simplest mechanisms for distributing work to the individual engines and collecting the results. Unlike the other methods we discussed, you can do this iteratively, which makes it easy to experiment with how you want to distribute and compute with the data. The Jupyter documentation has many more examples of parallel programming styles that include running the engines on cloud resources, supercomputer clusters, and across disparate networked computing resources. Although there are many other specialized parallel programming packages, Jupyter provides the best trade-off for generality against complexity across all of the major platforms.

1.13 Other Resources

The Python community is filled with supersmart and amazingly helpful people. One of the best places to get help with Scientific Python is the www.stackoverflow.com site, which hosts a competitive Q&A forum that is particularly welcoming for Python newbies. Several of the key Python developers regularly participate there, and the quality of the answers is very high. The mailing lists for any of the key tools (e.g., Numpy, Jupyter, Matplotlib) are also great for keeping up with the newest developments. Anything written by Hans Petter Langtangen [23] is excellent, especially if you have a physics background. The Scientific Python conference held annually in Austin is also a great place to see your favorite developers in person, ask questions, and participate in the many interesting subgroups organized around niche topics. The PyData workshop is a semiannual meeting focused on Python for large-scale data-intensive processing. Also, [44] may be helpful if you liked the level of presentation in this chapter.

Chapter 2
Probability

2.1 Introduction

This chapter takes a geometric view of probability theory and relates it to familiar concepts in linear algebra and geometry. This approach connects your natural geometric intuition to the key abstractions in probability that can help guide your reasoning. This is particularly important in probability, because it is easy to be misled. We need a bit of rigor and some intuition to guide us.

In grade school, you were introduced to the natural numbers (i.e., $1, 2, 3, ..$), and you learned how to manipulate them by operations like addition, subtraction, and multiplication. Later, you were introduced to positive and negative numbers and were again taught how to manipulate them. Ultimately, you were introduced to the calculus of the real line and learned how to differentiate, take limits, and so on. This progression not only provided more abstractions but also widened the field of problems you could successfully tackle. The same is true of probability. One way to think about probability is as a new number concept that allows you to tackle problems that have a special kind of *uncertainty* built into them. Thus, the key idea is that there is some number, say x, with a traveling companion, say, $f(x)$, and this companion represents the uncertainties about the value of x, as if looking at the number x through a frosted window. The degree of opacity of the window is represented by $f(x)$. If we want to manipulate x, then we have to figure out what to do with $f(x)$. For example, if we want $y = 2x$, then we have to understand how $f(x)$ generates $f(y)$.

Where is the *random* part? To conceptualize this, we need still another analogy: think about a beehive with the swarm around it representing $f(x)$ and the hive itself, which you can barely see through the swarm, as x. The random piece is you don't know *which* bee in particular is going to sting you! Once this happens, the uncertainty evaporates. Up until that happens, all we have is a concept of a swarm (i.e., density of bees), which represents a *potentiality* of which bee will ultimately sting. In summary, one way to think about probability is as a way of carrying through

J. Unpingco, *Python for Probability, Statistics, and Machine Learning*,
https://doi.org/10.1007/978-3-031-04648-3_2

mathematical reasoning (e.g., adding, subtracting, taking limits) with a notion of potentiality that is so-transformed by these operations.

2.1.1 Understanding Probability Density

In order to understand the heart of modern probability, which is built on the Lesbesgue theory of integration, we need to extend the concept of integration from basic calculus. To begin, let us consider the following piecewise function:

$$f(x) = \begin{cases} \frac{1}{2} & \text{if } 0 < x \le \frac{1}{4} \\ \frac{7}{6} & \text{if } \frac{1}{4} < x \le 1 \\ 0 & \text{otherwise} \end{cases}$$

as shown in Fig. 2.1. In calculus, you learned Riemann integration, which you can apply here as

$$\int_0^1 f(x)dx = \frac{1}{8} + \frac{21}{24} = 1$$

which has the usual interpretation as the area of the two rectangles that make up $f(x)$. So far, so good.

With Lesbesgue integration, the idea is very similar except that we focus on the y-axis instead of moving along the x-axis. The question is given $f(x) = \frac{1}{2}$; what is the set of x values for which this is true? For our example, this is true whenever $x \in (0, \frac{1}{4}]$. So now, we have a correspondence between the values of the

Fig. 2.1 Simple piecewise-constant function

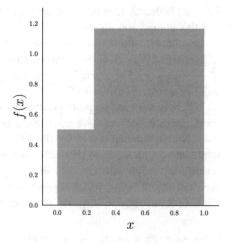

function (namely, $1/2$ and $7/6$) and the sets of x values for which this is true, namely, $\{(0, \frac{1}{4}]\}$ and $\{(\frac{1}{4}, 1]\}$, respectively. To compute the integral, we simply take the function values (i.e., $1/2, 7/6$) and some way of measuring the size of the corresponding interval (i.e., μ) as in the following:

$$\int_0^1 f d\mu = \frac{1}{2}\mu(\{(0, \frac{1}{4}]\}) + \frac{7}{6}\mu(\{(\frac{1}{4}, 1]\})$$

We have suppressed some of the notation above to emphasize generality. Note that we obtain the same value of the integral as in the Riemann case when $\mu((0, \frac{1}{4}]) = \frac{1}{4}$ and $\mu((\frac{1}{4}, 1]) = \frac{3}{4}$. By introducing the μ function as a way of measuring the intervals above, we have introduced another degree of freedom in our integration. This accommodates many weird functions that are not tractable using the usual Riemann theory, but we refer you to a proper introduction to Lesbesgue integration for further study [19]. Nonetheless, the key step in the above discussion is the introduction of the μ function, which we will encounter again as the so-called probability density function.

2.1.2 Random Variables

Most introductions to probability jump straight into *random variables* and then explain how to compute complicated integrals. The problem with this approach is that it skips over some of the important subtleties that we will now consider. Unfortunately, the term *random variable* is not very descriptive. A better term is *measurable function*. To understand why this is a better term, we have to dive into the formal constructions of probability by way of a simple example.

Consider tossing a fair six-sided die. There are only six outcomes possible,

$$\Omega = \{1, 2, 3, 4, 5, 6\}$$

As we know, if the die is fair, then the probability of each outcome is $1/6$. To say this formally, the measure of each set (i.e., $\{1\}, \{2\}, \ldots, \{6\}$) is $\mu(\{1\}) = \mu(\{2\}) \ldots = \mu(\{6\}) = 1/6$. In this case, the μ function we discussed earlier is the usual *probability* mass function, denoted by $\mathbb{P}$. The measurable function maps a set into a number on the real line. For example, $\{1\} \mapsto 1$ is one such function.

Now, here's where things get interesting. Suppose you were asked to construct a fair coin from the fair die. In other words, we want to throw the die and then record the outcomes as if we had just tossed a fair coin. How could we do this? One way would be to define a measurable function that says if the die comes up 3 or less, then we declare *heads* and otherwise declare *tails*. This has some strong intuition behind it, but let us articulate it in terms of formal theory. This strategy creates two different nonoverlapping sets $\{1, 2, 3\}$ and $\{4, 5, 6\}$. Each set has the same probability *measure*,

$$\mathbb{P}(\{1, 2, 3\}) = 1/2$$

$$\mathbb{P}(\{4, 5, 6\}) = 1/2$$

And the problem is solved. Every time the die comes up $\{1, 2, 3\}$, we record heads and record tails otherwise.

Is this the only way to construct a fair coin experiment from a fair die? Alternatively, we can define the sets as $\{1\}$, $\{2\}$, $\{3, 4, 5, 6\}$. If we define the corresponding measure for each set as the following:

$$\mathbb{P}(\{1\}) = 1/2$$

$$\mathbb{P}(\{2\}) = 1/2$$

$$\mathbb{P}(\{3, 4, 5, 6\}) = 0$$

then, we have another solution to the fair coin problem. To implement this, all we do is ignore every time the die shows 3 , 4 , 5 , 6 and throw again. This is wasteful, but it solves the problem. Nonetheless, we hope you can see how the interlocking pieces of the theory provide a framework for carrying the notion of uncertainty/potentiality from one problem to the next (e.g., from the fair die to the fair coin).

Let us consider a slightly more interesting problem where we toss two dice. We assume that each throw is *independent*, meaning that the outcome of one does not influence the other. What are the sets in this case? They are all pairs of possible outcomes from two throws as shown below,

$$\Omega = \{(1, 1), (1, 2), \ldots, (5, 6), (6, 6)\}$$

What are the measures of each of these sets? By virtue of the independence claim, the measure of each is the product of the respective measures of each element. For instance,

$$\mathbb{P}((1, 2)) = \mathbb{P}(\{1\})\mathbb{P}(\{2\}) = \frac{1}{6^2}$$

With all that established, we can ask the following question: what is the probability that the sum of the dice equals seven? As before, the first thing to do is characterize the measurable function for this as $X : (a, b) \mapsto (a + b)$. Next, we associate all of the (a, b) pairs with their sum. We can create a Python dictionary for this as shown,

```python
>>> d={ (i,j) :i+j for i in range(1,7) for j in range(1,7)}
```

The next step is to collect all of the (a, b) pairs that sum to each of the possible values from 2 to 12.

```python
>>> from collections import defaultdict
>>> dinv = defaultdict(list)
>>> for i,j in d.items():
...         dinv[j].append(i)
...
```

For example, `dinv[7]` contains the following list of pairs that sum to seven:

```
>>> dinv[7]
[(1, 6), (2, 5), (3, 4), (4, 3), (5, 2), (6, 1)]
```

The next step is to compute the probability measured for each of these items. Using the independence assumption, this means we have to compute the sum of the products of the individual item probabilities in `dinv`. Because we know that each outcome is equally likely, the probability of every term in the sum equals $1/36$. Thus, all we have to do is count the number of items in the corresponding list for each key in `dinv` and divide by `36`. For example, `dinv[11]` contains `[(5, 6), (6, 5)]`. The probability of $5+6=6+5=11$ is the probability of this set, which is composed of the sum of the probabilities of the individual elements $\{(5,6),(6,5)\}$. In this case, we have $\mathbb{P}(11) = \mathbb{P}(\{(5,6)\}) + \mathbb{P}(\{(6,5)\}) = 1/36 + 1/36 = 2/36$. Repeating this procedure for all the elements, we derive the probability mass function as shown below:

```
>>> X={i:len(j)/36. for i,j in dinv.items()}
>>> print(X)
{2: 0.027777777777777776, 3: 0.05555555555555555,
 4: 0.08333333333333333, 5: 0.1111111111111111,
 6: 0.1388888888888889, 7: 0.16666666666666666,
 8: 0.1388888888888889, 9: 0.1111111111111111,
 10: 0.08333333333333333, 11: 0.05555555555555555,
 12: 0.027777777777777776}
```

The above example exposes the elements of probability theory that are in play for this simple problem while deliberately suppressing some of the gory technical details. With this framework, we can ask other questions like the following: what is the probability that half the product of three dice will exceed their sum? We can solve this using the same method as in the following. First, let us create the first mapping:

```
>>> d={(i,j,k):((i*j*k)/2>i+j+k) for i in range(1,7)
...                               for j in range(1,7)
...                               for k in range(1,7)}
```

The keys of this dictionary are the triples, and the values are the logical values of whether or not half the product of three dice exceeds their sum. Now, we do the inverse mapping to collect the corresponding lists:

```
>>> dinv = defaultdict(list)
>>> for i,j in d.items():
...     dinv[j].append(i)
...
```

Note that `dinv` contains only two keys, `True` and `False`. Again, because the dice are independent, the probability of any triple is $1/6^3$. Finally, we collect this for each outcome as in the following:

```
>>> X={i:len(j)/6.0**3 for i,j in dinv.items()}
>>> print(X)
{False: 0.37037037037037035, True: 0.6296296296296297}
```

Thus, the probability of half the product of three dice exceeding their sum is `136/(6.0**3) = 0.63`. The set that is induced by the random variable has only two elements in it, `True` and `False`, with $\mathbb{P}(\text{True}) = 136/216$ and $\mathbb{P}(\text{False}) = 1 - 136/216$.

As a final example to exercise another layer of generality, let is consider the first problem with the two dice, where we want the probability of a seven, but this time one of the dice is no longer fair. The distribution for the unfair die is the following:

$$\mathbb{P}(\{1\}) = \mathbb{P}(\{2\}) = \mathbb{P}(\{3\}) = \frac{1}{9}$$

$$\mathbb{P}(\{4\}) = \mathbb{P}(\{5\}) = \mathbb{P}(\{6\}) = \frac{2}{9}$$

From our earlier work, we know the elements corresponding to the sum of seven are the following:

$$\{(1, 6), (2, 5), (3, 4), (4, 3), (5, 2), (6, 1)\}$$

Because we still have the independence assumption, all we need to change is the probability computation of each of elements. For example, given that the first die is the unfair one, we have

$$\mathbb{P}((1, 6)) = \mathbb{P}(1)\mathbb{P}(6) = \frac{1}{9} \times \frac{1}{6}$$

and likewise for (2, 5) we have the following:

$$\mathbb{P}((2, 5)) = \mathbb{P}(2)\mathbb{P}(5) = \frac{1}{9} \times \frac{1}{6}$$

and so forth. Summing all of these gives the following:

$$\mathbb{P}_X(7) = \frac{1}{9} \times \frac{1}{6} + \frac{1}{9} \times \frac{1}{6} + \frac{1}{9} \times \frac{1}{6} + \frac{2}{9} \times \frac{1}{6} + \frac{2}{9} \times \frac{1}{6} + \frac{2}{9} \times \frac{1}{6} = \frac{1}{6}$$

Let us try computing this using Pandas instead of Python dictionaries. First, we construct a `DataFrame` object with an index of tuples consisting of all pairs of possible dice outcomes.

```
>>> from pandas import DataFrame
>>> d=DataFrame(index=[(i,j) for i in range(1,7) for j in
    range(1,7)],
...             columns=['sm','d1','d2','pd1','pd2','p'])
```

Now, we can populate the columns that we set up above, where the outcome of the first die is the `d1` column and the outcome of the second die is `d2`:

```
>>> d.d1=[i[0] for i in d.index]
>>> d.d2=[i[1] for i in d.index]
```

Next, we compute the sum of the dices in the `sm` column:

```
>>> d.sm=list(map(sum,d.index))
```

With that established, the DataFrame now looks like the following:

```
>>> d.head(5) # show first five lines
        sm  d1  d2   pd1   pd2    p
(1, 1)   2   1   1   NaN   NaN  NaN
(1, 2)   3   1   2   NaN   NaN  NaN
(1, 3)   4   1   3   NaN   NaN  NaN
(1, 4)   5   1   4   NaN   NaN  NaN
(1, 5)   6   1   5   NaN   NaN  NaN
```

Next, we fill out the probabilities for each face of the unfair die (`d1`) and the fair die (`d2`):

```
>>> d.loc[d.d1<=3,'pd1']=1/9.
>>> d.loc[d.d1 > 3,'pd1']=2/9.
>>> d.pd2=1/6.
>>> d.head(10)
        sm  d1  d2       pd1       pd2    p
(1, 1)   2   1   1  0.111111  0.166667  NaN
(1, 2)   3   1   2  0.111111  0.166667  NaN
(1, 3)   4   1   3  0.111111  0.166667  NaN
(1, 4)   5   1   4  0.111111  0.166667  NaN
(1, 5)   6   1   5  0.111111  0.166667  NaN
(1, 6)   7   1   6  0.111111  0.166667  NaN
(2, 1)   3   2   1  0.111111  0.166667  NaN
(2, 2)   4   2   2  0.111111  0.166667  NaN
(2, 3)   5   2   3  0.111111  0.166667  NaN
(2, 4)   6   2   4  0.111111  0.166667  NaN
```

Finally, we can compute the joint probabilities for the sum of the shown faces as the following:

```
>>> d.p = d.pd1 * d.pd2
>>> d.head(5)
         sm  d1  d2       pd1       pd2         p
(1, 1)    2   1   1  0.111111  0.166667  0.018519
(1, 2)    3   1   2  0.111111  0.166667  0.018519
(1, 3)    4   1   3  0.111111  0.166667  0.018519
(1, 4)    5   1   4  0.111111  0.166667  0.018519
(1, 5)    6   1   5  0.111111  0.166667  0.018519
```

With all that established, we can compute the density of all the dice outcomes by using groupby as in the following:

```
>>> d.groupby('sm')['p'].sum()
sm
2         0.018519
3         0.037037
4         0.055556
5         0.092593
6          0.12963
7         0.166667
8         0.148148
9          0.12963
10        0.111111
11        0.074074
12        0.037037
Name: p, dtype: object
```

These examples have shown how the theory of probability breaks down sets and measurements of those sets and how these can be combined to develop the probability mass functions for new random variables.

2.1.3   Continuous Random Variables

The same ideas work with continuous variables, but managing the sets becomes trickier, because the real line, unlike discrete sets, has many limiting properties already built into it that have to be handled carefully. Nonetheless, let us start with an example that should illustrate the analogous ideas. Suppose a random variable X is uniformly distributed on the unit interval. What is the probability that the variable takes on values less than 1/2?

In order to build intuition onto the discrete case, let us go back to our dice-throwing experiment with the fair dice. The sum of the values of the dice is a measurable function:

$$Y: \{1, 2, \ldots, 6\}^2 \mapsto \{2, 3, \ldots, 12\}$$

That is, Y is a mapping of the cartesian product of sets to a discrete set of outcomes. In order to compute probabilities of the set of outcomes, we need to derive the probability measure for Y, $\mathbb{P}_Y$, from the corresponding probability measures for each die. Our previous discussion went through the mechanics of that. This means that

$$\mathbb{P}_Y : \{2, 3, \ldots, 12\} \mapsto [0, 1]$$

Note there is a separation between the function definition and where the target items of the function are measured in probability. More bluntly,

$$Y : A \mapsto B$$

with

$$\mathbb{P}_Y : B \mapsto [0, 1]$$

Thus, to compute $\mathbb{P}_Y$, which is derived from other random variables, we have to express the equivalence classes in B in terms of their progenitor A sets.

The situation for continuous variables follows the same pattern, but with many more deep technicalities that we are going to skip. For the continuous case, the random variable is now

$$X : \mathbb{R} \mapsto \mathbb{R}$$

with corresponding probability measure

$$\mathbb{P}_X : \mathbb{R} \mapsto [0, 1]$$

But where are the corresponding sets here? Technically, these are the *Borel* sets, but we can just think of them as intervals. Returning to our question, what is the probability that a uniformly distributed random variable on the unit interval takes values less than $1/2$? Rephrasing this question according to the framework, we have the following:

$$X : [0, 1] \mapsto [0, 1]$$

with corresponding

$$\mathbb{P}_X : [0, 1] \mapsto [0, 1]$$

To answer the question, by the definition of the uniform random variable on the unit interval, we compute the following integral:

$$\mathbb{P}_X([0, 1/2]) = \mathbb{P}_X(0 < X < 1/2) = \int_0^{1/2} dx = 1/2$$

where the above integral's dx sweeps through intervals of the B-type. The measure of any dx interval (i.e., A-type set) is equal to dx, by definition of the uniform random variable. To get all the moving parts into one notationally rich integral, we can also write this as

$$\mathbb{P}_X(0 < X < 1/2) = \int_0^{1/2} d\mathbb{P}_X(dx) = 1/2$$

Now, let us consider a slightly more complicated and interesting example. As before, suppose we have a uniform random variable, X, and let us introduce another random variable defined

$$Y = 2X$$

Now, what is the probability that $0 < Y < \frac{1}{2}$? To express this in our framework, we write:

$$Y : [0, 1] \mapsto [0, 2]$$

with corresponding

$$\mathbb{P}_Y : [0, 2] \mapsto [0, 1]$$

To answer the question, we need to measure the set $[0, 1/2]$, with the probability measure for Y, $\mathbb{P}_Y([0, 1/2])$. How can we do this? Because Y is derived from the X random variable, as with the fair-dice throwing experiment, we have to create a set of equivalences in the target space (i.e., B-type sets) that reflect back on the input space (i.e., A-type sets). That is, what is the interval $[0, 1/2]$ equivalent to in terms of the X random variable? Because, functionally, $Y = 2X$, then the B-type interval $[0, 1/2]$ corresponds to the A-type interval $[0, 1/4]$. From the probability measure of X, we compute this with the integral

$$\mathbb{P}_Y([0, 1/2]) = \mathbb{P}_X([0, 1/4]) = \int_0^{1/4} dx = 1/4$$

Now, let us up the ante and consider the following random variable:

$$Y = X^2$$

where now X is still uniformly distributed but now over the interval $[-1/2, 1/2]$. We can express this in our framework as

$$Y : [-1/2, 1/2] \mapsto [0, 1/4]$$

with corresponding

$$\mathbb{P}_Y : [0, 1/4] \mapsto [0, 1]$$

What is the $\mathbb{P}_Y(Y < 1/8)$? In other words, what is the measure of the set $B_Y = [0, 1/8]$? As before, because X is derived from our uniformly distributed random variable, we have to reflect the B_Y set onto sets of the A-type. The thing to recognize is that because X^2 is symmetric about zero, all B_Y sets reflect back into two sets. This means that for any set B_Y, we have the correspondence $B_Y = A_X^+ \cup A_X^-$. So, we have

$$B_Y = \left\{ 0 < Y < \frac{1}{8} \right\} = \left\{ 0 < X < \frac{1}{\sqrt{8}} \right\} \bigcup \left\{ -\frac{1}{\sqrt{8}} < X < 0 \right\}$$

From this perspective, we have the following solution:

$$\mathbb{P}_Y(B_Y) = \mathbb{P}(A_X^+) + \mathbb{P}(A_X^-)$$

Also,

$$A_X^+ = \left\{ 0 < X < \frac{1}{\sqrt{8}} \right\}$$

$$A_X^- = \left\{ -\frac{1}{\sqrt{8}} < X < 0 \right\}$$

Therefore,

$$\mathbb{P}_Y(B_Y) = \frac{1}{\sqrt{8}} + \frac{1}{\sqrt{8}}$$

because $\mathbb{P}(A_X^+) = \mathbb{P}(A_X^-) = 1/\sqrt{8}$. Let us see if this comes out using the usual transformation of variables method from calculus. Using this method, the density $f_Y(y) = \frac{1}{\sqrt{y}}$. Then, we obtain

$$\int_0^{\frac{1}{8}} \frac{1}{\sqrt{y}} dy = \frac{1}{\sqrt{2}}$$

which is what we got using the sets method. Note that you would favor the calculus method in practice, but it is important to understand the deeper mechanics, because sometimes the usual calculus method fails, as the next problem shows.

2.1.4 Transformation of Variables Beyond Calculus

Suppose X and Y are uniformly distributed in the unit interval, and we define Z as

$$Z = \frac{X}{Y - X}$$

What is the $f_Z(z)$? If you try this using the usual calculus method, you will fail (try it!). The problem is one of the technical prerequisites for the calculus method is not in force.

The key observation is that $Z \notin (-1, 0]$. If this were possible, the X and Y would have different signs, which cannot happen, given that X and Y are uniformly distributed over $(0, 1]$. Now, let us consider when $Z > 0$. In this case, $Y > X$ because Z cannot be positive otherwise. For the density function, we are interested in the set $\{0 < Z < z\}$. We want to compute

$$\mathbb{P}(Z < z) = \int \int B_1 dX dY$$

with

$$B_1 = \{0 < Z < z\}$$

Now, we have to translate that interval into an interval relevant to X and Y. For $0 < Z$, we have $Y > X$. For $Z < z$, we have $Y > X(1/z + 1)$. Putting this together gives

$$A_1 = \{\max(X, X(1/z + 1)) < Y < 1\}$$

Integrating this over Y as follows:

$$\int_0^1 \{\max(X, X(1/z + 1)) < Y < 1\} dY = \frac{z - X - Xz}{z} \text{ where } z > \frac{X}{1 - X}$$

and integrating this one more time over X gives

$$\int_0^{\frac{z}{1+z}} \frac{-X + z - Xz}{z} dX = \frac{z}{2(z + 1)} \text{ where } z > 0$$

Note that this is the computation for the *probability* itself, not the probability density function. To get that, all we have to do is differentiate the last expression to obtain

$$f_Z(z) = \frac{1}{(z + 1)^2} \text{ where } z > 0$$

Now, we need to compute this density using the same process for when $z < -1$. We want the interval $Z < z$ for when $z < -1$. For a fixed z, this is equivalent to $X(1 + 1/z) < Y$. Because z is negative, this also means that $Y < X$. Under these terms, we have the following integral:

$$\int_0^1 \{X(1/z + 1) < Y < X\} dY = -\frac{X}{z} \text{ where } z < -1$$

and integrating this one more time over X gives the following:

$$-\frac{1}{2z} \text{ where } z < -1$$

To get the density for $z < -1$, we differentiate this with respect to z to obtain the following:

$$f_Z(z) = \frac{1}{2z^2} \text{ where } z < -1$$

Putting this all together, we obtain

$$f_Z(z) = \begin{cases} \frac{1}{(z+1)^2} & \text{if } z > 0 \\ \frac{1}{2z^2} & \text{if } z < -1 \\ 0 & \text{otherwise} \end{cases}$$

We will leave it as an exercise to show that this integrates out to one.

2.1.5 Independent Random Variables

Independence is a standard assumption. Mathematically, the necessary and sufficient condition for independence between two random variables X and Y is the following:

$$\mathbb{P}(X, Y) = \mathbb{P}(X)\mathbb{P}(Y)$$

Two random variables X and Y are *uncorrelated* if

$$\mathbb{E}\left((X - \overline{X})(Y - \overline{Y})\right) = 0$$

where $\overline{X} = \mathbb{E}(X)$. Note that uncorrelated random variables are sometimes called *orthogonal* random variables. Uncorrelatedness is a weaker property than independence, however. For example, consider the discrete random variables X and Y uniformly distributed over the set $\{1, 2, 3\}$, where

$$X = \begin{cases} 1 & \text{if } \omega = 1 \\ 0 & \text{if } \omega = 2 \\ -1 & \text{if } \omega = 3 \end{cases}$$

and also

$$Y = \begin{cases} 0 & \text{if } \omega = 1 \\ 1 & \text{if } \omega = 2 \\ 0 & \text{if } \omega = 3 \end{cases}$$

Thus, $\mathbb{E}(X) = 0$ and $\mathbb{E}(XY) = 0$, so X and Y are uncorrelated. However, we have

$$\mathbb{P}(X = 1, Y = 1) = 0 \neq \mathbb{P}(X = 1)\mathbb{P}(Y = 1) = \frac{1}{9}$$

So, these two random variables are *not* independent. Thus, uncorrelatedness does not imply independence, generally, but there is the important case of Gaussian random variables for which it does. To see this, consider the probability density function for two zero-mean, unit-variance Gaussian random variables X and Y:

$$f_{X,Y}(x, y) = \frac{e^{\frac{x^2 - 2\rho xy + y^2}{2(\rho^2 - 1)}}}{2\pi\sqrt{1 - \rho^2}}$$

where $\rho := \mathbb{E}(XY)$ is the correlation coefficient. In the uncorrelated case, where $\rho = 0$, the probability density function factors into the following:

$$f_{X,Y}(x, y) = \frac{e^{-\frac{1}{2}(x^2 + y^2)}}{2\pi} = \frac{e^{-\frac{x^2}{2}}}{\sqrt{2\pi}} \frac{e^{-\frac{y^2}{2}}}{\sqrt{2\pi}} = f_X(x) f_Y(y)$$

which means that X and Y are independent.

Independence and conditional independence are closely related, as in the following:

$$\mathbb{P}(X, Y | Z) = \mathbb{P}(X | Z)\mathbb{P}(Y | Z)$$

which says that X and Y and independent conditioned on Z. Conditioning independent random variables can break their independence. For example, consider two independent Bernoulli-distributed random variables, $X_1, X_2 \in \{0, 1\}$. We define $Z = X_1 + X_2$. Note that $Z \in \{0, 1, 2\}$. In the case where $Z = 1$, we have

$$\mathbb{P}(X_1 | Z = 1) > 0$$
$$\mathbb{P}(X_2 | Z = 1) > 0$$

Even though X_1, X_2 are independent, after conditioning on Z, we have the following:

$$\mathbb{P}(X_1 = 1, X_2 = 1 | Z = 1) = 0 \neq \mathbb{P}(X_1 = 1 | Z = 1)\mathbb{P}(X_2 = 1 | Z = 1)$$

Thus, conditioning on Z breaks the independence of X_1, X_2. This also works in the opposite direction—conditioning can make dependent random variables independent. Define $Z_n = \sum_i^n X_i$ with X_i independent, integer-valued random variables. The Z_n variables are dependent because they stack the same telescoping set of X_i variables. Consider the following:

$$\mathbb{P}(Z_1 = i, Z_3 = j | Z_2 = k) = \frac{\mathbb{P}(Z_1 = i, Z_2 = k, Z_3 = j)}{\mathbb{P}(Z_2 = k)} \tag{2.1}$$

$$= \frac{\mathbb{P}(X_1 = i)\mathbb{P}(X_2 = k - i)\mathbb{P}(X_3 = j - k)}{\mathbb{P}(Z_2 = k)} \tag{2.2}$$

where the factorization comes from the independence of the X_i variables. Using the definition of conditional probability,

$$\mathbb{P}(Z_1 = i | Z_2) = \frac{\mathbb{P}(Z_1 = i, Z_2 = k)}{\mathbb{P}(Z_2 = k)}$$

We can continue to expand Eq. 2.1

$$\mathbb{P}(Z_1 = i, Z_3 = j | Z_2 = k) = \mathbb{P}(Z_1 = i | Z_2) \frac{\mathbb{P}(X_3 = j - k)\mathbb{P}(Z_2 = k)}{\mathbb{P}(Z_2 = k)}$$

$$= \mathbb{P}(Z_1 = i | Z_2)\mathbb{P}(Z_3 = j | Z_2)$$

where $\mathbb{P}(X_3 = j - k)\mathbb{P}(Z_2 = k) = \mathbb{P}(Z_3 = j, Z_2)$. Thus, we see that dependence between random variables can be broken by conditioning to create conditionally independent random variables. As we have just witnessed, understanding how conditioning influences independence is important and is the main topic of study in probabilistic graphical models, a field with many algorithms and concepts to extract these notions of conditional independence from graph-based representations of random variables.

2.1.6   Classic Broken Rod Example

Let us do one last example to exercise fluency in our methods by considering the following classic problem: given a rod of unit-length, broken independently and randomly at two places, what is the probability that you can assemble the three remaining pieces into a triangle? The first task is to find a representation of a triangle as an easy-to-apply constraint. What we want is something like the following:

$$\mathbb{P}(\text{ triangle exists }) = \int_0^1 \int_0^1 \{\text{ triangle exists }\} dX dY$$

where X and Y are independent and uniformly distributed in the unit interval. Heron's formula for the area of the triangle

$$\text{area} = \sqrt{(s-a)(s-b)(s-c)s}$$

where $s = (a+b+c)/2$ is what we need. The idea is that this yields a valid area only when each of the terms under the square root is greater than or equal to zero. Thus, suppose that we have

$$a = X$$

$$b = Y - X$$

$$c = 1 - Y$$

assuming that $Y > X$. Thus, the criterion for a valid triangle boils down to

$$\{(s > a) \wedge (s > b) \wedge (s > c) \wedge (X < Y)\}$$

After a bit of manipulation, this consolidates into the following:

$$\left\{ \frac{1}{2} < Y < 1 \bigwedge \frac{1}{2}(2Y - 1) < X < \frac{1}{2} \right\}$$

which we integrate out by dX first to obtain

$$\mathbb{P}(\text{ triangle exists }) = \int_0^1 \int_0^1 \left\{ \frac{1}{2} < Y < 1 \bigwedge \frac{1}{2}(2Y - 1) < X < \frac{1}{2} \right\} dX dY$$

$$\mathbb{P}(\text{ triangle exists }) = \int_{\frac{1}{2}}^1 (1 - Y) dY$$

and then by dY to obtain finally

$$\mathbb{P}(\text{ triangle exists }) = \frac{1}{8}$$

when $Y > X$. By symmetry, we get the same result for $X > Y$. Thus, the final result is the following:

$$\mathbb{P}(\text{ triangle exists }) = \frac{1}{8} + \frac{1}{8} = \frac{1}{4}$$

We can quickly check using this result using Python for the case $Y > X$ using the following code:

```
>>> import numpy as np
>>> x,y = np.random.rand(2,1000) # uniform rv
>>> a,b,c = x,(y-x),1-y # 3 sides
>>> s = (a+b+c)/2
>>> np.mean((s>a) & (s>b) & (s>c) & (y>x)) # approx 1/8=0.125
0.121
```

Programming Tip

The chained logical & symbols above tell Numpy that the logical operation should be considered element-wise.

2.2 Projection Methods

The concept of projection is key to developing an intuition about conditional probability. We already have a natural intuition of projection from looking at the shadows of objects on a sunny day. As we will see, this simple idea consolidates many abstract ideas in optimization and mathematics. Consider Fig. 2.2, where we want to find a point along the blue line (namely, $\mathbf{x}$) that is closest to the black square (namely, $\mathbf{y}$). In other words, we want to inflate the gray circle until it just touches the black line. Recall that the circle boundary is the set of points for which

$$\sqrt{(\mathbf{y} - \mathbf{x})^T (\mathbf{y} - \mathbf{x})} = \|\mathbf{y} - \mathbf{x}\| = \epsilon$$

for some value of ϵ. So we want a point $\mathbf{x}$ along the line that satisfies this for the smallest ϵ. Then, that point will be the closest point on the black line to the black square. It may be obvious from the diagram, but the closest point on the line occurs, where the line segment from the black square to the black line is perpendicular to the line. At this point, the gray circle just touches the black line. This is illustrated in Fig. 2.3.

Programming Tip

Figure 2.2 uses the matplotlib.patches module. This module contains primitive shapes like circles, ellipses, and rectangles that can be assembled into complex graphics. After importing a particular shape, you can apply that shape to an existing axis using the add_patch method. The patches themselves can by styled using the usual formatting keywords like color and alpha.

Fig. 2.2 Given the point **y** (black square), we want to find the **x** along the line that is closest to it. The gray circle is the locus of points within a fixed distance from **y**

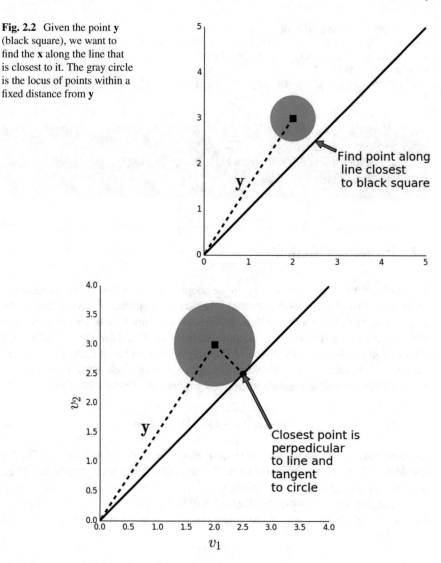

Find point along line closest to black square

Fig. 2.3 The closest point on the line occurs when the line is tangent to the circle. When this happens, the black line and the line (minimum distance) are perpendicular

Now that we can see what's going on, we can construct the solution analytically. We can represent an arbitrary point along the black line as follows:

$$\mathbf{x} = \alpha \mathbf{v}$$

where $\alpha \in \mathbb{R}$ slides the point up and down the line with

$$\mathbf{v} = [1, 1]^T$$

Formally, $\mathbf{v}$ is the *subspace* onto which we want to *project* $\mathbf{y}$. At the closest point, the vector between $\mathbf{y}$ and $\mathbf{x}$ (the *error* vector above) is perpedicular to the line. This means that

$$(\mathbf{y} - \mathbf{x})^T \mathbf{v} = 0$$

and by substituting and working out the terms, we obtain

$$\alpha = \frac{\mathbf{y}^T \mathbf{v}}{\|\mathbf{v}\|^2}$$

The *error* is the distance between $\alpha \mathbf{v}$ and $\mathbf{y}$. This is a right triangle, and we can use the Pythagorean theorem to compute the squared length of this error as

$$\epsilon^2 = \|(\mathbf{y} - \mathbf{x})\|^2 = \|\mathbf{y}\|^2 - \alpha^2 \|\mathbf{v}\|^2 = \|\mathbf{y}\|^2 - \frac{\|\mathbf{y}^T \mathbf{v}\|^2}{\|\mathbf{v}\|^2}$$

where $\|\mathbf{v}\|^2 = \mathbf{v}^T \mathbf{v}$. Note that since $\epsilon^2 \geq 0$, this also shows that

$$\|\mathbf{y}^T \mathbf{v}\| \leq \|\mathbf{y}\| \|\mathbf{v}\|$$

which is the famous and useful Cauchy-Schwarz inequality which we will exploit later. Finally, we can assemble all of this into the *projection* operator

$$\mathbf{P}_v = \frac{1}{\|\mathbf{v}\|^2} \mathbf{v} \mathbf{v}^T$$

With this operator, we can take any $\mathbf{y}$ and find the closest point on $\mathbf{v}$ by doing

$$\mathbf{P}_v \mathbf{y} = \mathbf{v} \left(\frac{\mathbf{v}^T \mathbf{y}}{\|\mathbf{v}\|^2} \right)$$

where we recognize the term in parenthesis as the α we computed earlier. It's called an *operator*, because it takes a vector ($\mathbf{y}$) and produces another vector ($\alpha \mathbf{v}$). Thus, projection unifies geometry and optimization.

2.2.1 Weighted Distance

We can easily extend this projection operator to cases, where the measure of distance between $\mathbf{y}$ and the subspace $\mathbf{v}$ is weighted. We can accommodate these weighted distances by rewriting the projection operator as

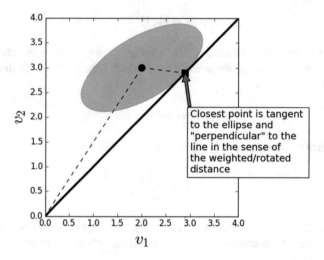

Fig. 2.4 In the weighted case, the closest point on the line is tangent to the ellipse and is still perpendicular in the sense of the weighted distance

$$P_v = v \frac{v^T Q^T}{v^T Q v} \qquad (2.3)$$

where Q is positive definite matrix. In the previous case, we started with a point y and inflated a circle centered at y until it just touched the line defined by v, and this point was closest point on the line to y. The same thing happens in the general case with a weighted distance, except now we inflate an ellipse, not a circle, until the ellipse touches the line.

Note that the error vector $(y - \alpha v)$ in Fig. 2.4 is still perpendicular to the line (subspace v) but in the space of the weighted distance. The difference between the first projection (with the uniform circular distance) and the general case (with the elliptical weighted distance) is the inner product between the two cases. For example, in the first case, we have $y^T v$, and in the weighted case, we have $y^T Q^T v$. To move from the uniform circular case to the weighted ellipsoidal case, all we had to do was change all of the vector inner products. Before we finish, we need a formal property of projections:

$$P_v P_v = P_v$$

known as the *idempotent* property, which basically says that once we have projected onto a subspace, subsequent projections leave us in the same subspace. You can verify this by computing Eq. 2.2.1.

Thus, projection ties a minimization problem (closest point to a line) to an algebraic concept (inner product). It turns out that these same geometric ideas from linear algebra [42] can be translated to the conditional expectation. How this works is the subject of our next section.

2.3 Conditional Expectation as Projection

Now that we understand projection methods geometrically, we can apply them to conditional probability. This is the *key* concept that ties probability to geometry, optimization, and linear algebra.

Inner Product for Random Variables From our previous work on projection for vectors in $\mathbb{R}^n$, we have a good geometric grasp on how projection is related to minimum mean squared error (MMSE). By one abstract step, we can carry all of our geometric interpretations to the space of random variables. For example, we previously noted that at the point of projection, we had the following orthogonal (i.e., perpendicular vectors) condition,

$$(\mathbf{y} - \mathbf{v}_{opt})^T \mathbf{v} = 0$$

which by noting the inner product slightly more abstractly as $\langle \mathbf{x}, \mathbf{y} \rangle = \mathbf{x}^T \mathbf{y}$, we can express as

$$\langle \mathbf{y} - \mathbf{v}_{opt}, \mathbf{v} \rangle = 0$$

and by defining the inner product for the random variables X and Y as

$$\langle X, Y \rangle = \mathbb{E}(XY)$$

we have the same relationship:

$$\langle X - h_{opt}(Y), Y \rangle = 0$$

which holds not for vectors in $\mathbb{R}^n$ but for random variables X and Y and functions of those random variables. Exactly why this is true is technical, but it turns out that one can build up the *entire theory of probability* this way [33], by using the expectation as an inner product.

Furthermore, by abstracting out the inner product concept, we have connected Minimum mean squared error (MMSE) optimization problems, geometry, and random variables. That's a lot of mileage to get out of an abstraction, and it enables us to shift between these interpretations to address real problems. Soon, we'll do this with some examples, but first we collect the most important result that flows naturally from this abstraction.

Conditional Expectation as Projection   The conditional expectation is the minimum mean squared error (MMSE) solution to the following problem:[1]

$$\min_{h} \int_{\mathbb{R}^2} (x - h(y))^2 f_{X,Y}(x, y) dx dy$$

with the minimizing $h_{opt}(Y)$ as

$$h_{opt}(Y) = \mathbb{E}(X|Y)$$

which is another way of saying that among all possible functions $h(Y)$, the one
that minimizes the MSE is $\mathbb{E}(X|Y)$. From our previous discussion on projection,
we noted that these MMSE solutions can be thought of as projections onto a
subspace that characterizes Y. For example, we previously noted that at the point
of projection, we have perpendicular terms,

$$\langle X - h_{opt}(Y), Y \rangle = 0 \tag{2.4}$$

but since we know that the MMSE solution

$$h_{opt}(Y) = \mathbb{E}(X|Y)$$

we have by direct substitution

$$\mathbb{E}(X - \mathbb{E}(X|Y), Y) = 0 \tag{2.5}$$

That last step seems pretty innocuous, but it ties MMSE to conditional expectation
to the inner project abstraction and, in so doing, reveals the conditional expectation
to be a projection operator for random variables. Before we develop this further,
let us grab some quick dividends. From the previous equation, by linearity of the
expectation, we obtain

$$\mathbb{E}(XY) = \mathbb{E}(Y\mathbb{E}(X|Y))$$

which is the so-called *tower* property of the expectation. Note that we could have
found this by using the formal definition of conditional expectation

$$\mathbb{E}(X|Y) = \int_{\mathbb{R}^2} x \frac{f_{X,Y}(x, y)}{f_Y(y)} dx dy$$

---

[1] See appendix for proof using the Cauchy-Schwarz inequality.

and brute-force direct integration

$$\mathbb{E}(Y\mathbb{E}(X|Y)) = \int_{\mathbb{R}} y \int_{\mathbb{R}} x \frac{f_{X,Y}(x, y)}{f_Y(y)} f_Y(y) dx dy$$

$$= \int_{\mathbb{R}^2} xy f_{X,Y}(x, y) dx dy$$

$$= \mathbb{E}(XY)$$

which is not very geometrically intuitive. This lack of geometric intuition makes it hard to apply these concepts and keep track of these relationships.

We can keep pursuing this analogy and obtain the length of the error term from the orthogonality property of the MMSE solution as

$$\langle X - h_{opt}(Y), X - h_{opt}(Y) \rangle = \langle X, X \rangle - \langle h_{opt}(Y), h_{opt}(Y) \rangle$$

and then by substituting all the notation, we obtain

$$\mathbb{E}(X - \mathbb{E}(X|Y))^2 = \mathbb{E}(X)^2 - \mathbb{E}(\mathbb{E}(X|Y))^2$$

which would be tough to compute by direct integration.

To formally establish that $\mathbb{E}(X|Y)$ *is* in fact *a projection operator*, we need to show idempotency. Recall that idempotency means that once we project something onto a subspace, further projections do nothing. In the space of random variables, $\mathbb{E}(X|\cdot)$ is the idempotent projection as we can show by noting that

$$h_{opt} = \mathbb{E}(X|Y)$$

is purely a function of Y, so that

$$\mathbb{E}(h_{opt}(Y)|Y) = h_{opt}(Y)$$

because Y is fixed, this verifies idempotency. Thus, conditional expectation is the corresponding projection operator for random variables. We can continue to carry over our geometric interpretations of projections for vectors (**v**) into random variables (X). With this important result, let us consider some examples of conditional expectations obtained by using brute force to find the optimal MMSE function h_{opt} as well as by using our new perspective on conditional expectation.

Example Suppose we have a random variable, X, then what constant is closest to X in the sense of the mean squared error (MSE)? In other words, which $c \in \mathbb{R}$ minimizes the following mean squared error:

$$\text{MSE} = \mathbb{E}(X - c)^2$$

we can work this out in many ways. First, use calculus-based optimization,

$$\mathbb{E}(X - c)^2 = \mathbb{E}(c^2 - 2cX + X^2) = c^2 - 2c\mathbb{E}(X) + \mathbb{E}(X^2)$$

and then take the first derivative with respect to c and solve

$$c_{opt} = \mathbb{E}(X)$$

Remember that X may potentially take on many values, but this says that the closest number to X in the MSE sense is $\mathbb{E}(X)$. This is intuitively pleasing. Coming at this same problem using our inner product, from Eq. 2.3, we know that at the point of projection

$$\mathbb{E}((X - c_{opt})1) = 0$$

where the 1 we are projecting onto. By linearity of the expectation, the following gives

$$c_{opt} = \mathbb{E}(X)$$

Using the projection approach, because $\mathbb{E}(X|Y)$ is the projection operator, with $Y = \Omega$ (the entire underlying probability space), we have, using the definition of conditional expectation:

$$\mathbb{E}(X|Y = \Omega) = \mathbb{E}(X)$$

This is because of the subtle fact that a random variable over the entire Ω space can only be a constant. Thus, we just worked the same problem three ways (optimization, orthogonal inner products, projection).

Example Let us consider the following example with probability density $f_{X,Y} = x + y$ where $(x, y) \in [0, 1]^2$ and compute the conditional expectation straight from the definition:

$$\mathbb{E}(X|Y) = \int_0^1 x \frac{f_{X,Y}(x, y)}{f_Y(y)} dx = \int_0^1 x \frac{x + y}{y + 1/2} dx = \frac{3y + 2}{6y + 3}$$

That was pretty easy because the density function was so simple. Now, let us do it the hard way by going directly for the MMSE solution $h(Y)$. Then,

$$\text{MSE} = \min_h \int_0^1 \int_0^1 (x - h(y))^2 f_{X,Y}(x, y) dx dy$$

$$= \min_h \int_0^1 yh^2(y) - yh(y) + \frac{1}{3}y + \frac{1}{2}h^2(y) - \frac{2}{3}h(y) + \frac{1}{4}dy$$

Now, we have to find a function h that is going to minimize this. Solving for a function, as opposed to solving for a number, is generally very, very hard, but because we are integrating over a finite interval, we can use the Euler-Lagrange method from variational calculus to take the derivative of the integrand with respect to the function $h(y)$ and set it to zero. Using Euler-Lagrange methods, we obtain the following result:

$$2yh(y) - y + h(y) - \frac{2}{3} = 0$$

Solving this gives

$$h_{opt}(y) = \frac{3y + 2}{6y + 3}$$

which is what we obtained before. Finally, we can solve this using our inner product in Eq. 2.3 as

$$\mathbb{E}((X - h(Y))Y) = 0$$

Writing this out gives,

$$\int_0^1 \int_0^1 (x - h(y))y(x + y)dxdy = \int_0^1 \frac{1}{6}y(-3(2y + 1)h(y) + 3y + 2)dy = 0$$

and the integrand must be zero,

$$2y + 3y^2 - 3yh(y) - 6y^2h(y) = 0$$

and solving this for $h(y)$ gives the same solution:

$$h_{opt}(y) = \frac{3y + 2}{6y + 3}$$

Thus, doing it by the brute-force integration from the definition, optimization, or inner product gives us the same answer, but, in general, no method is necessarily easiest, because they both involve potentially difficult or impossible integration, optimization, or functional equation solving. The point is that now that we have a deep toolbox, we can pick and choose which tools we want to apply for different problems.

Before we leave this example, let us use Sympy to verify the length of the error function we found earlier for this example:

$$\mathbb{E}(X - \mathbb{E}(X|Y))^2 = \mathbb{E}(X)^2 - \mathbb{E}(\mathbb{E}(X|Y))^2$$

that is based on the Pythagorean theorem. First, we need to compute the marginal densities:

```
>>> from sympy.abc import y,x
>>> from sympy import integrate, simplify
>>> fxy = x + y                      # joint density
>>> fy = integrate(fxy,(x,0,1))  # marginal density
>>> fx = integrate(fxy,(y,0,1))  # marginal density
```

Then, we need to write out the conditional expectation:

```
>>> EXY = (3*y+2)/(6*y+3) # conditional expectation
```

Next, we can compute the left side, $\mathbb{E}(X - \mathbb{E}(X|Y))^2$, as the following:

```
>>> # from the definition
>>> LHS=integrate((x-EXY)**2*fxy,(x,0,1),(y,0,1))
>>> LHS # left-hand-side
1/12 - log(3)/144
```

We can similarly compute the right side, $\mathbb{E}(X)^2 - \mathbb{E}(\mathbb{E}(X|Y))^2$, as the following:

```
>>> # using Pythagorean theorem
>>> RHS=integrate((x)**2*fx,(x,0,1))-integrate((EXY)**2*fy,
    (y,0,1))
>>> RHS # right-hand-side
1/12 - log(3)/144
```

Finally, we can verify that the left and right sides match:

```
>>> print(simplify(LHS-RHS)==0)
True
```

In this section, we have pulled together all the projection and least-squares optimization ideas from the previous sections to connect geometric notions of projection from vectors in $\mathbb{R}^n$ to random variables. This resulted in the remarkable realization that the conditional expectation is in fact a projection operator for random variables. Knowing this allows to approach difficult problems in multiple ways, depending on which way is more intuitive or tractable in a particular situation. Indeed, finding the right problem to solve is the hardest part, so having many ways of looking at the same concepts is crucial.

For much more detailed development, the book by Mikosch [31] has some excellent sections covering much of this material with a similar geometric interpretation. Kobayashi [21] does too. Nelson [33] also has a similar presentation based on hyperreal numbers.

2.3.1   Appendix

We want to prove that the conditional expectation is the minimum mean squared error minimizer of the following:

$$J = \min_{h} \int_{\mathbb{R}^2} |X - h(Y)|^2 f_{X,Y}(x, y) dx dy$$

We can expand this as follows:

$$J = \min_{h} \int_{\mathbb{R}^2} |X|^2 f_{X,Y}(x, y) dx dy + \int_{\mathbb{R}^2} |h(Y)|^2 f_{X,Y}(x, y) dx dy$$

$$- \int_{\mathbb{R}^2} 2Xh(Y) f_{X,Y}(x, y) dx dy$$

To minimize this, we have to maximize the following:

$$A = \max_{h} \int_{\mathbb{R}^2} Xh(Y) f_{X,Y}(x, y) dx dy$$

Breaking up the integral using the definition of conditional expectation

$$A = \max_{h} \int_{\mathbb{R}} \left(\int_{\mathbb{R}} X f_{X|Y}(x|y) dx \right) h(Y) f_Y(y) dy \qquad (2.6)$$

$$= \max_{h} \int_{\mathbb{R}} \mathbb{E}(X|Y) h(Y) f_Y(Y) dy \qquad (2.7)$$

From the properties of the Cauchy-Schwarz inequality, we know that the maximum happens when $h_{opt}(Y) = \mathbb{E}(X|Y)$, so we have found the optimal $h(Y)$ function as

$$h_{opt}(Y) = \mathbb{E}(X|Y)$$

which shows that the optimal function is the conditional expectation.

2.4 Conditional Expectation and Mean Squared Error

In this section, we work through a detailed example using conditional expectation and optimization methods. Suppose we have two fair six-sided dice (X and Y) and we want to measure the sum of the two variables as $Z = X + Y$. Further, let us suppose that given Z, we want the best estimate of X in the mean-squared-sense. Thus, we want to minimize the following:

$$J(\alpha) = \sum (x - \alpha z)^2 \mathbb{P}(x, z)$$

where $\mathbb{P}$ is the probability mass function for this problem. The idea is that when we have solved this problem, we will have a function of Z that is going to be the minimum MSE estimate of X. We can substitute in for Z in J and get:

$$J(\alpha) = \sum (x - \alpha(x + y))^2 \mathbb{P}(x, y)$$

Let us work out the steps in Sympy in the following:

```
>>> import sympy as S
>>> from sympy.stats import density, E, Die
>>> x = Die('D1',6)       # 1st six sided die
>>> y = Die('D2',6)       # 2nd six sides die
>>> a = S.symbols('a')
>>> z = x+y               # sum of 1st and 2nd die
>>> J = E((x-a*(x+y))**2) # expectation
>>> S.simplify(J)
329*a**2/6 - 329*a/6 + 91/6
```

With all that setup, we can now use basic calculus to minimize the objective function J:

```
>>> sol,=S.solve(S.diff(J,a),a)  # using calculus to minimize
>>> sol # solution is 1/2
1/2
```

> **Programming Tip**
>
> Sympy has a `stats` module that can do some basic work with expressions involving probability densities and expectations. The above code uses its E function to compute the expectation.

This says that $z/2$ is the MSE estimate of X given Z, which means geometrically (interpreting the MSE as a squared distance weighted by the probability mass function) that $z/2$ is as *close* to x as we are going to get for a given z.

Let us look at the same problem using the conditional expectation operator $\mathbb{E}(\cdot|z)$ and apply it to our definition of Z. Then

$$\mathbb{E}(z|z) = \mathbb{E}(x + y|z) = \mathbb{E}(x|z) + \mathbb{E}(y|z) = z$$

using the linearity of the expectation. Now, since by the symmetry of the problem (i.e., two identical die), we have

$$\mathbb{E}(x|z) = \mathbb{E}(y|z)$$

we can plug this in and solve

$$2\mathbb{E}(x|z) = z$$

which once again gives

$$\mathbb{E}(x|z) = \frac{z}{2}$$

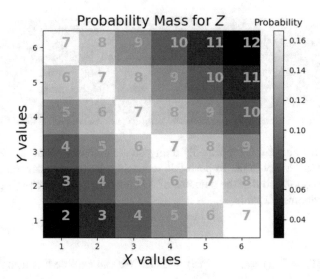

Fig. 2.5 The values of Z are in yellow with the corresponding values for X and Y on the axes. The gray scale colors indicate the underlying joint probability density

which is equal to the estimate we just found by minimizing the MSE. Let us explore this further with Fig. 2.5. Figure 2.5 shows the values of Z in yellow with the corresponding values for X and Y on the axes. Suppose $z = 2$, then the closest X to this is $X = 1$, which is what $\mathbb{E}(x|z) = z/2 = 1$ gives. What happens when $Z = 7$? In this case, this value is spread out diagonally along the X axis; so if $X = 1$, then Z is 6 units away; if $X = 2$, then Z is 5 units away; and so on.

Now, back to the original question, if we had $Z = 7$ and we wanted to get as close as we could to this using X, then why not choose $X = 6$, which is only one unit away from Z? The problem with doing that is $X = 6$ only occurs 1/6 of the time, so we are not likely to get it right the other 5/6 of the time. So, 1/6 of the time we are one unit away but 5/6 of the time, we are much more than one unit away. This means that the MSE score is going to be worse. Since each value of X from 1 to 6 is equally likely, to play it safe, we choose 7/2 as the estimate, which is what the conditional expectation suggests.

We can check this claim with samples using Sympy below:

```
>>> from sympy import stats
>>> # Eq constrains Z
>>> samples_z7 = lambda : stats.sample(x, S.Eq(z,7))
>>> # using 6 as an estimate
>>> mn = np.mean([(6-samples_z7())**2 for i in range(100)])
>>> #7/2 is the MSE estimate
>>> mn0=np.mean([(7/2.-samples_z7())**2 for i in range(100)])
>>> print('MSE=%3.2f using 6 vs MSE=%3.2f using 7/2 '
    % (mn,mn0))
MSE=9.91 using 6 vs MSE=3.47 using 7/2
```

Programming Tip
The `stats.sample(x, S.Eq(z,7))` function call samples the x variable subject to a condition on the z variable. In other words, it generates random samples of x die, given that the sum of the outcomes of that die and the y die add up to `z==7`.

Please run the above code repeatedly until you are convinced that the $\mathbb{E}(x|z)$ gives the lower MSE every time. To push this reasoning, let us consider the case where the die is so biased so that the outcome of 6 is ten times more probable than any of the other outcomes. That is,

$$\mathbb{P}(6) = 2/3$$

whereas $\mathbb{P}(1) = \mathbb{P}(2) = \ldots = \mathbb{P}(5) = 1/15$. We can explore this using Sympy as in the following:

```
>>> # here 6 is ten times more probable than any other outcome
>>> x=stats.FiniteRV('D3',{1:1/15., 2:1/15.,
...                        3:1/15., 4:1/15.,
...                        5:1/15., 6:2/3.})
```

As before, we construct the sum of the two dice and plot the corresponding probability mass function in Fig. 2.6. As compared with Fig. 2.5, the probability mass has been shifted away from the smaller numbers.

Let us see what the conditional expectation says about how we can estimate X from Z.

Fig. 2.6 The values of Z are in yellow with the corresponding values for X and Y on the axes

```
>>> E(x, S.Eq(z,7)) # conditional expectation E(x|z=7)
5.00000000000000
```

Now that we have $\mathbb{E}(x|z = 7) = 5$, we can generate samples as before and see if this gives the minimum MSE.

```
>>> samples_z7 = lambda : stats.sample(x, S.Eq(z,7))
>>> #using 6 as an estimate
>>> mn= np.mean([(6-samples_z7())**2 for i in range(100)])
>>> #5 is the MSE estimate
>>> mn0= np.mean([(5-samples_z7())**2 for i in range(100)])
>>> print('MSE=%3.2f using 6 vs MSE=%3.2f using 5 ' % (mn,mn0))
MSE=4.30 using 6 vs MSE=1.94 using 5
```

Using a simple example, we have emphasized the connection between minimum mean squared error problems and conditional expectation. Hopefully, the last two figures helped expose the role of the probability density. Next, we'll continue revealing the true power of the conditional expectation as we continue to develop corresponding geometric intuition.

2.5   Worked Examples of Conditional Expectation and Mean Square Error Optimization

Brzezniak [6] is a great book because it approaches conditional expectation through a sequence of exercises, which is what we are trying to do here. The main difference is that Brzezniak takes a more abstract measure-theoretic approach to the same problems. Note that you *do* need to grasp measure theory for advanced areas in probability, but for what we have covered so far, working the same problems in his text using our methods is illuminating. It always helps to have more than one way to solve *any* problem. I have numbered the examples corresponding to the book and tried to follow its notation.

Example This is Example 2.1 from Brzezniak. Three coins, 10p, 20p, and 50p, are tossed. The values of the coins that land heads-up are totaled. What is the expected total given that two coins have landed heads-up? In this case, we have we want to compute $\mathbb{E}(\xi|\eta)$, where

$$\xi := 10X_{10} + 20X_{20} + 50X_{50}$$

where $X_i \in \{0, 1\}$, and where X_{10} is the Bernoulli-distributed random variable corresponding to the 10p coin (and so on). Thus, ξ represents the total value of the heads-up coins. The η represents the condition that only two of the three coins are heads-up

$$\eta := X_{10}X_{20}(1 - X_{50}) + (1 - X_{10})X_{20}X_{50} + X_{10}(1 - X_{20})X_{50}$$

and is a function that is nonzero *only*, when two of the three coins lands heads-up. Each triple term catches each of these three possibilities. For example, the first term equals one, when the 10p and 20p are heads-up and the 50p is heads down. The remaining terms are zero.

To compute the conditional expectation, we want to find a function h of η that minimizes the mean-squared-error (MSE)

$$\text{MSE} = \sum_{X \in \{0,1\}^3} \frac{1}{2^3} (\xi - h(\eta))^2$$

where the sum is taken over all possible triples of outcomes for $\{X_{10}, X_{20}, X_{50}\}$, because each of the three coins has a $\frac{1}{2}$ chance of coming up heads.

Now, the question boils down to how can we characterize the function $h(\eta)$. Note that $\eta \mapsto \{0, 1\}$, so h takes on only two values. So, the orthogonal inner product condition is the following:

$$\langle \xi - h(\eta), \eta \rangle = 0$$

Because are only interested in $\eta = 1$, this simplifies to

$$\langle \xi - h(1), 1 \rangle = 0$$
$$\langle \xi, 1 \rangle = \langle h(1), 1 \rangle$$

This doesn't look so hard to evaluate, but we have to compute the integral over the set where $\eta = 1$. In other words, we need the set of triples $\{X_{10}, X_{20}, X_{50}\}$, where $\eta = 1$. That is, we can compute

$$\int_{\{\eta=1\}} \xi \, dX = h(1) \int_{\{\eta=1\}} dX$$

which is what Brzezniak does. Instead, we can define $h(\eta) = \alpha\eta$ and then find α. Rewriting the orthogonal condition gives

$$\langle \xi - \eta, \alpha\eta \rangle = 0$$
$$\langle \xi, \eta \rangle = \alpha \langle \eta, \eta \rangle$$
$$\alpha = \frac{\langle \xi, \eta \rangle}{\langle \eta, \eta \rangle}$$

where

$$\langle \xi, \eta \rangle = \sum_{X \in \{0,1\}^3} \frac{1}{2^3} (\xi\eta)$$

Note that we can just sweep over all triples $\{X_{10}, X_{20}, X_{50}\}$, because the definition of $h(\eta)$ zeros out when $\eta = 0$ anyway. All we have to do is plug everything in and solve. This tedious job is perfect for Sympy.

```
>>> import sympy as S
>>> X10,X20,X50 = S.symbols('X10,X20,X50',real=True)
>>> xi  = 10*X10+20*X20+50*X50
>>> eta = X10*X20*(1-X50)+X10*(1-X20)*(X50)+(1-X10)*X20*(X50)
>>> num = S.summation(xi*eta, (X10,0,1),(X20,0,1),(X50,0,1))
>>> den = S.summation(eta*eta, (X10,0,1),(X20,0,1),(X50,0,1))
>>> alpha = num/den
>>> alpha # alpha=160/3
160/3
```

This means that

$$\mathbb{E}(\xi|\eta) = \frac{160}{3}\eta$$

which we can check with a quick simulation

```
>>> import pandas as pd
>>> d = pd.DataFrame(columns=['X10','X20','X50'])
>>> d.X10 = np.random.randint(0,2,1000)
>>> d.X10 = np.random.randint(0,2,1000)
>>> d.X20 = np.random.randint(0,2,1000)
>>> d.X50 = np.random.randint(0,2,1000)
```

> **Programming Tip**
>
> The code above creates an empty Pandas dataframe with the named columns. The next four lines assigns values to each of the columns.

The code above simulates flipping the three coins 1000 times. Each column of the dataframe is either 0 or 1, corresponding to heads-down or heads-up, respectively. The condition is that two of the three coins have landed heads-up. Next, we can group the columns according to their sums. Note that the sum can only be in $\{0, 1, 2, 3\}$ corresponding to 0 heads-up, 1 heads-up, and so on.

```
>>> grp=d.groupby(d.eval('X10+X20+X50'))
```

> **Programming Tip**
>
> The eval function of the Pandas dataframe takes the named columns and evaluates the given formula. At the time of this writing, only simple formulas involving primitive operations are possible.

Next, we can get the 2 group, which corresponds to exactly two coins having landed heads-up, and then evaluate the sum of the values of the coins. Finally, we can take the mean of these sums.

```
>>> grp.get_group(2).eval('10*X10+20*X20+50*X50').mean()
52.60162601626016
```

The result is close to `160/3=53.33`, which supports the analytic result. The following code shows that we can accomplish the same simulation using pure Numpy.

```
>>> import numpy as np
>>> from numpy import array
>>> x=np.random.randint(0,2,(3,1000))
>>> print(np.dot(x[:,x.sum(axis=0)==2].T,array([10,20,50]))
    .mean())
52.860759493670884
```

In this case, we used the Numpy dot product to compute the value of the heads-up coins. The `sum(axis=0)==2` part selects the columns that correspond to two heads-up coins.

Still another way to get at the same problem is to forego the random sampling part and just consider all possibilities exhaustively using the `itertools` module in Python's standard library.

```
>>> import itertools as it
>>> list(it.product((0,1),(0,1),(0,1)))
[(0, 0, 0), (0, 0, 1), (0, 1, 0), (0, 1, 1), (1, 0, 0),
 (1, 0, 1), (1, 1, 0), (1, 1, 1)]
```

Note that we need to call `list` above in order to trigger the iteration in `it.product`. This is because the `itertools` module is generator-based so it does not actually *do* the iteration until it is iterated over (by `list` in this case). This shows all possible triples (X_{10}, X_{20}, X_{50}), where 0 and 1 indicate heads-down and heads-up, respectively. The next step is to filter out the cases that correspond to two heads-up coins.

```
>>> list(filter(lambda i:sum(i)==2,it.product((0,1),(0,1),
    (0,1))))
[(0, 1, 1), (1, 0, 1), (1, 1, 0)]
```

Next, we need to compute the sum of the coins and combine the prior code.

```
>>> list(map(lambda k:10*k[0]+20*k[1]+50*k[2],
...          filter(lambda i:sum(i)==2,
...          it.product((0,1),(0,1),(0,1)))))
[70, 60, 30]
```

The mean of the output is `53.33`, which is yet another way to get the same result. For this example, we demonstrated the full spectrum of approaches made possible using Sympy, Numpy, and Pandas. It is always valuable to have multiple ways of approaching the same problem and cross-checking the result.

2.5.1 Example

This is Example 2.2 from Brzezniak. Three coins, 10p, 20p, and 50p, are tossed as before. What is the conditional expectation of the total amount shown by the three coins, given the total amount shown by the 10p and 20p coins only? For this problem,

$$\xi :=10X_{10} + 20X_{20} + 50X_{50}$$

$$\eta :=30X_{10}X_{20} + 20(1 - X_{10})X_{20} + 10X_{10}(1 - X_{20})$$

which takes on four values $\eta \mapsto \{0, 10, 20, 30\}$ and only considers the 10p and 20p coins. In contrast to the last problem, here we are interested in $h(\eta)$ for all of the values of η. Naturally, there are only four values for $h(\eta)$ corresponding to each of these four values. Let us first consider $\eta = 10$. The orthogonal condition is then

$$\langle \xi - h(10), 10 \rangle = 0$$

The domain for $\eta = 10$ is $\{X_{10} = 1, X_{20} = 0, X_{50}\}$, which we can integrate out of the expectation below:

$$\mathbb{E}_{\{X_{10}=1, X_{20}=0, X_{50}\}}(\xi - h(10))10 = 0$$

$$\mathbb{E}_{\{X_{50}\}}(10 - h(10) + 50X_{50}) = 0$$

$$10 - h(10) + 25 = 0$$

which gives $h(10) = 35$. Repeating the same process for $\eta \in \{20, 30\}$ gives $h(20) = 45$ and $h(30) = 55$, respectively. This is the approach Brzezniak takes. On the other hand, we can just look at affine functions, $h(\eta) = a\eta + b$ and use brute-force calculus.

```
>>> from sympy.abc import a,b
>>> h = a*eta + b
>>> eta = X10*X20*30 + X10*(1-X20)*(10)+ (1-X10)*X20*(20)
>>> MSE=S.summation((xi-h)**2*S.Rational(1,8),(X10,0,1),
...                 (X20,0,1),
...                 (X50,0,1))
>>> sol=S.solve([S.diff(MSE,a),S.diff(MSE,b)],(a,b))
>>> print(sol)
{a: 64/3, b: 32}
```

Programming Tip

The `Rational` function from Sympy code expresses a rational number that Sympy is able to manipulate as such. This is different that specifying a fraction like `1/8.`, which Python would automatically compute as a floating point number (i.e., `0.125`). The advantage of using `Rational` is that Sympy can later produce rational numbers as output, which are sometimes easier to make sense of.

This means that

$$\mathbb{E}(\xi|\eta) = 25 + \eta \tag{2.8}$$

Since η takes on only four values, $\{0, 10, 20, 30\}$, we can write this out explicitly as

$$\mathbb{E}(\xi|\eta) = \begin{cases} 25 & \text{for } \eta = 0 \\ 35 & \text{for } \eta = 10 \\ 45 & \text{for } \eta = 20 \\ 55 & \text{for } \eta = 30 \end{cases} \tag{2.9}$$

Alternatively, we can use orthogonal inner products to write out the following conditions for the postulated affine function:

$$\langle \xi - h(\eta), \eta \rangle = 0 \tag{2.10}$$

$$\langle \xi - h(\eta), 1 \rangle = 0 \tag{2.11}$$

Writing these out and solving for a and b is tedious and a perfect job for Sympy. Starting with Eq. 2.10,

```
>>> expr=S.expand((xi-h)*eta)
>>> print(expr)
30*X10**2*X20*X50*a - 10*X10**2*X20*a - 10*X10**2*X50*a +
   100*X10**2 + 60*X10*X20**2*X50*a - 20*X10*X20**2*a -
   30*X10*X20*X50*a + 400*X10*X20 + 500*X10*X50 - 10*X10*b -
   20*X20**2*X50*a + 400*X20**2 + 1000*X20*X50 - 20*X20*b
```

and then because $\mathbb{E}(X_i^2) = 1/2 = \mathbb{E}(X_i)$, we make the following substitutions:

```
>>> expr.xreplace({X10**2:0.5, X20**2:0.5,X10:0.5,X20:0.5,
   X50:0.5})
-7.5*a - 15.0*b + 725.0
```

We can do this for the other orthogonal inner product in Eq. 2.11 as follows:

> **Programming Tip**
> Because Sympy symbols are hashable, they can be used as keys in Python dictionaries as in the `xreplace` function above.

```
>>> S.expand((xi-h)*1).xreplace({X10**2:0.5,
...                              X20**2:0.5,
...                              X10:0.5,
...                              X20:0.5,
...                              X50:0.5})
-0.375*a - b + 40.0
```

Then, combining this result with the previous one and solving for a and b gives

```
>>> S.solve([-350.0*a-15.0*b+725.0,-15.0*a-b+40.0])
{a: 1.00000000000000, b: 25.0000000000000}
```

which again gives us the final solution:

$$\mathbb{E}(\xi|\eta) = 25 + \eta$$

The following is a quick simulation to demonstrate this. We can build on the Pandas dataframe we used for the last example and create a new column for the sum of the 10p and 20p coins, as shown below.

```
>>> d['sm'] = d.eval('X10*10+X20*20')
```

We can group this by the values of this sum:

```
>>> d.groupby('sm').mean()
      X10   X20       X50
sm
0     0.0   0.0   0.502024
10    1.0   0.0   0.531646
20    0.0   1.0   0.457831
30    1.0   1.0   0.516854
```

But we want the expectation of the value of the coins:

```
>>> d.groupby('sm').mean().eval('10*X10+20*X20+50*X50')
sm
0      25.101215
10     36.582278
20     42.891566
30     55.842697
dtype: float64
```

which is very close to our analytical result in Eq. 2.5.1.

2.5.2   *Example*

This is Example 2.3 paraphrased from Brzezniak. Given X uniformly distributed on [0, 1], find $\mathbb{E}(\xi|\eta)$, where

$$\xi(x) = 2x^2$$

$$\eta(x) = \begin{cases} 1 & \text{if } x \in [0, 1/3] \\ 2 & \text{if } x \in (1/3, 2/3) \\ 0 & \text{if } x \in (2/3, 1] \end{cases}$$

Note that this problem is different from the previous two, because the sets that characterize η are intervals instead of discrete points. Nonetheless, we will eventually have three values for $h(\eta)$ because $\eta \mapsto \{0, 1, 2\}$. For $\eta = 1$, we have the orthogonal conditions:

$$\langle \xi - h(1), 1 \rangle = 0$$

which boils down to

$$\mathbb{E}_{\{x \in [0,1/3]\}}(\xi - h(1)) = 0$$

$$\int_0^{\frac{1}{3}} (2x^2 - h(1))dx = 0$$

and then by solving this for $h(1)$ gives $h(1) = 2/24$. This is the way Brzezniak works this problem. Alternatively, we can use $h(\eta) = a + b\eta + c\eta^2$ and brute-force calculus.

```
>>> x,c,b,a = S.symbols('x,c,b,a')
>>> xi = 2*x**2
>>> eta=S.Piecewise((1,S.And(S.Gt(x,0),
...                          S.Lt(x,S.Rational(1,3)))), #  0 < x < 1/3
...                  (2,S.And(S.Gt(x,S.Rational(1,3)),
...                          S.Lt(x,S.Rational(2,3)))),# 1/3 < x < 2/3,
...                  (0,S.And(S.Gt(x,S.Rational(2,3)),
...                          S.Lt(x,1))))              # 1/3 < x < 2/3
>>> h = a + b*eta + c*eta**2
>>> J = S.integrate((xi-h)**2, (x,0,1))
>>> sol = S.solve([S.diff(J,a),
...                S.diff(J,b),
...                S.diff(J,c),
...                ],
...                (a,b,c))
>>> print(sol)
{a: 38/27, b: -20/9, c: 8/9}
>>> print(S.piecewise_fold(h.subs(sol)))
Piecewise((2/27, (x > 0) & (x < 1/3)), (14/27, (x > 1/3) &
 (x < 2/3)), (38/27, (x > 2/3) & (x < 1)))
```

Thus, collecting this result gives

$$\mathbb{E}(\xi | \eta) = \frac{38}{27} - \frac{20}{9}\eta + \frac{8}{9}\eta^2$$

which can be rewritten as a piecewise function of x:

$$\mathbb{E}(\xi | \eta(x)) = \begin{cases} \frac{2}{27} & \text{for } 0 < x < \frac{1}{3} \\ \frac{14}{27} & \text{for } \frac{1}{3} < x < \frac{2}{3} \\ \frac{38}{27} & \text{for } \frac{2}{3} < x < 1 \end{cases} \tag{2.12}$$

Alternatively, we can use the orthogonal inner product conditions directly by choosing $h(\eta) = c + \eta b + \eta^2 a$:

$$\langle \xi - h(\eta), 1 \rangle = 0$$

$$\langle \xi - h(\eta), \eta \rangle = 0$$

$$\langle \xi - h(\eta), \eta^2 \rangle = 0$$

and then solving for *a*,*b*, and *c*.

```
>>> x,a,b,c,eta = S.symbols('x,a,b,c,eta',real=True)
>>> xi   = 2*x**2
>>> eta=S.Piecewise((1,S.And(S.Gt(x,0),
...                        S.Lt(x,S.Rational(1,3)))),   #
  0 < x < 1/3
...                     (2,S.And(S.Gt(x,S.Rational(1,3)),
...                        S.Lt(x,S.Rational(2,3)))),  #
  1/3 < x < 2/3,
...                     (0,S.And(S.Gt(x,S.Rational(2,3)),
...                        S.Lt(x,1))))  # 1/3 < x < 2/3
>>> h = c+b*eta+a*eta**2
```

Then, the orthogonal conditions become

```
>>> S.integrate((xi-h)*1,(x,0,1))
-5*a/3 - b - c + 2/3
>>> S.integrate((xi-h)*eta,(x,0,1))
-3*a - 5*b/3 - c + 10/27
>>> S.integrate((xi-h)*eta**2,(x,0,1))
-17*a/3 - 3*b - 5*c/3 + 58/81
```

Now, we just combine the three equations and solve for the parameters:

```
>>> eqs=[ -5*a/3 - b - c + 2/3,
...    -3*a - 5*b/3 - c + 10/27,
...    -17*a/3 - 3*b - 5*c/3 + 58/81]
>>> sol=S.solve(eqs)
>>> print(sol)
{a: 0.888888888888889, b: -2.22222222222222, c:
  1.40740740740741}
```

We can assemble the final result by substituting in the solution

```
>>> print(S.piecewise_fold(h.subs(sol)))
Piecewise((0.074074074074074, (x > 0) & (x < 1/3)),
  (0.518518518518518, (x > 1/3) & (x < 2/3)),
  (1.40740740740741, (x > 2/3) & (x < 1)))
```

which is the same as our analytic result in Eq. 2.5.2, just in decimal format.

Programming Tip

The definition of Sympy's piecewise function is verbose because of the way Python parses inequality statements. As of this writing, this has not been reconciled in Sympy, so we have to use the verbose declaration.

To reinforce our result, let us do a quick simulation using Pandas.

```
>>> d = pd.DataFrame(columns=['x','eta','xi'])
>>> d.x = np.random.rand(1000)
>>> d.xi = 2*d.x**2
>>> d.xi.head()
0      0.649201
1      1.213763
2      1.225751
3      0.005203
4      0.216274
Name: xi, dtype: float64
```

Now, we can use the `pd.cut` function to group the x values in the following:

```
>>> pd.cut(d.x,[0,1/3,2/3,1]).head()
0      (0.333, 0.667]
1      (0.667, 1.0]
2      (0.667, 1.0]
3      (0.0, 0.333]
4      (0.0, 0.333]
Name: x, dtype: category
Categories (3, interval[float64, right]): [(0.0, 0.333]
 < (0.333, 0.667] < (0.667, 1.0]]
```

Note that the `head()` call above is only to limit the printout shown. The categories listed are each of the intervals for `eta` that we specified using the `[0,1/3,2/3,1]` list. Now that we know how to use `pd.cut`, we can just compute the mean on each group as shown below:

```
>>> d.groupby(pd.cut(d.x,[0,1/3,2/3,1])).mean()['xi']
x
(0.0, 0.333]        0.073048
(0.333, 0.667]      0.524023
(0.667, 1.0]        1.397096
Name: xi, dtype: float64
```

which is pretty close to our analytic result in Eq. 2.5.2. Alternatively, `sympy.stats` has some limited tools for the same calculation.

```
>>> from sympy.stats import E, Uniform
>>> x=Uniform('x',0,1)
>>> E(2*x**2,S.And(x < S.Rational(1,3), x > 0))
2/27
>>> E(2*x**2,S.And(x < S.Rational(2,3), x > S.Rational(1,3)))
14/27
>>> E(2*x**2,S.And(x < 1, x > S.Rational(2,3)))
38/27
```

which again gives the same result still another way.

2.5.3 Example

This is Example 2.4 from Brzezniak. Find $\mathbb{E}(\xi|\eta)$ for

$$\xi(x) = 2x^2$$

$$\eta = \begin{cases} 2 & \text{if } 0 \le x < \frac{1}{2} \\ x & \text{if } \frac{1}{2} < x \le 1 \end{cases}$$

Once again, X is uniformly distributed on the unit interval. Note that η is no longer discrete for every domain. For the domain $0 < x < 1/2$, $h(2)$ takes on only one value, say, h_0. For this domain, the orthogonal condition becomes

$$\mathbb{E}_{\{\eta=2\}}((\xi(x) - h_0)2) = 0$$

which simplifies to

$$\int_0^{1/2} 2x^2 - h_0 dx = 0$$

$$\int_0^{1/2} 2x^2 dx = \int_0^{1/2} h_0 dx$$

$$h_0 = 2 \int_0^{1/2} 2x^2 dx$$

$$h_0 = \frac{1}{6}$$

For the other domain, where $\{\eta = x\}$ in Eq. 2.5.3, we again use the orthogonal condition:

$$\mathbb{E}_{\{\eta=x\}}((\xi(x) - h(x))x) = 0$$

$$\int_{1/2}^1 (2x^2 - h(x))x dx = 0$$

$$h(x) = 2x^2$$

Assembling the solution gives

$$\mathbb{E}(\xi|\eta(x)) = \begin{cases} \frac{1}{6} & \text{for } 0 \le x < \frac{1}{2} \\ 2x^2 & \text{for } \frac{1}{2} < x \le 1 \end{cases}$$

although this result is not explicitly written as a function of η.

2.5.4   *Example*

This is Exercise 2.6 in Brzezniak. Find $\mathbb{E}(\xi|\eta)$, where

$$\xi(x) = 2x^2$$

$$\eta(x) = 1 - |2x - 1|$$

and X is uniformly distributed in the unit interval. We can write this out as a piecewise function in the following:

$$\eta = \begin{cases} 2x & \text{for } 0 \le x < \frac{1}{2} \\ 2 - 2x & \text{for } \frac{1}{2} < x \le 1 \end{cases}$$

The discontinuity is at $x = 1/2$. Let us start with the $\{\eta = 2x\}$ domain.

$$\mathbb{E}_{\{\eta=2x\}}((2x^2 - h(2x))2x) = 0$$

$$\int_0^{1/2} (2x^2 - h(2x))2x\,dx = 0$$

We can make this explicitly a function of η by a change of variables ($\eta = 2x$), which gives

$$\int_0^1 (\eta^2/2 - h(\eta))\frac{\eta}{2}d\eta = 0$$

Thus, for this domain, $h(\eta) = \eta^2/2$. Note that due to the change of variables, $h(\eta)$ is valid defined over $\eta \in [0, 1]$.

For the other domain where $\{\eta = 2 - 2x\}$, we have

$$\mathbb{E}_{\{\eta=2-2x\}}((2x^2 - h(2 - 2x))(2 - 2x)) = 0$$

$$\int_{1/2}^1 (2x^2 - h(2 - 2x))(2 - 2x)dx = 0$$

Once again, a change of variables makes the η dependency explicit using $\eta = 2 - 2x$, which gives

$$\int_0^1 ((2 - \eta)^2/2 - h(\eta))\frac{\eta}{2}d\eta = 0$$

$$h(\eta) = (2 - \eta)^2/2$$

Once again, the change of variables means this solution is valid over $\eta \in [0, 1]$. Thus, because both pieces are valid over the same domain ($\eta \in [0, 1]$), we can just add them to get the final solution:

$$h(\eta) = \eta^2 - 2\eta + 2$$

A quick simulation can help bear this out.

```
>>> from pandas import DataFrame
>>> d = DataFrame(columns=['xi','eta','x','h','h1','h2'])
```

```
>>> # 100 random samples
>>> d.x = np.random.rand(100)
>>> d.xi = d.eval('2*x**2')
>>> d.eta =1-abs(2*d.x-1)
>>> d.h1=d[(d.x<0.5)].eval('eta**2/2')
>>> d.h2=d[(d.x>=0.5)].eval('(2-eta)**2/2')
>>> d.fillna(0,inplace=True)
>>> d.h = d.h1+d.h2
>>> d.head()
          xi        eta         x          h         h1         h2
0   1.102459   0.515104   0.742448   1.102459   0.000000   1.102459
1   0.239610   0.692257   0.346128   0.239610   0.239610   0.000000
2   1.811868   0.096389   0.951806   1.811868   0.000000   1.811868
3   0.000271   0.023268   0.011634   0.000271   0.000271   0.000000
4   0.284240   0.753977   0.376988   0.284240   0.284240   0.000000
```

Note that we have to be careful, where we apply the individual solutions using the slice (d.x<0.5) index. The fillna part ensures that the default NaN that fills out the empty row-etries is replaced with zero before combining the individual solutions. Otherwise, the NaN values would circulate through the rest of the computation. The following is the essential code that draws Fig. 2.7.

```
from matplotlib.pyplot import subplots
fig,ax=subplots()
ax.plot(d.xi,d.eta,'.',alpha=.3,label='$\eta$')
ax.plot(d.xi,d.h,'k.',label='$h(\eta)$')
ax.legend(loc=0,fontsize=18)
ax.set_xlabel('$2 x^2$',fontsize=18)
ax.set_ylabel('$h(\eta)$',fontsize=18)
```

> **Programming Tip**
>
> Basic formatting works for the labels in Fig. 2.7. The loc=0 in the legend function is the code for the *best* placement for the labels in the legend. The individual labels should be specified when the elements are drawn individually, otherwise they will be hard to separate out later. This is accomplished using the label keyword in the plot commands.

Figure 2.7 shows the ξ data plotted against η and $h(\eta) = \mathbb{E}(\xi|\eta)$. Points on the diagonal are points where ξ and $\mathbb{E}(\xi|\eta)$ match. As shown by the dots, there is no agreement between the raw η data and ξ. Thus, one way to think about the conditional expectation is as a functional transform that bends the curve onto the diagonal line. The black dots plot ξ versus $\mathbb{E}(\xi|\eta)$ and the two match everywhere along the diagonal line. This is to be expected because the conditional expectation is the MSE best estimate for ξ among all functions of η.

Fig. 2.7 The diagonal line shows where the conditional expectation equals the ξ function

$$\xi = 2x^2$$

2.5.5 Example

This is Exercise 2.14 from Brzezniak. Find $\mathbb{E}(\xi|\eta)$ where

$$\xi(x) = 2x^2$$

$$\eta = \begin{cases} 2x & \text{if } 0 \le x < \frac{1}{2} \\ 2x - 1 & \text{if } \frac{1}{2} < x \le 1 \end{cases}$$

and X is uniformly distributed in the unit interval. This is the same as the last example, and the only difference here is that η is not continuous at $x = \frac{1}{2}$, as before. The first part is exactly the same as the first part of the prior example, so we will skip it here. The second part follows the same reasoning as the last example, so we will just write the answer for the $\{\eta = 2x - 1\}$ case as the following:

$$h(\eta) = \frac{(1 + \eta)^2}{2}, \quad \forall \eta \in [0, 1]$$

and then adding these up as before gives the full solution:

$$h(\eta) = \frac{1}{2} + \eta + \eta^2$$

The interesting part about this example is shown in Fig. 2.8. The dots show where η is discontinuous, yet the $h(\eta) = \mathbb{E}(\xi|\eta)$ solution is equal to ξ (i.e., matches the diagonal). This illustrates the power of the orthogonal inner product technique, which does not need continuity or complex set-theoretic arguments to calculate

Fig. 2.8 The diagonal line shows where the conditional expectation equals the ξ function

solutions. By contrast, I urge you to consider Brzezniak's solution to this problem which requires such methods.

Extending projection methods to random variables provides multiple ways for calculating solutions to conditional expectation problems. In this section, we also worked out corresponding simulations using a variety of Python modules. It is always advisable to have more than one technique at hand to cross-check potential solutions. We worked out some of the examples in Brzezniak's book using our methods as a way to show multiple ways to solve the same problem. Comparing Brzezniak's measure-theoretic methods to our less abstract techniques is a great way to get a handle on both concepts, which are important for advanced study in stochastic process.

2.6 Useful Distributions

2.6.1 Normal Distribution

Without a doubt, the normal (Gaussian) distribution is the most important and foundational probability distribution. The one-dimensional form is the following:

$$f(x) = \frac{e^{-\frac{(x-\mu)^2}{2\sigma^2}}}{\sqrt{2\pi\sigma^2}}$$

where $\mathbb{E}(x) = \mu$ and $\mathbb{V}(x) = \sigma^2$. The multidimensional version for $\mathbf{x} \in \mathbb{R}^n$ is the following:

$$f(\mathbf{x}) = \frac{1}{\det(2\pi\mathbf{R})^{\frac{1}{2}}} e^{-\frac{1}{2}(\mathbf{x}-\boldsymbol{\mu})^T \mathbf{R}^{-1}(\mathbf{x}-\boldsymbol{\mu})}$$

where $\mathbf{R}$ is the covariance matrix with entries

$$R_{i,j} = \mathbb{E}\left[(x_i - \bar{x}_i)(x_j - \bar{x}_j)\right]$$

A key property of the normal distribution is that it is completely specified by its first two moments. Another key property is that the normal distribution is preserved under linear transformations. For example:

$$\mathbf{y} = \mathbf{A}\mathbf{x}$$

means $\mathbf{y} \sim \mathcal{N}(\mathbf{A}\mathbf{x}, \mathbf{A}\mathbf{R_x}\mathbf{A}^T)$. This means that it is easy to do linear algebra and matrix operations with normal distributed random variables. There are many intuitive geometric relationships that are preserved with normal distributed random variables, as discussed in the Gauss-Markov chapter.

2.6.2 Multinomial Distribution

The multinomial distribution generalizes the binomial distribution. Recall that the binomial distribution characterizes the number of heads obtained in n trials. Consider the problem of n balls to be divided among r available bins, where each bin may accommodate more than one ball. For example, suppose $n = 10$ and and $r = 3$, then one possible valid configuration is $\mathbf{N}_{10} = [3, 3, 4]$. The probability that a ball lands in the ith bin is p_i, where $\sum p_i = 1$. The multinomial distribution characterizes the probability distribution of $\mathbf{N}_n$. The binomial distribution is a special case of the multinomial distribution with $n = 2$. The multinomial distribution is implemented in the `scipy.stats` module as shown below:

```
>>> from scipy.stats import multinomial
>>> rv = multinomial(10,[1/3]*3)
>>> rv.rvs(4)
array([[2, 2, 6],
       [4, 2, 4],
       [2, 4, 4],
       [2, 6, 2]])
```

Note that the sum across the columns is always n:

```
>>> rv.rvs(10).sum(axis=1)
array([10, 10, 10, 10, 10, 10, 10, 10, 10, 10])
```

To derive the probability mass function, we define the *occupancy vector*, $\mathbf{e}_i \in \mathbb{R}^r$, which is a binary vector with exactly one nonzero component (i.e., a unit vector).

Then, the $\mathbf{N}_n$ vector can be written as the sum of n vectors $\mathbf{X}$, each drawn from the set $\{\mathbf{e}_j\}_{j=1}^r$:

$$\mathbf{N}_n = \sum_{i=1}^{n} \mathbf{X}_i$$

where the probability $\mathbb{P}(\mathbf{X} = \mathbf{e}_j) = p_j$. Thus, $\mathbf{N}_n$ has a discrete distribution over the set of vectors with nonnegative components that sum to n. Because the $\mathbf{X}$ vectors are independent and identically distributed, the probability of any particular $\mathbf{N}_n = [x_1, x_2, \cdots, x_r]^\top = \mathbf{x}$ is

$$\mathbb{P}(\mathbf{N}_n = x) = C_n p_1^{x_1} p_2^{x_2} \cdots p_r^{x_r}$$

where C_n is a combinatorial factor that accounts for all the ways a component can sum to x_j. Consider that there are $\binom{n}{x_1}$ ways that the first component can be chosen. This leaves $n - x_1$ balls left for the rest of the vector components. Thus, the second component has $\binom{n-x_1}{x_2}$ ways to pick a ball. Following the same pattern, the third component has $\binom{n-x_1-x_2}{x_3}$ ways and so forth

$$C_n = \binom{n}{x_1}\binom{n - x_1}{x_2}\binom{n - x_1 - x_2}{x_3}\cdots\binom{n - x_1 - x_2 - \cdots - x_{r-1}}{x_r}$$

simplifies to the following:

$$C_n = \frac{n!}{x_1! \cdots x_r!}$$

Thus, the probability mass function for the multinomial distribution is the following:

$$\mathbb{P}(\mathbf{N}_n = x) = \frac{n!}{x_1! \cdots x_r!} p_1^{x_1} p_2^{x_2} \cdots p_r^{x_r}$$

The expectation of this distribution is the following:

$$\mathbb{E}(\mathbf{N}_n) = \sum_{i=1}^{n} \mathbb{E}(X_i)$$

by the linearity of the expectation. Then,

$$\mathbb{E}(X_i) = \sum_{j=1}^{r} p_j \mathbf{e}_j = I\mathbf{p} = \mathbf{p}$$

where p_j are the components of the vector $\mathbf{p}$ and $\mathbf{I}$ is the identity matrix. Then, because this is the same for any X_i, we have

$$\mathbb{E}(\mathbf{N}_n) = n\mathbf{p}$$

For the covariance of $\mathbf{N}_n$, we need to compute the following:

$$\text{Cov}(\mathbf{N}_n) = \mathbb{E}\left(\mathbf{N}_n\mathbf{N}_n^\top\right) - \mathbb{E}(\mathbf{N}_n)\mathbb{E}(\mathbf{N}_n)^\top$$

For the first term on the right, we have

$$\mathbb{E}\left(\mathbf{N}_n\mathbf{N}_n^\top\right) = \mathbb{E}\left((\sum_{i=1}^n X_i)(\sum_{j=1}^n X_j^\top)\right)$$

and for $i = j$, we have

$$\mathbb{E}(X_i X_i^\top) = \text{diag}(\mathbf{p})$$

and for $i \neq j$, we have

$$\mathbb{E}(X_i X_j^\top) = \mathbf{p}\mathbf{p}^\top$$

Note that this term has elements on the diagonal. Then, combining the above two equations gives the following:

$$\mathbb{E}(\mathbf{N}_n\mathbf{N}_n^\top) = n\text{diag}(\mathbf{p}) + (n^2 - n)\mathbf{p}\mathbf{p}^\top$$

Now, we can assemble the covariance matrix:

$$\text{Cov}(\mathbf{N}_n) = n\text{diag}(\mathbf{p}) + (n^2 - n)\mathbf{p}\mathbf{p}^\top - n^2\mathbf{p}\mathbf{p}^\top = n\text{diag}(\mathbf{p}) - n\mathbf{p}\mathbf{p}^\top$$

Specifically, the off-diagonal terms are $np_i p_j$ and the diagonal terms are $np_i(1-p_i)$.

2.6.3 Chi-Square Distribution

The χ^2 distribution appears in many different contexts so it's worth understanding. Suppose we have n independent random variables X_i such that $X_i \sim \mathcal{N}(0, 1)$. We are interested in the following random variable $R = \sqrt{\sum_i X_i^2}$. The joint probability density of X_i is the following:

$$f_{\mathbf{X}}(X) = \frac{e^{-\frac{1}{2}\sum_i X_i^2}}{(2\pi)^{\frac{n}{2}}}$$

where the $\mathbf{X}$ represents a vector of X_i random variables. You can think of R as the radius of an n-dimensional sphere. The volume of this sphere is given by the following formula:

$$V_n(R) = \frac{\pi^{\frac{n}{2}}}{\Gamma(\frac{n}{2} + 1)} R^n$$

To reduce the amount of notation we define,

$$A := \frac{\pi^{\frac{n}{2}}}{\Gamma(\frac{n}{2} + 1)}$$

The differential of this volume is the following:

$$dV_n(R) = nAR^{n-1}dR$$

In terms of the X_i coordinates, the probability (as always) integrates out to one.

$$\int f_{\mathbf{X}}(\mathbf{X})dV_n(\mathbf{X}) = 1$$

In terms of R, the change of variable provides

$$\int f_{\mathbf{X}}(R)nAR^{n-1}dR$$

Thus,

$$f_R(R) := f_{\mathbf{X}}(R) = nAR^{n-1}\frac{e^{-\frac{1}{2}R^2}}{(2\pi)^{\frac{n}{2}}}$$

But we are interested in the distribution $Y = R^2$. Using the same technique again,

$$\int f_R(R)dR = \int f_R(\sqrt{Y})\frac{dY}{2\sqrt{Y}}$$

Finally,

$$f_Y(Y) := nAY^{\frac{n-1}{2}}\frac{e^{-\frac{1}{2}Y}}{(2\pi)^{\frac{n}{2}}}\frac{1}{2\sqrt{Y}}$$

Table 2.1 Diagnosis table

Mild infection	Strong infection	No infection	Total
128	136	420	684

Then, finally substituting back in A gives the χ^2 distribution with n degrees of freedom:

$$f_Y(Y) = n\frac{\pi^{\frac{n}{2}}}{\Gamma(\frac{n}{2}+1)}Y^{n/2-1}\frac{e^{-\frac{1}{2}Y}}{(2\pi)^{\frac{n}{2}}}\frac{1}{2} = \frac{2^{-\frac{n}{2}-1}n}{\Gamma\left(\frac{n}{2}+1\right)}e^{-Y/2}Y^{\frac{n}{2}-1}$$

with corresponding mean and variance

$$\overline{Y} = n, \ \mathbb{V}(Y) = 2n$$

Example Hypothesis testing is a common application of the χ^2 distribution. Consider Table 2.1, which tabulates the infection status of a certain population. The hypothesis is that these data are distributed according to the multinomial distribution with the following rates for each group, $p_1 = 1/4$ (mild infection), $p_2 = 1/4$ (strong infection), and $p_3 = 1/2$ (no infection). Suppose n_i is the count of persons in the ith column and $\sum_i n_i = n = 684$. Let k denote the number of columns. Then, in order to apply the central limit theorem, we want to sum the n_i random variables, but these all sum to n, a constant, which prohibits using the theorem. Instead, suppose we sum the n_i variables up to $k-1$ terms. Then,

$$z = \sum_{i=1}^{k-1} n_i$$

is asymptotically normally distributed by the theorem with mean $\mathbb{E}(z) = \sum_{i=1}^{k-1} np_i$. Using our previous results and notation for multinomial random variables, we can write this as

$$z = [\mathbf{1}_{k-1}^\top, 0]\mathbf{N}_n$$

where $\mathbf{1}_{k-1}$ is a vector of all ones of length $k-1$ and $\mathbf{N}_n \in \mathbb{R}^k$. With this notation, we have

$$\mathbb{E}(z) = n[\mathbf{1}_{k-1}^\top, 0]\mathbf{p} = \sum_{i=1}^{k-1} np_i = n(1 - p_k)$$

We can get the variance of z using the same method,

$$\mathbb{V}(z) = [\mathbf{1}_{k-1}^\top, 0]\text{Cov}(\mathbf{N}_n)[\mathbf{1}_{k-1}^\top, 0]^\top$$

which gives

$$\mathbb{V}(z) = [\mathbf{1}_{k-1}^\top, 0](n\text{diag}(\mathbf{p}) - n\mathbf{pp}^\top)[\mathbf{1}_{k-1}^\top, 0]^\top$$

The variance is then

$$\mathbb{V}(z) = n(1 - p_k)p_k$$

With the mean and variance established, we can subtract the hypothesized mean for each column under the hypothesis and create the transformed variable,

$$z' = \sum_{i=1}^{k-1} \frac{n_i - np_i}{\sqrt{n(1 - p_k)p_k}} \sim \mathcal{N}(0, 1)$$

by the central limit theorem. Likewise,

$$\sum_{i=1}^{k-1} \frac{(n_i - np_i)^2}{n(1 - p_k)p_k} \sim \chi_{k-1}^2$$

With all that established, we can test if the data in the table follows the hypothesized multinomial distribution.

```
>>> from scipy import stats
>>> p1 = p2 = 1/4
>>> p3, n = 1/2, 684
>>> v = n*p3*(1-p3)
>>> z = (128-n*p1)**2/v + (136-n*p2)**2/v
>>> degrees_freedom = 2
>>> 1-stats.chi2(degrees_freedom).cdf(z)
0.00012486166748693073
```

This value is very low and suggests that the hypothesized multinomial distribution is not a good one for this data. Note that this approximation only works when n is large in comparison to the number of columns in the table or when there are few columns but a large number of counts in each entry of the table. The latter case arises because each table entry converges to a normally distribution. The z statistic we computed above is known more generally as the Pearson chi-square test statistic.

2.6.4 Poisson and Exponential Distributions

The Poisson distribution for a random variable X represents a number of outcomes occurring in a given time interval (t).

$$p(x; \lambda t) = \frac{e^{-\lambda t} (\lambda t)^x}{x!}$$

The Poisson distribution is closely related to the binomial distribution, $b(k; n, p)$ where p is small and n is large. That is, when there is a low-probability event but many trials, n. Recall that the binomial distribution is the following:

$$b(k; n, p) = \binom{n}{k} p^k (1 - p)^{n-k}$$

for $k = 0$ and taking the logarithm of both sides, we obtain

$$\log b(0; n, p) = (1 - p)^n = \left(1 - \frac{\lambda}{n}\right)^n$$

Then, the Taylor expansion of this gives the following:

$$\log b(0; n, p) \approx -\lambda - \frac{\lambda^2}{2n} - \cdots$$

For large n, this results in

$$b(0; n, p) \approx e^{-\lambda}$$

A similar argument for k leads to the Poisson distribution. Conveniently, we have $\mathbb{E}(X) = \mathbb{V}(X) = \lambda$. For example, suppose that the average number of vehicles passing under a tollgate per hour is 3. Then, the probability that six vehicles pass under the gate in a given hour is $p(x = 6; \lambda t = 3) = \frac{81}{30e^3} \approx 0.05$.

The Poisson distribution is available from the scipy.stats module. The following code computes the last result:

```
>>> from scipy.stats import poisson
>>> x = poisson(3)
>>> print(x.pmf(6))
0.05040940672246224
```

The Poisson distribution is important for applications involving reliability and queueing and is closely related to the exponential distribution in this case. The Poisson distribution is used to compute the probability of specific numbers of events during a particular time period. In many cases, the time period (X) itself is the random variable. For example, we might be interested in understanding the time X between arrivals of vehicles at a checkpoint. With the Poisson distribution, the probability of *no* events occurring in the span of time up to time t is given by the following:

$$p(0; \lambda t) = e^{-\lambda t}$$

Now, suppose X is the time to the first event. The probability that the length of time until the first event will exceed x is given by the following:

$$\mathbb{P}(X > x) = e^{-\lambda x}$$

Then, the cumulative distribution function is given by the following:

$$\mathbb{P}(0 \leq X \leq x) = F_X(x) = 1 - e^{-\lambda x}$$

Taking the derivative gives the *exponential* distribution:

$$f_X(x) = \lambda e^{-\lambda x}$$

where $\mathbb{E}(X) = 1/\lambda$ and $\mathbb{V}(X) = \frac{1}{\lambda^2}$. For example, suppose we want to know the probability of a certain component lasting beyond $T = 10$ years, where T is modeled as a an exponential random variable with $1/\lambda = 5$ years. Then, we have $1 - F_X(10) = e^{-2} \approx 0.135$.

The exponential distribution is available in the `scipy.stats` module. The following code computes the result of the example above. Note that the parameters are described in slightly different terms as above, as described in the corresponding documentation for `expon`.

```
>>> from scipy.stats import expon
>>> x = expon(0,5) # create random variable object
>>> print(1 - x.cdf(10))
0.1353352832366127
```

Connection Between Poisson and Multinomial Distributions Suppose we observe a sample of K independent Poisson random variables with $t = 1$:

$$X_i \sim \text{Poisson}(\mu_i)$$

The total count of Poisson events

$$X_T = \sum_{i=1}^{K} X_i$$

is another random variable. By properties of the Poisson distribution, $X_T \sim \text{Poisson}(\sum_{i=1}^{n} \mu_i)$. Consider the following conditional probability:

$$\mathbb{P}(X_1 = x_1, \cdots, X_K = x_K | X_T = n) = \frac{\mathbb{P}(X_1 = x_1, \cdots, X_K = x_K \wedge X_T = n)}{\mathbb{P}(X_T = n)}$$

$$= \frac{\Pi_i \exp(-\mu_i)\mu_i^{x_i}/x_i!}{\exp(-\sum_{i=1}^{K} \mu_i)(\sum_{i=1}^{K} \mu_i)^n/n!}$$

$$= \left(\frac{n!}{\Pi_i x_i!}\right) \prod_{i=1}^{K} p_i^{x_i}$$

where $p_i = \mu_i/(\sum_{i=j}^{K} \mu_i)$. This is the multinomial distribution with parameters p_i, so this means that the set of independent Poisson distributed random variables conditioned on a fixed sum is multinomially distributed.

2.6.5 Gamma Distribution

We have previously discussed how the exponential distribution can be created from the Poisson events. The exponential distribution has the *memoryless* property, namely,

$$\mathbb{P}(T > t_0 + t | T > t_0) = \mathbb{P}(T > t)$$

For example, given T as the random variable representing the time until failure, this means that a component that has survived up through t_0 has the same failure probability of lasting t units beyond that point. To derive this result, it is easier to compute the complementary event:

$$\mathbb{P}(t_0 < T < t_0 + t | T > t_0) = \mathbb{P}(t_0 < T < t_0 + t) = e^{-\lambda t}\left(e^{\lambda t} - 1\right)$$

Then, one minus this result shows the memoryless property, which, unrealistically, does not account for wear over the first t hours. The *gamma* distribution can remedy this.

Recall that the exponential distribution describes the time until the occurrence of a Poisson event. The random variable X for the time until a specified number of Poisson events (α) is described by the *gamma* distribution. Thus, the exponential distribution is a special case of the gamma distribution when $\alpha = 1$ and $\beta = 1/\lambda$. For $x > 0$, the gamma distribution is the following:

$$f(x; \alpha, \beta) = \frac{\beta^{-\alpha} x^{\alpha-1} e^{-\frac{x}{\beta}}}{\Gamma(\alpha)}$$

and $f(x; \alpha, \beta) = 0$ when $x \leq 0$ and Γ is the gamma function. For example, suppose that vehicles passing under a gate follows a Poisson process, with an average of five vehicles passing per hour, what is the probability that at most an hour will have passed before two vehicles pass the gate? If X is time in hours that transpires before the two vehicles pass, then we have $\beta = 1/5$ and $\alpha = 2$. The required probability $\mathbb{P}(X < 1) \approx 0.96$. The gamma distribution has $\mathbb{E}(X) = \alpha\beta$ and $\mathbb{V}(X) = \alpha\beta^2$.

The following code computes the result of the example above. Note that the parameters are described in slightly different terms as above, as described in the corresponding documentation for gamma.

```
>>> from scipy.stats import gamma
>>> x = gamma(2,scale=1/5) # create random variable object
>>> print(x.cdf(1))
0.9595723180054873
```

2.6.6   Beta Distribution

The uniform distribution assigns a single constant value over the unit interval. The beta distribution generalizes this to a function over the unit interval. The probability density function of the beta distribution is the following:

$$f(x) = \frac{1}{\beta(a, b)} x^{a-1}(1 - x)^{b-1}$$

where

$$\beta(a, b) = \int_0^1 x^{a-1}(1 - x)^{b-1} dx$$

Note that $a = b = 1$ yields the uniform distribution. In the special case for integers where $0 \leq k \leq n$, we have

$$\int_0^1 \binom{n}{k} x^k (1 - x)^{n-k} dx = \frac{1}{n+1}$$

To get this result without calculus, we can use an experiment by Thomas Bayes. Start with n white balls and one gray ball. Uniformly at random, toss them onto the unit interval. Let X be the number of white balls to the left of the gray ball. Thus, $X \in \{0, 1, \ldots, n\}$. To compute $\mathbb{P}(X = k)$, we condition on the probability of the position B of the gray ball, which is uniformly distributed over the unit interval ($f(p) = 1$). Thus, we have

$$\mathbb{P}(X = k) = \int_0^1 \mathbb{P}(X = k | B = p) f(p) dp = \int_0^1 \binom{n}{k} p^k (1 - p)^{n-k} dp$$

Now, consider a slight variation on the experiment, where we start with $n + 1$ white balls, and again toss them onto the unit interval and then later choose one ball at random to color gray. Using the same X as before, by symmetry, because any one of the $n + 1$ balls is equally likely to be chosen, we have

$$\mathbb{P}(X = k) = \frac{1}{n+1}$$

for $k \in \{0, 1, \ldots, n\}$. Both situations describe the same problem, because it does not matter whether we paint the ball before or after we throw it. Setting the last two equations equal gives the desired result without using calculus.

$$\int_0^1 \binom{n}{k} p^k (1 - p)^{n-k} dp = \frac{1}{n+1}$$

The following code shows where to get the beta distribution from the `scipy` module.

```
>>> from scipy.stats import beta
>>> x = beta(1,1) # create random variable object
>>> print(x.cdf(1))
1.0
```

Given this experiment, it is not too surprising that there is an intimate relationship between the beta distribution and binomial random variables. Suppose we want to estimate the probability of heads for coin-tosses using Bayesian inference. Using this approach, all unknown quantities are treated as random variables. In this case, the probability of heads (p) is the unknown quantity that requires a *prior* distribution. Let us choose the beta distribution as the prior distribution, `Beta(a, b)`. Then, conditioning on p, we have

$$X|p \sim \texttt{binom}(n, p)$$

which says that X is conditionally distributed as a binomial. To get the posterior probability, $f(p|X = k)$, we have the following Bayes rule:

$$f(p|X = k) = \frac{\mathbb{P}(X = k|p)f(p)}{\mathbb{P}(X = k)}$$

with the corresponding denominator:

$$\mathbb{P}(X = k) = \int_0^1 \binom{n}{k} p^k (1 - p)^{n-k} f(p)dp$$

Note that unlike with our experiment before, $f(p)$ is not constant. Without substituting in all of the distributions, we observe that the posterior is a function of p which means that everything else that is not a function of p is a constant. This gives

$$f(p|X = k) \propto p^{a+k-1}(1 - p)^{b+n-k-1}$$

which is another beta distribution with parameters $a + k, b + n - k$. This special relationship in which the beta prior probability distribution on p on data that are conditionally binomial distributed yields the posterior that is also binomial distributed is known as *conjugacy*. We say that the beta distribution is the conjugate prior of the binomial distribution.

2.6.7 Dirichlet-Multinomial Distribution

The Dirichlet-multinomial distribution is a discrete multivariate distribution, also known as the multivariate Polya distribution. The Dirichlet-multinomial distribution arises in situations, where the usual multinomial distribution is inadequate. For example, if a multinomial distribution is used to model the number of balls that land in a set of bins and the multinomial parameter vector (i.e., probabilities of balls landing in particular bins) varies from trial to trial, then the Dirichlet distribution can be used to include variation in those probabilities, because the Dirichlet distribution is defined over a simplex that describes the multinomial parameter vector.

Specifically, suppose we have K rival events, each with probability μ_k. Then, the probability of the vector $\boldsymbol{\mu}$, given that each event has been observed α_k times, is the following:

$$\mathbb{P}(\boldsymbol{\mu}|\boldsymbol{\alpha}) \propto \prod_{k=1}^{K} \mu_k^{\alpha_k - 1}$$

where $0 \le \mu_k \le 1$ and $\sum \mu_k = 1$. Note that this last sum is a constraint that makes the distribution $K - 1$ dimensional. The normalizing constant for this distribution is the multinomial beta function:

$$\text{Beta}(\boldsymbol{\alpha}) = \frac{\prod_{k=1}^{K} \Gamma(\alpha_k)}{\Gamma(\sum_{k=1}^{K} \alpha_k)}$$

The elements of the $\boldsymbol{\alpha}$ vector are also called *concentration* parameters. As before, the Dirichlet distribution can be found in the `scipy.stats` module:

```
>>> from scipy.stats import dirichlet
>>> d = dirichlet([ 1,1,1 ])
>>> d.rvs(3) # get samples from distribution
array([[0.33938968, 0.62186914, 0.03874119],
       [0.21593733, 0.54123298, 0.24282969],
       [0.37483713, 0.07830673, 0.54685613]])
```

Note that each of the rows sums to one. This is because of the $\sum \mu_k = 1$ constraint. We can generate more samples and plot this using `Axes3D` in Matplotlib in Fig. 2.9.

Notice that the generated samples lie on the triangular simplex shown. The corners of the triangle correspond to each of the components in the $\boldsymbol{\mu}$. Using, a nonuniform $\boldsymbol{\alpha} = [2, 3, 4]$ vector, we can visualize the probability density function using the `pdf` method on the `dirichlet` object as shown in Fig. 2.10. By choosing the $\boldsymbol{\alpha} \in \mathbb{R}^3$, the peak of the density function can be moved within the corresponding triangular simplex.

We have seen that the beta distribution generalizes the uniform distribution over the unit interval. Likewise, the Dirichlet distribution generalizes the beta distribution over a vector with components in the unit interval. Recall that the

Fig. 2.9 One thousand
samples from a Dirichlet
distribution with $\boldsymbol{\alpha} = [1, 1, 1]$

Fig. 2.10 Probability density
function for the Dirichlet
distribution with $\boldsymbol{\alpha} = [2, 3, 4]$

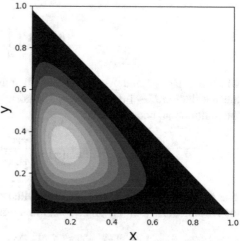

binomial distribution and the beta distribution form a conjugate pair for Bayesian inference, because with $p \sim$ Beta,

$$X|p \sim \text{Binomial}(n, p)$$

That is, the data conditioned on p is binomial distributed. Analogously, the multinomial distribution and the Dirichlet distribution also form such a conjugate pair with multinomial parameter $p \sim$ Dirichlet:

$$X|p \sim \text{multinomial}(n, p)$$

For this reason, the Dirichlet-multinomial distribution is popular in machine learning text processing, because nonzero probabilities can be assigned to words absent in specific documents, which helps generalization performance.

2.6.8   Negative Binomial Distribution

The negative binomial distribution is used to characterize the number of trials until a specified number of failures (r) occur. For example, suppose 1 indicates failure and 0 indicates success. Then, the negative binomial distribution characterizes the probability of n=2 successes with four (k=4) prior failures, with the sequence terminating in a success (e.g., 001001) with success probability $\mathbb{P}(1) = 1/3$. For the negative binomial distribution, $\mathbb{P}(k) = \frac{80}{729}$.

The probability mass function is the following:

$$\mathbb{P}(k) = \binom{n+k-1}{n-1} p^n (1-p)^k$$

which describes the distribution of k failures in a sequence of trials with p success probability before n successes occur. The mean and variance of this distribution is the following:

$$\mathbb{E}(k) = \frac{n(1-p)}{p}$$

$$\mathbb{V}(k) = \frac{n(1-p)}{p^2}$$

The following simulation shows an example sequence generated for the negative binomial distribution.

```
>>> import random
>>> n=2      # num of failures
>>> p=1/3    # prob of failure
>>> nc = 0 # counter
>>> seq= []
>>> while nc< n:
...         v,=random.choices([0,1],[1-p,p])
...         seq.append(v)
...         nc += (v == 1)
...
>>> seq,len(seq)
([1, 0, 0, 0, 1], 5)
```

Keep in mind that the negative binomial distribution characterizes the family of such sequences with the specified number of failures.

2.6.9   Negative Multinomial Distribution

The discrete negative multinomial distribution is an extension of the negative binomial distribution to account for more than two possible outcomes. That is, there are other alternatives whose respective probabilities sum to one less the failure probability, $p_f = 1 - \sum_{k=1}^n p_i$. For example, a random sample from this distribution with parameters $n = 2$ (number of observed failures) and with $p_a = \frac{1}{3}$, $p_b = \frac{1}{2}$ means that the failure probability, $p_f = \frac{1}{6}$. Thus, a sample from this distribution like [2, 9] means that 2 of the a objects were observed in the sequence, 9 of the b objects were observed, and there were two failure symbols (say, F) with one of them at the end of the sequence.

The probability mass function is the following:

$$\mathbb{P}(\mathbf{k}) = (n)_{\sum_{i=0}^m k_i} p_f^n \prod_{i=1}^m \frac{p_i^{k_i}}{k_i!}$$

where p_f is the probability of failure and the other p_i terms are the probabilities of the other alternatives in the sequence. The $(a)_n$ notation is the rising factorial function (e.g., $a_3 = a(a+1)(a+2)$). The mean and variance of this distribution is the following:

$$\mathbb{E}(\mathbf{k}) = \frac{n}{p_f}\mathbf{p}$$

$$\mathbb{V}(k) = \frac{n}{p_f^2}\mathbf{p}\mathbf{p}^T + \frac{n}{p_f}\,\text{diag}(\mathbf{p})$$

The following simulation shows the sequences generated for the negative multinomial distribution.

```
>>> import random
>>> from collections import Counter
>>> n=2                          # num of failure items
>>> p=[1/3,1/2]                  # prob of other non-failure items
>>> items = ['a','b','F']        # F marks failure item
>>> nc = 0                       # counter
>>> seq= []
>>> while nc< n:
...         v,=random.choices(items,p+[1-sum(p)])
...         seq.append(v)
...         nc += (v == 'F')
...
>>> c=Counter(seq)
>>> print(c)
Counter({'a': 5, 'b': 2, 'F': 2})
```

The values of the `Counter` dictionary above are the $\mathbf{k}$ vectors in the probability mass function for the negative multinomial distribution. Importantly, these are not the probabilities of a particular sequence but of a family of sequences with the same corresponding `Counter` values. The probability mass function implemented in Python is the following:

```
>>> from scipy.special import factorial
>>> import numpy as np
>>> def negative_multinom_pdf(p,n):
...      assert len(n) == len(p)
...      term = [i**j for i,j in zip(p,n)]
...      num=np.prod(term)*(1-sum(p))*factorial(sum(n))
...      den = np.prod([factorial(i) for i in n])
...      return num/den
...
```

Evaluating this with the prior `Counter` result,

```
>>> negative_multinom_pdf([1/3,1/2],[c['a'],c['b']])
0.00360082304526749
```

2.7 Information Entropy

We are in a position to discuss information entropy. This will give us a powerful perspective on how information passes between experiments and will prove important in certain machine learning algorithms.

There used to be a TV game show where the host would hide a prize behind one of three doors and the contestant would have to pick one of the doors. However, before opening the door of the contestant's choice, the host would open one of the other doors and ask the contestant if she wanted to change her selection. This is the classic *Monty Hall* problem. The question is should the contestant stay with her original choice or switch after seeing what the host has revealed? From the information theory perspective, does the information environment change when the host reveals what is behind one of the doors? The important detail here is that the host *never* opens the door with the prize behind it, regardless of the contestant's choice. That is, the host *knows* where the prize is, but he does not reveal that information directly to the contestant. This is the fundamental problem information theory addresses—how to aggregate and reason about partial information. We need a concept of information that can accommodate this kind of question.

2.7.1 Information Theory Concepts

The Shannon *information content* of an outcome x is defined as

$$h(x) = \log_2 \frac{1}{P(x)}$$

where $P(x)$ is the probability of x. The *entropy* of the ensemble X is defined to be the Shannon information content of

$$H(X) = \sum_x P(x) \log_2 \frac{1}{P(x)}$$

It is no accident that the entropy has this functional form as the expectation of $h(x)$. It leads to a deep and powerful theory of information.

To get some intuition about what information entropy means, consider a sequence of three-bit numbers, where each individual bit is equally likely. Thus, the individual information content of a single bit is $h(x) = \log_2(2) = 1$. The units of entropy are *bits*, so this says that information content of a single bit is one bit. Because the three-bit number has elements that are mutually independent and equally likely, the information entropy of the three-bit number is $h(X) = 2^3 \times \log_2(2^3)/8 = 3$. Thus, the basic idea of information content at least makes sense at this level.

A better way to interpret this question is as follows: how much information would I have to provide in order to uniquely encode an arbitrary three-bit number? In this case, you would have to answer three questions: *Is the first bit zero or one? Is the second bit zero or one? Is the third bit zero or one?* Answering these questions uniquely specifies the unknown three-bit number. Because the bits are mutually independent, knowing the state of any of the bits does not inform the remainder.

Next, let us consider a situation that lacks this mutual independence. Suppose in a group of nine otherwise identical balls there is a heavier one. Furthermore, we also have a measuring scale that indicates whether one side is heavier, lighter, or equal to the other. How could we identify the heavier ball? At the outset, the information content, which measures the uncertainty of the situation, is $\log_2(9)$, because one of the nine balls is heavier. Figure 2.11 shows one strategy. We could arbitrarily select out one of the balls (shown by the square), leaving the remaining eight to be balanced. The thick, black horizontal line indicates the scale. The items below and above this line indicate the counterbalanced sides of the scale.

If we get lucky, the scale will report that the group of four walls on either side of the balance are equal in weight. This means that the ball that was omitted is the heavier one. This is indicated by the hashed left-pointing arrow. In this case, all the uncertainty has evaporated, and the *informational value* of that one weighing is equal to $\log_2(9)$. In other words, the scale has reduced the uncertainty to zero (i.e., found the heavy ball). On the other hand, the scale could report that the upper group of four balls is heavier (black, upward-pointing arrow) or lighter (gray, downward-pointing arrow). In this case, we cannot isolate the heavier ball until we perform all of the indicated weighings, moving from left to right. Specifically, the four balls on the heavier side have to be split by a subsequent weighing into two balls and then to one ball before the heavy ball can be identified. Thus, this process takes three weighings. The first one has information content $\log_2(9/8)$, the next has $\log_2(4)$, and the final one has $\log_2(2)$. Adding all these up sums to $\log_2(9)$. Thus, whether or not the heavier ball is isolated in the first weighing, the strategy consumes $\log_2(9)$ bits, as it must, to find the heavy ball.

However, this is not the only strategy. Figure 2.12 shows another. In this approach, the nine balls are split up into three groups of three balls apiece. Two

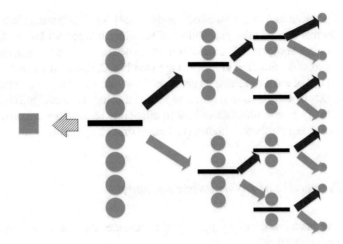

Fig. 2.11 One heavy ball is hidden among eight identical balls. By weighing groups sequentially, we can determine the heavy ball

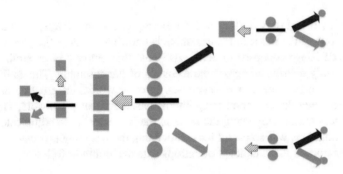

Fig. 2.12 For this strategy, the balls are broken up into three groups of equal size and subsequently weighed

groups are weighed. If they are of equal weight, then this means the heavier ball is in the group that was left out (dashed arrow). Then, this group is split into two groups, with one element left out. If the two balls on the scale weigh the same, then it means the excluded one is the heavy one. Otherwise, it is one of the balls on the scale. The same process follows if one of the initially weighed groups is heavier (black upward-facing arrow) or lighter (gray lower-facing arrow). As before the information content of the situation is $\log_2(9)$. The first weighing reduces the uncertainty of the situation by $\log_2(3)$ and the subsequent weighing reduces it by another $\log_2(3)$. As before, these sum to $\log_2(9)$, but here we only need two weighings, whereas the first strategy in Fig. 2.11 takes an average of $1/9 + 3 * 8/9 \approx 2.78$ weighings, which is more than two from the second strategy in Fig. 2.12.

Why does the second strategy use fewer weighings? To reduce weighings, we need each weighing to adjudicate equally probable situations as many times as possible. Choosing one of the nine balls at the outset (i.e, first strategy in Fig. 2.11) does not do this because the probability of selecting the correct ball is 1/9. This does not create a equiprobable situation in the process. The second strategy leaves an equally probable situation at every stage (see Fig. 2.12), so it extracts the most information out of each weighing as possible. Thus, the information content tells us how many bits of information have to be resolved using *any* strategy (i.e., $\log_2(9)$ in this example). It also illuminates how to efficiently remove uncertainty, namely, by adjudicating equiprobable situations as many times as possible.

2.7.2   Properties of Information Entropy

Now that we have the flavor of the concepts, consider the following properties of the information entropy:

$$H(X) \geq 0$$

with equality if and only if $P(x) = 1$ for exactly one x. Intuitively, this means that when just one of the items in the ensemble is known absolutely (i.e., with $P(x) = 1$), the uncertainty collapses to zero. Also note that entropy is maximized when P is uniformly distributed across the elements of the ensemble. This is illustrated in Fig. 2.13 for the case of two outcomes. In other words, information entropy is maximized when the two conflicting alternatives are equally probable. This is the mathematical reason why using the scale in the last example to adjudicate equally probable situations was so useful for abbreviating the weighing process.

Most importantly, the concept of entropy extends jointly as follows:

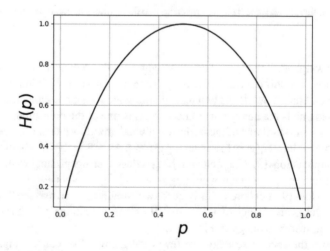

Fig. 2.13 The information entropy is maximized when $p = 1/2$

$$H(X, Y) = \sum_{x,y} P(x, y) \log_2 \frac{1}{P(x, y)}$$

If and only if X and Y are independent, entropy becomes additive:

$$H(X, Y) = H(X) + H(Y)$$

2.7.3 Kullback-Leibler Divergence

Notions of information entropy lead to notions of distance between probability distributions that will become important for machine learning methods. The Kullback-Leibler divergence between two probability distributions P and Q that are defined over the same set is defined as

$$D_{KL}(P, Q) = \sum_{x} P(x) \log_2 \frac{P(x)}{Q(x)}$$

Note that $D_{KL}(P, Q) \geq 0$ with equality if and only if $P = Q$. Sometimes the Kullback-Leibler divergence is called the Kullback-Leibler distance, but it is not formally a distance metric, because it is asymmetrical in P and Q. The Kullback-Leibler divergence defines a relative entropy as the loss of information if P is modeled in terms of Q. There is an intuitive way to interpret the Kullback-Leibler divergence and understand its lack of symmetry. Suppose we have a set of messages to transmit, each with a corresponding probability $\{(x_1, P(x_1)), (x_2, P(x_2)), \ldots, (x_n, P(x_n))\}$. Based on what we know about information entropy, it makes sense to encode the length of the message by $\log_2 \frac{1}{p(x)}$ bits. This parsimonious strategy means that more frequent messages are encoded with fewer bits. Thus, we can rewrite the entropy of the situation as before:

$$H(X) = \sum_{k} P(x_k) \log_2 \frac{1}{P(x_k)}$$

Now, suppose we want to transmit the same set of messages, but with a different set of probability weights, $\{(x_1, Q(x_1)), (x_2, Q(x_2)), \ldots, (x_n, Q(x_n))\}$. In this situation, we can define the cross-entropy as

$$H_q(X) = \sum_{k} P(x_k) \log_2 \frac{1}{Q(x_k)}$$

Note that only the purported length of the encoded message has changed, not the probability of that message. The difference between these two is the Kullback-Leibler divergence:

$$D_{KL}(P, Q) = H_q(X) - H(X) = \sum_x P(x) \log_2 \frac{P(x)}{Q(x)}$$

In this light, the Kullback-Leibler divergence is the average difference in the encoded lengths of the same set of messages under two different probability regimes. This should help explain the lack of symmetry of the Kullback-Leibler divergence up because left to themselves, P and Q would provide the optimal-length encodings separately, but there can be no necessary symmetry in how each regime would rate the informational value of each message ($Q(x_i)$ versus $P(x_i)$). Given that each encoding is optimal length in its own regime means that it must therefore be at least suboptimal in another, thus giving rise to the Kullback-Leibler divergence. In the case where the encoding length of all messages remains the same for the two regimes, then the Kullback-Leibler divergence is zero.[2]

2.7.4 Conditional Entropy and Mutual Information

Given the joint density $p(X, Y)$, the *conditional entropy* is defined as the following:

$$H(Y|X) = -\mathbb{E}_{X,Y} \log p(Y|X)$$

This leads to the following intuitive (chain rule) relationship:

$$H(X, Y) = H(X) + H(Y|X)$$

That is, the joint information in X, Y is the sum of the information in X plus the new information in Y, conditioned on X. The conditioning on X is important, because we already have all of the information about X from $H(X)$, so we just need the additional nonredundant information from Y. To derive this result, note that $p(X, Y) = p(Y|X)p(X)$, so that taking the following expectation of the logarithm, we obtain the following:

$$\mathbb{E} \log p(X, Y) = \mathbb{E} \log p(Y|X) + \mathbb{E} \log p(X)$$
$$H(X, Y) = H(Y|X) + H(X)$$

Importantly, $H(Y|X) \neq H(X|Y)$ but $H(X) - H(X|Y) = H(Y) - H(Y|X)$.

Mutual Information The Kullback-Leibler Divergence between $p(X, Y)$ and $p(X)p(Y)$ is the *mutual information*:

---

[2] The best, easy-to-understand presentation of this material is Chapter 4 of Mackay's text [28]. Another good reference is Chapter 4 of [14].

$$I(X; Y) = D_K L(p(X, Y), p(X)p(Y))$$

$$= \sum p(X, Y) \log_2 \frac{p(X, Y)}{p(X)p(Y)}$$

By writing out the conditional distributions, it is straightforward to show that

$$I(X; Y) = H(X) - H(X|Y) = H(Y) - H(Y|X)$$

where the last equality comes from symmetry. The mutual information represents the reduction in the uncertainty of X, given the knowledge of Y. Because we have $H(X, Y) = H(X) + H(Y|X)$, we have

$$I(X; Y) = H(X) + H(Y) - H(X, Y)$$

Interestingly, taking $Y = X$ gives the so-called *self-information*:

$$I(X; X) = H(X)$$

which is just the entropy of the random variable.

2.7.5 Cross-Entropy as Maximum Likelihood

Reconsidering maximum likelihood from our statistics chapter in more general terms, we have

$$\theta_{ML} = \arg \max_{\theta} \sum_{i=1}^{n} \log p_{model}(x_i; \theta)$$

where p_{model} is the assumed underlying probability density function parameterized by θ for the x_i data elements. Dividing the above summation by n does not change the derived optimal values, but it allows us to rewrite this using the empirical density function for x as the following:

$$\theta_{ML} = \arg \max_{\theta} \mathbb{E}_{x \sim \hat{p}_{data}} (\log p_{model}(x_i; \theta))$$

Note that we have the distinction between p_{data} and $\hat{p}_{data}$, where the former is the unknown distribution of the data and the latter is the estimated distribution of the data we have on hand.

The cross-entropy can be written as the following:

$$D_{KL}(P, Q) = \mathbb{E}_{X \sim P}(\log P(x)) - \mathbb{E}_{X \sim P}(\log Q(x))$$

where $X \sim P$ means the random variable X has distribution P. Thus, we have

$$\theta_{ML} = \arg\max_{\theta} D_{KL}(\hat{p}_{data}, p_{model})$$

That is, we can interpret maximum likelihood as the cross-entropy between the p_{model} and the $\hat{p}_{data}$ distributions. The first term has nothing to do with the estimated θ, so maximizing this is the same as minimizing the following:

$$\mathbb{E}_{x \sim \hat{p}_{data}}(\log p_{model}(x_i; \theta))$$

because information entropy is always nonnegative. The important interpretation is that maximum likelihood is an attempt to choose θ model parameters that make the empirical distribution of the data match the model distribution.

2.8   Moment Generating Functions

Generating moments usually involves integrals that are extremely difficult to compute. Moment generating functions make this much, much easier. The moment generating function is defined as

$$M(t) = \mathbb{E}(\exp(tX))$$

The first moment is the mean, which we can easily compute from $M(t)$ as

$$\frac{dM(t)}{dt} = \frac{d}{dt}\mathbb{E}(\exp(tX)) = \mathbb{E}\frac{d}{dt}(\exp(tX))$$
$$= \mathbb{E}(X\exp(tX))$$

Now, we have to set $t = 0$ and we have the mean:

$$M^{(1)}(0) = \mathbb{E}(X)$$

continuing this derivative process again, we obtain the second moment as

$$M^{(2)}(t) = \mathbb{E}(X^2\exp(tX))$$
$$M^{(2)}(0) = \mathbb{E}(X^2)$$

With this in hand, we can easily compute the variance as

$$\mathbb{V}(X) = \mathbb{E}(X^2) - \mathbb{E}(X)^2 = M^{(2)}(0) - M^{(1)}(0)^2$$

Example Returning to our favorite binomial distribution, let us compute some moments using Sympy.

```
>>> import sympy as S
>>> from sympy import stats
>>> p,t = S.symbols('p t',positive=True)
>>> x=stats.Binomial('x',10,p)
>>> mgf = stats.E(S.exp(t*x))
```

Now, let us compute the first moment (aka, mean) using the usual integration method and using moment generating functions:

```
>>> print(S.simplify(stats.E(x)))
10*p
>>> print(S.simplify(S.diff(mgf,t).subs(t,0)))
10*p
```

Otherwise, we can compute this directly as follows:

```
>>> print(S.simplify(stats.moment(x,1)))  # mean
10*p
>>> print(S.simplify(stats.moment(x,2)))  # 2nd moment
10*p*(9*p + 1)
```

In general, the moment generating function for the binomial distribution is the following:

$$M_X(t) = (p(e^t - 1) + 1)^n$$

A key aspect of moment generating functions is that they are unique identifiers of probability distributions. By the uniqueness theorem, given two random variables X and Y, if their respective moment generating functions are equal, then the corresponding probability distribution functions are equal.

Example Let us use the uniqueness theorem to consider the following problem. Suppose we know that the probability distribution of X given $U = p$ is binomial with parameters n and p. For example, suppose X represents the number of heads in n coin flips, given the probability of heads is p. We want to find the unconditional distribution of X. Writing out the moment generating function as the following:

$$\mathbb{E}(e^{tX}|U = p) = (pe^t + 1 - p)^n$$

Because U is uniform over the unit interval, we can integrate this part out

$$\mathbb{E}(e^{tX}) = \int_0^1 (pe^t + 1 - p)^n dp$$

$$= \frac{1}{n+1} \frac{e^{t(n+1)-1}}{e^t - 1}$$

$$= \frac{1}{n+1}(1 + e^t + e^{2t} + e^{3t} + \ldots + e^{nt})$$

Thus, the moment generating function of X corresponds to that of a random variable that is equally likely to be any of the values $0, 1, \ldots, n$. This is another way of saying that the distribution of X is discrete uniform over $\{0, 1, \ldots, n\}$. Concretely, suppose we have a box of coins whose individual probability of heads is unknown and that we dump the box on the floor, spilling all of the coins. If we then count the number of coins facing heads-up, that distribution is uniform.

Moment generating functions are useful for deriving distributions of sums of independent random variables. Suppose X_1 and X_2 are independent and $Y = X_1 + X_2$. Then, the moment generating function of Y follows from the properties of the expectation:

$$M_Y(t) = \mathbb{E}(e^{tY}) = \mathbb{E}(e^{tX_1 + tX_2})$$

$$= \mathbb{E}(e^{tX_1} e^{tX_2}) = \mathbb{E}(e^{tX_1})\mathbb{E}(e^{tX_2})$$

$$= M_{X_1}(t) M_{X_2}(t)$$

Example Suppose we have two normally distributed random variables, $X_1 \sim \mathcal{N}(\mu_1, \sigma_1)$ and $X_2 \sim \mathcal{N}(\mu_2, \sigma_2)$ with $Y = X_1 + X_2$. We can save some tedium by exploring this in Sympy:

```
>>> S.var('x:2',real=True)
(x0, x1)
>>> S.var('mu:2',real=True)
(mu0, mu1)
>>> S.var('sigma:2',positive=True)
(sigma0, sigma1)
>>> S.var('t',positive=True)
t
>>> x0=stats.Normal(x0,mu0,sigma0)
>>> x1=stats.Normal(x1,mu1,sigma1)
```

Programming Tip

The `S.var` function defines the variable and injects it into the global namespace. This is sheer laziness. It is more expressive to define variables explicitly as in `x = S.symbols('x')`. Also notice that we used the Greek names for the `mu` and `sigma` variables. This will come in handy later when we want to render the equations in the Jupyter notebook which understands how to typeset these symbols in . The `var('x:2')` creates two symbols, `x0` and `x1`. Using the colon this way makes it easy to generate array-like sequences of symbols.

In the next block, we compute the moment generating functions:

```
>>> mgf0=S.simplify(stats.E(S.exp(t*x0)))
>>> mgf1=S.simplify(stats.E(S.exp(t*x1)))
>>> mgfY=S.simplify(mgf0*mgf1)
```

The moment generating functions of a normally distributed random variable is the following:

$$e^{\mu_0 t + \frac{\sigma_0^2 t^2}{2}}$$

Note the coefficients of t. To show that Y is normally distributed, we want to match the moment generating function of Y to this format. The following is the form of the moment generating function of Y:

$$M_Y(t) = e^{\frac{t}{2}(2\mu_0 + 2\mu_1 + \sigma_0^2 t + \sigma_1^2 t)}$$

We can extract the exponent using Sympy and collect on the t variable using the following code:

```
>>> S.collect(S.expand(S.log(mgfY)),t)
t**2*(sigma0**2/2 + sigma1**2/2) + t*(mu0 + mu1)
```

Thus, by the uniqueness theorem, Y is normally distributed with $\mu_Y = \mu_0 + \mu_1$ and $\sigma_Y^2 = \sigma_0^2 + \sigma_1^2$.

> **Programming Tip**
>
> When using the Jupyter notebook, you can do `S.init_printing` to get the mathematical typesetting to work in the browser. Otherwise, if you want to keep the raw expression and to selectively render to , then you can `from IPython.display import Math`, and then use `Math(S.latex(expr))` to see the typeset version of the expression.

2.9 Monte Carlo Sampling Methods

So far, we have studied analytical ways to transform random variables and how to augment these methods using Python. In spite of all this, we frequently must resort to purely numerical methods to solve real-world problems. Hopefully, now that we have seen the deeper theory, these numerical methods will feel more concrete. Suppose we want to generate samples of a given density, $f(x)$, given we already can generate samples from a uniform distribution, $\mathcal{U}[0, 1]$. How do we know a random sample v comes from the $f(x)$ distribution? One approach is to look at how a histogram of samples of v approximates $f(x)$. Specifically,

Fig. 2.14 The histogram
approximates the target
probability density

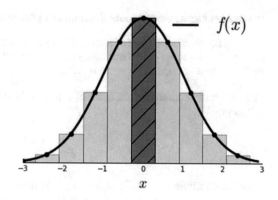

$$\mathbb{P}(v \in N_\Delta(x)) = f(x)\Delta x \tag{2.13}$$

which says that the probability that a sample is in some N_Δ neighborhood of x
is approximately $f(x)\Delta x$. Figure 2.14 shows the target probability density function
$f(x)$ and a histogram that approximates it. The histogram is generated from samples
v. The hatched rectangle in the center illustrates Eq. 2.9. The area of this rectangle
is approximately $f(x)\Delta x$ where $x = 0$, in this case. The width of the rectangle is
$N_\Delta(x)$ The quality of the approximation may be clear visually, but to know that v
samples are characterized by $f(x)$, we need the statement of Eq. 2.9, which says that
the proportion of samples v that fill the hatched rectangle is approximately equal to
$f(x)\Delta x$.

Now that we know how to evaluate samples v that are characterized by the
density $f(x)$, let us consider how to create these samples for both discrete and
continuous random variables.

2.9.1 Inverse CDF Method for Discrete Variables

Suppose we want to generate samples from a fair six-sided die. Our workhouse
uniform random variable is defined continuously over the unit interval, and the fair
six-sided die is discrete. We must first create a mapping between the continuous
random variable u and the discrete outcomes of the die. This mapping is shown in
Fig. 2.15, where the unit interval is broken up into segments, each of length $1/6$.
Each individual segment is assigned to one of the die outcomes. For example, if
$u \in [1/6, 2/6)$, then the outcome for the die is 2. Because the die is fair, all segments
on the unit interval are the same length. Thus, our new random variable v is derived
from u by this assignment.

For example, for $v = 2$, we have

$$\mathbb{P}(v = 2) = \mathbb{P}(u \in [1/6, 2/6)) = 1/6$$

Fig. 2.15 A uniform distribution random variable on the unit interval is assigned to the six outcomes of a fair die using these segments

where, in the language of Eq. 2.9, $f(x) = 1$ (uniform distribution), $\Delta x = 1/6$, and $N_\Delta(2) = [1/6, 2/6)$. Naturally, this pattern holds for all the other die outcomes in $\{1, 2, 3, .., 6\}$. Let us consider a quick simulation to make this concrete. The following code generates uniform random samples and stacks them in a Pandas dataframe.

```
>>> import pandas as pd
>>> import numpy as np
>>> from pandas import DataFrame
>>> u= np.random.rand(100)
>>> df = DataFrame(data=u,columns=['u'])
```

The next block uses pd.cut to map the individual samples to the set $\{1, 2, \ldots, 6\}$ labeled v.

```
>>> labels = [1,2,3,4,5,6]
>>> df['v']=pd.cut(df.u,np.linspace(0,1,7),
...                include_lowest=True,labels=labels)
```

The v column contains the samples drawn from the fair die.

```
>>> df.head()
          u  v
0  0.356225  3
1  0.466557  3
2  0.776817  5
3  0.836790  6
4  0.037928  1
```

The following is a count of the number of samples in each group. There should be roughly the same number of samples in each group because the die is fair.

```
>>> df.groupby('v').count()
    u
v
1  17
2  15
3  18
4  20
5  14
6  16
```

So far, so good. We now have a way to simulate a fair die from a uniformly distributed random variable.

To extend this to unfair die, we need only make some small adjustments to this code. For example, suppose that we want an unfair die so that $\mathbb{P}(1) = \mathbb{P}(2) = $

$\mathbb{P}(3) = 1/12$ and $\mathbb{P}(4) = \mathbb{P}(5) = \mathbb{P}(6) = 1/4$. The only change we have to make is with pd.cut as follows:

```
>>> df['v']=pd.cut(df.u, [0,1/12,2/12,3/12,2/4,3/4,1],
...                     include_lowest=True,labels=labels)
>>> df.groupby('v').count()/df.shape[0]
        u
v
1    0.10
2    0.07
3    0.05
4    0.28
5    0.29
6    0.21
```

where now these are the individual probabilities of each digit. You can take more than 100 samples to get a clearer view of the individual probabilities, but the mechanism for generating them is the same. The method is called the inverse CDF [3] method, because the CDF (namely, [0,1/12,2/12,3/12,2/4,3/4,1]) in the last example has been inverted (using the pd.cut method) to generate the samples. The inversion is easier to see for continuous variables, which we consider next.

2.9.2   Inverse CDF Method for Continuous Variables

The method above applies to continuous random variables, but now we have to squeeze the intervals down to individual points. In the example above, our inverse function was a piecewise function that operated on uniform random samples. In this case, the piecewise function collapses to a continuous inverse function. We want to generate random samples v for an invertible CDF. Consider $u \sim \mathcal{U}[0, 1]$ and the following:

$$\mathbb{P}_u(F_v(x) < u < F_v(x + \Delta x)) = F_v(x + \Delta x) - F_v(x)$$

$$= \int_x^{x+\Delta x} f(v)dv \approx f(x)\Delta x$$

which says that the probability that the sample u is contained in the interval $[F_v(x), F_v(x+\Delta x)]$ is approximately equal to $f_v(x)\Delta x$. For invertible F_v, we have the following:

$$\mathbb{P}_u(x < F_v^{-1}(u) < x + \Delta x) = \mathbb{P}_u(F_v(x) < u < F_v(x + \Delta x))$$

$$= F_v(x + \Delta x) - F_v(x)$$

$$= \int_x^{x+\Delta x} f_v(v)dv \approx f_v(x)\Delta x$$

---

[3] Cumulative density function. Namely, $F(x) = \mathbb{P}(X < x)$.

This means that $v = F_v^{-1}(u)$ is distributed according to $f_v(x)$, which is what we want.

Let us try this to generate samples from the exponential distribution:

$$f_\alpha(x) = \alpha e^{-\alpha x}$$

which has the following CDF:

$$F(x) = 1 - e^{-\alpha x}$$

and corresponding inverse:

$$F^{-1}(u) = \frac{1}{\alpha} \ln \frac{1}{(1-u)}$$

Now, all we have to do is generate some uniformly distributed random samples and then feed them into F^{-1}.

```
>>> from numpy import array, log
>>> import scipy.stats
>>> alpha = 1.    # distribution parameter
>>> nsamp = 1000 # num of samples
>>> # define uniform random variable
>>> u=scipy.stats.uniform(0,1)
>>> # define inverse function
>>> Finv=lambda u: 1/alpha*log(1/(1-u))
>>> # apply inverse function to samples
>>> v = array(list(map(Finv,u.rvs(nsamp))))
```

Now, we have the samples from the exponential distribution, but how do we know the method is correct with samples distributed accordingly? Fortunately, scipy.stats already has an exponential distribution, so we can check our work against the reference using a *probability plot* (i.e., also known as a *quantile-quantile* plot). The following code sets up the probability plot from scipy.stats.

```
fig,ax=subplots()
scipy.stats.probplot(v,(1,),dist='expon',plot=ax)
```

Note that we have to supply an axes object (ax) for it to draw on. The result is Fig. 2.16. The more the samples line match the diagonal line, the more they match the reference distribution (i.e., exponential distribution in this case). You may also want to try dist=norm in the code above to see what happens when the normal distribution is the reference distribution.

Fig. 2.16 The samples created using the inverse cdf method match the exponential reference distribution

2.9.3 Rejection Method

In some cases, inverting the CDF may be impossible. The *rejection* method can handle this situation. The idea is to pick two uniform random variables $u_1 \sim \mathcal{U}[a, b]$ and $u_2 \sim \mathcal{U}[0, 1]$ so that

$$\mathbb{P}\left(u_1 \in N_\Delta(x) \bigwedge u_2 < \frac{f(u_1)}{M}\right) \approx \frac{\Delta x}{b - a} \frac{f(u_1)}{M}$$

where we take $x = u_1$ and $f(x) < M$. This is a two-step process. First, draw u_1 and u_2. Second, feed u_1 into $f(x)$, and if $u_2 < f(u_1)/M$, then u_1 is a valid sample for $f(x)$. Thus, u_1 is the proposed sample from f that may or may not be rejected, depending on u_2. The only job of the M constant is to scale down the $f(x)$, so that the u_2 variable can span the range. The *efficiency* of this method is the probability of accepting u_1, which comes from integrating out the above approximation:

$$\int \frac{f(x)}{M(b - a)} dx = \frac{1}{M(b - a)} \int f(x) dx = \frac{1}{M(b - a)}$$

This means that we don't want an unnecessarily large M, because that makes it more likely that samples will be discarded.

Fig. 2.17 The rejection method generate samples in the histogram that nicely match the target distribution. Unfortunately, the efficiency is not so good

Let us try this method for a density that does not have a continuous inverse.[4]

$$f(x) = \exp\left(-\frac{(x-1)^2}{2x}\right)(x+1)/12$$

where $x > 0$. The following code implements the rejection plan.

```
>>> x = np.linspace(0.001,15,100)
>>> f= lambda x: np.exp(-(x-1)**2/2./x)*(x+1)/12.
>>> fx = f(x)
>>> M=0.3                          # scale factor
>>> u1 = np.random.rand(10000)*15  # uniform random samples
        scaled out
>>> u2 = np.random.rand(10000)     # uniform random samples
>>> idx,= np.where(u2<=f(u1)/M)    # rejection criterion
>>> v = u1[idx]
```

Figure 2.17 shows a histogram of the so-generated samples that nicely fits the probability density function. The title in the figure shows the efficiency (the number of *rejected* samples), which is poor. It means that we threw away most of the proposed samples. Thus, even though there is nothing conceptually wrong with this result, the low efficiency must be fixed, as a practical matter. Figure 2.18 shows where the proposed samples were rejected. Samples under the curve were retained (i.e., $u_2 < \frac{f(u_1)}{M}$), but the vast majority of the samples are outside this umbrella.

The rejection method uses u_1 to select along the domain of $f(x)$, and the other u_2 uniform random variable decides whether to accept or not. One idea would be to choose u_1 so that x values are coincidentally those that are near the peak of $f(x)$, instead of uniformly anywhere in the domain, especially near the tails, which are

---
[4] Note that this example density does not *exactly* integrate out to one like a probability density function should, but the normalization constant for this is distracting for our purposes here.

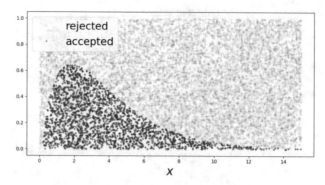

Fig. 2.18 The proposed samples under the curve were accepted and the others were not. This shows the majority of samples were rejected

low probability anyway. Now, the trick is to find a new density function $g(x)$ to sample from that has a similar concentration of probability density. One way to find such a $g(x)$ is to familiarize oneself with the probability density functions that have adjustable parameters and fast random sample generators already (e.g., see `scipy.stats`). For densities with bounded support, the family of β densities is a good place to start.

To be explicit, what we want is $u_1 \sim g(x)$ so that, returning to our earlier argument,

$$\mathbb{P}\left(u_1 \in N_\Delta(x) \bigwedge u_2 < \frac{f(u_1)}{M}\right) \approx g(x)\Delta x \frac{f(u_1)}{M}$$

but this is *not* what we need here. The problem is with the second part of the logical $\bigwedge$ conjunction, $u < \frac{f(u_1)}{M}$. We need to put a condition there that will give us something proportional to $f(x)$. Let us define the following:

$$h(x) = \frac{f(x)}{g(x)} \tag{2.14}$$

with corresponding maximum on the domain as h_{max} and then go back and construct the second part of the clause as

$$\mathbb{P}\left(u_1 \in N_\Delta(x) \bigwedge u_2 < \frac{h(u_1)}{h_{max}}\right) \approx g(x)\Delta x \frac{h(u_1)}{h_{max}} = f(x)/h_{max}$$

Recall that satisfying this criterion means that $u_1 = x$. As before, we can estimate the probability of acceptance of the u_1 as $1/h_{max}$.

Now, how to construct the $g(x)$ function in the denominator of Eq. 2.9.3? Here's where familarity with some standard probability densities pays off. For this case, we choose the χ^2 distribution. The following plots the $g(x)$ and $f(x)$ (left plot) and the

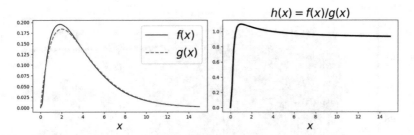

Fig. 2.19 The plot on the right shows $h(x) = f(x)/g(x)$ and the one on the left shows $f(x)$ and $g(x)$ separately

Fig. 2.20 Using the updated method, the histogram matches the target probability density function with high efficiency

corresponding $h(x) = f(x)/g(x)$ (right plot). Note that $g(x)$ and $f(x)$ have peaks that almost coincide, which is what we are looking for (Fig. 2.19).

```
>>> ch=scipy.stats.chi2(4) # chi-squared
>>> h = lambda x: f(x)/ch.pdf(x) # h-function
```

Now, let us generate some samples from this χ^2 distribution with the rejection method.

```
>>> hmax=h(x).max()
>>> u1 = ch.rvs(5000)          # samples from chi-square
         distribution
>>> u2 = np.random.rand(5000) # uniform random samples
>>> idx = (u2 <= h(u1)/hmax)  # rejection criterion
>>> v = u1[idx]               # keep these only
```

Using the χ^2 distribution with the rejection method results in throwing away less than 10% of the generated samples compared with our prior example, where we threw out at least 80%. This is dramatically more efficient! Figure 2.20 shows that the histogram and the probability density function match. For completeness, Fig. 2.21 shows the samples with the corresponding threshold $h(x)/h_{max}$ that was used to select them.

Fig. 2.21 Fewer proposed points were rejected in this case, which means better efficiency

2.10　Sampling Importance Resampling

An alternative to the rejection method that does not involve rejecting samples or coming up with M bounds or bounding functions is the sampling importance resampling (SIR) method. Choose a tractable g probability density function, and draw a n samples from it, $\{x_i\}_{i=1}^n$. Our objective is to derive samples f. Next, compute the following:

$$q_i = \frac{w_i}{\sum w_i}$$

where

$$w_i = \frac{f(x_i)}{g(x_i)}$$

The q_i defines a probability mass function whose samples approximate samples from f. To see this, consider

$$\mathbb{P}(X \le a) = \sum_{i=1}^n q_i \mathbb{I}_{(-\infty,a]}(x_i)$$

$$= \frac{\sum_{i=1}^n w_i \mathbb{I}_{(-\infty,a]}(x_i)}{\sum_{i=1}^n w_i}$$

$$= \frac{\frac{1}{n}\sum_{i=1}^n \frac{f(x_i)}{g(x_i)} \mathbb{I}_{(-\infty,a]}(x_i)}{\frac{1}{n}\sum_{i=1}^n \frac{f(x_i)}{g(x_i)}}$$

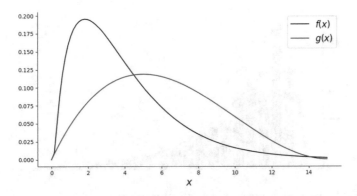

Fig. 2.22 Two densities to generate samples using SIR

Because the x_i samples are generated from the g probability distribution, the numerator is approximately

$$\mathbb{E}_g\left(\frac{f(x)}{g(x)}\right) = \int_{-\infty}^{a} f(x)dx$$

which gives, because the denominator is just for normalization, the following:

$$\mathbb{P}(X \leq a) = \int_{-\infty}^{a} f(x)dx$$

which shows that the samples generated this way are f-distributed. Note more samples have to be generated from this probability mass function the further away g is from the desired function f. Further, because there is no rejection step, we no longer have the issue of efficiency.

For example, let us choose a beta distribution for g, as in the following code:

```
>>> g = scipy.stats.beta(2,3)
```

This distribution does not bear a strong resemblance to our desired f function from last section. as shown in Fig. 2.22 below. Note that we scaled the domain of the beta distribution to get it close to the support of f.

In the next block, we sample from the g distribution and compute the weights as described above. The final step is to sample from this new probability mass function. The resulting normalized histogram is shown compared to the target f probability density function in Fig. 2.23.

```
>>> xi = g.rvs(500)
>>> w = np.array([f(i*15)/g.pdf(i) for i in xi])
>>> fsamples=np.random.choice(xi*15,5000,p = w/w.sum())
```

In this section, we investigated how to generate random samples from a given distribution, beit discrete or continuous. For the continuous case, the key issue was

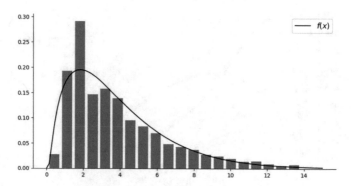

Fig. 2.23 Histogram and probability density function using SIR

whether or not the cumulative density function had a continuous inverse. If not, we had to turn to the rejection method and find an appropriate related density that we could easily sample from to use as part of a rejection threshold. Finding such a function is an art, but many families of probability densities have been studied over the years that already have fast random number generators.

The rejection method has many complicated extensions that involve careful partitioning of the domains and lots of special methods for corner cases. Nonetheless, all of these advanced techniques are still variations on the same fundamental theme we illustrated here [12], [18].

2.11   Useful Inequalities

In practice, few quantities can be analytically calculated. Some knowledge of bounding inequalities helps find the ballpark for potential solutions. This sections discusses three key inequalities that are important for probability, statistics, and machine learning.

2.11.1   Markov's Inequality

Let X be a nonnegative random variable and suppose that $\mathbb{E}(X) < \infty$. Then, for any $t > 0$,

$$\mathbb{P}(X > t) \leq \frac{\mathbb{E}(X)}{t}$$

This is a foundational inequality that is used as a stepping stone to other inequalities. It is easy to prove. Because $X > 0$, we have the following:

Fig. 2.24 The χ_1^2 density has much of its weight on the left, which is excluded in the establishment of the Markov inequality

Fig. 2.25 The shaded area shows the region between the curves on either side of the Markov inequality

$$\mathbb{E}(X) = \int_0^\infty x f_x(x) dx = \underbrace{\int_0^t x f_x(x) dx}_{\text{omit this}} + \int_t^\infty x f_x(x) dx$$

$$\geq \int_t^\infty x f_x(x) dx \geq t \int_t^\infty f_x(x) dx = t \mathbb{P}(X > t)$$

The step that establishes the inequality is the part where the $\int_0^t x f_x(x) dx$ is omitted. For a particular $f_x(x)$ that may be concentrated around the $[0, t]$ interval, this could be a lot to throw out. For that reason, the Markov inequality is considered a *loose* inequality, meaning that there is a substantial gap between both sides of the inequality. For example, as shown in Fig. 2.24, the χ^2 distribution has a lot of its mass on the left, which would be omitted in the Markov inequality. Figure 2.25 shows the two curves established by the Markov inequality. The gray-shaded region is the gap between the two terms and indicates that looseness of the bound (fatter-shaded region) for this case.

2.11.2 Chebyshev's Inequality

Chebyshev's inequality drops out directly from the Markov inequality. Let $\mu = \mathbb{E}(X)$ and $\sigma^2 = \mathbb{V}(X)$. Then, we have

$$\mathbb{P}(|X - \mu| \geq t) \leq \frac{\sigma^2}{t^2}$$

Note that if we normalize so that $Z = (X - \mu)/\sigma$, we have $\mathbb{P}(|Z| \geq k) \leq 1/k^2$. In particular, $\mathbb{P}(|Z| \geq 2) \leq 1/4$. We can illustrate this inequality using Sympy statistics module:

```
>>> import sympy
>>> import sympy.stats as ss
>>> t=sympy.symbols('t')
>>> x=ss.ChiSquared('x',1)
```

To get the left side of the Chebyshev inequality, we have to write this out as the following conditional probability:

```
>>> r = ss.P((x-1) > t,x>1)+ss.P(-(x-1) > t,x<1)
```

We could take the above expression, which is a function of t and attempt to compute the integral, but that would take a very long time (the expression is very long and complicated, which is why we did not print it out above). In this situation, it's better to use the built-in cumulative density function as in the following (after some rearrangement of the terms):

```
>>> w=(1-ss.cdf(x)(t+1))+ss.cdf(x)(1-t)
```

To plot this, we can evaluate at a variety of t values by using the `.subs` substitution method, but it is more convenient to use the `lambdify` method to convert the expression to a function.

```
>>> fw=sympy.lambdify(t,w)
```

Then, we can evaluate this function using something like

```
>>> [fw(i) for i in [0,1,2,3,4,5]]
[1.0,0.157299207050285,0.08326451666355039,0.045500263896358
42,0.0253473186774682,0.014305878435429631]
```

to produce the following Fig. 2.26.

Programming Tip

Note that we cannot use vectorized inputs for the `lambdify` function because it contains embedded functions that are only available in Sympy. Otherwise, we could have used `lambdify(t,fw,numpy)` to specify the corresponding functions in Numpy to use for the expression.

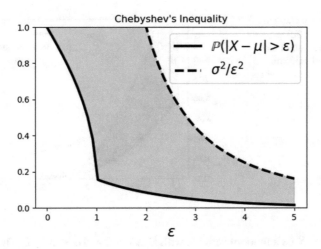

Fig. 2.26 The shaded area shows the region between the curves on either side of the Chebyshev inequality

2.11.3   Hoeffding's Inequality

Hoeffding's inequality is similar, but less loose, than Markov's inequality. Let $X_1, \ldots, X_n$ be iid observations such that $\mathbb{E}(X_i) = \mu$ and $a \leq X_i \leq b$. Then, for any $\epsilon > 0$, we have

$$\mathbb{P}(|\overline{X}_n - \mu| \geq \epsilon) \leq 2\exp(-2n\epsilon^2/(b-a)^2)$$

where $\overline{X}_n = \frac{1}{n}\sum_i^n X_i$. Note that we further assume that the individual random variables are bounded.

Corollary If $X_1, \ldots, X_n$ are independent with $\mathbb{P}(a \leq X_i \leq b) = 1$ and all with $\mathbb{E}(X_i) = \mu$. Then, we have

$$|\overline{X}_n - \mu| \leq \sqrt{\frac{c}{2n}\log\frac{2}{\delta}}$$

where $c = (b-a)^2$. We will see this inequality again in the machine learning chapter. Figure 2.27 shows the Markov and Hoeffding bounds for the case of ten identically and uniformly distributed random variables, $X_i \sim \mathcal{U}[0, 1]$. The solid line shows $\mathbb{P}(|\overline{X}_n - 1/2| > \epsilon)$. Note that the Hoeffding inequality is tighter than the Markov inequality and that both of them merge when ϵ gets big enough.

Proof of Hoeffding's Inequality We will need the following lemma to prove Hoeffding's inequality.

Fig. 2.27 This shows the Markov and Hoeffding bounds for the case of ten identically and uniformly distributed random variables

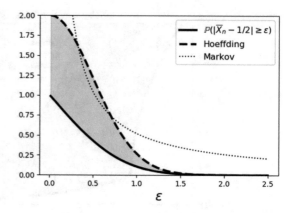

Lemma Let X be a random variable with $\mathbb{E}(X) = 0$ and $a \leq X \leq b$. Then, for any $s > 0$, we have the following:

$$\mathbb{E}(e^{sX}) \leq e^{s^2(b-a)^2/8} \tag{2.15}$$

Because X is contained in the closed interval $[a, b]$, we can write it as a convex combination of the endpoints of the interval.

$$X = \alpha_1 a + \alpha_2 b$$

where $\alpha_1 + \alpha_2 = 1$. Solving for the α_i terms, we have

$$\alpha_1 = \frac{x-a}{b-a}$$

$$\alpha_2 = \frac{b-x}{b-a}$$

From Jensen's inequality, for a convex functions f, we know that

$$f\left(\sum \alpha_i x_i\right) \leq \sum \alpha_i f(x_i)$$

Given the convexity of e^X, we therefore have

$$e^{sX} \leq \alpha_1 e^{sa} + \alpha_2 e^{sb}$$

With $\mathbb{E}(X) = 0$, we can write the expectation of both sides

$$\mathbb{E}(e^{sX}) \leq \mathbb{E}(\alpha_1)e^{sa} + \mathbb{E}(\alpha_2)e^{sb}$$

with $\mathbb{E}(\alpha_1) = \frac{b}{b-a}$ and $\mathbb{E}(\alpha_2) = \frac{-a}{b-a}$. Thus, we have

$$\mathbb{E}(e^{sX}) \le \frac{b}{b-a}e^{sa} - \frac{a}{b-a}e^{sb}$$

Using $p := \frac{-a}{b-a}$, we can rewrite the following:

$$\frac{b}{b-a}e^{sa} - \frac{a}{b-a}e^{sb} = (1-p)e^{sa} + pe^{sb} =: e^{\phi(u)}$$

where

$$\phi(u) = -pu + \log(1 - p + pe^u)$$

and $u = s(b-a)$. Note that $\phi(0) = \phi'(0) = 0$. Also, $\phi''(0) = p(1-p) \le 1/4$.
Thus, the Taylor expansion of $\phi(u) \approx \frac{u^2}{2}\phi''(t) \le \frac{u^2}{8}$ for $t \in [0, u]$■.

   To prove Hoeffding's inequality, we start with Markov's inequality:

$$\mathbb{P}(X \ge \epsilon) \le \frac{\mathbb{E}(X)}{\epsilon}$$

Then, given $s > 0$, we have the following,

$$\mathbb{P}(X \ge \epsilon) = \mathbb{P}(e^{sX} \ge e^{s\epsilon}) \le \frac{\mathbb{E}(e^{sX})}{e^{s\epsilon}}$$

We can write the one-sided Hoeffding inequality as the following:

$$\mathbb{P}(\overline{X}_n - \mu \ge \epsilon) \le e^{-s\epsilon}\mathbb{E}(\exp(\frac{s}{n}\sum_{i=1}^{n}(X_i - \mathbb{E}(X_i))))$$

$$= e^{-s\epsilon}\prod_{i=1}^{n}\mathbb{E}(e^{\frac{s}{n}(X_i - \mathbb{E}(X_i))})$$

$$\le e^{-s\epsilon}\prod_{i=1}^{n}e^{\frac{s^2}{n^2}(b-a)^2/8}$$

$$= e^{-s\epsilon}e^{\frac{s^2}{n}(b-a)^2/8}$$

Now, we want to pick $s > 0$ to minimize this upper bound. Then, with $s = \frac{4n\epsilon}{(b-a)^2}$

$$\mathbb{P}(\overline{X}_n - \mu \ge \epsilon) \le e^{-\frac{2n\epsilon^2}{(b-a)^2}}$$

The other side of the inequality follows similarly to obtain Hoeffding's inequality
■.

2.11.4   Jensen's Inequality

If f is a convex function with random variable v, then

$$\mathbb{E}(f(v)) \geq f(\mathbb{E}(v))$$

The proof of this is straightforward. Define $L(v) = av + b$ with $a, b \in \mathbb{R}$. Choose a and b so that $L(\mathbb{E}(v)) = f(\mathbb{E}(v))$, which makes L tangent to f at $\mathbb{E}(v)$. By the convexity of f, we have $f(v) \geq L(v)$. We can take the expectation of both sides of this:

$$
\begin{aligned}
\mathbb{E}(f(v)) &\geq \mathbb{E}(L(v)) \\
&= \mathbb{E}(av + b) \\
&= a\mathbb{E}(v) + b \\
&= L(\mathbb{E}(v)) \\
&= f(\mathbb{E}(v))
\end{aligned}
$$

equality holds when f is linear. For a concave function f, the sense of the inequality is reversed.

Chapter 3
Statistics

3.1 Introduction

Statistics starts with the data and then, by careful argument, draws conclusions about the underlying probability structure that generated the data. In this way, statistics is the inverse problem to probability, where we started with the probability density function and then reasoned about transformations of that function. Statistics is about developing a model about the data that has both explanatory and predictive powers. That model must emphasize aspects of the data that are important to our motives. Any model must necessarily be a simplification of reality, and there are always many models to choose from, so a big part of statistics is choosing well. For example, here's a practical problem:

- You have already taken the entrance exam twice, and you want to know if it's worth it to take it a third time in the hopes that your score will improve. Because only the last score is reported, you are worried that you may do worse the third time. How do you decide whether or not to take the test again?

For this problem, we have relative few data points, and we want to choose a statistical model that has good predictive power as opposed to explanatory power. We need to quantify the quality of our prediction, so we can make the decision to take the exam again or not. Statistics provides structured methods to address our concerns.

Statistics, as an inverse problem, is seldom well-posed, meaning that there is not only one model for each dataset. When we start with just the data, we lack the underlying probability density that we discussed in the last chapter. In the following, we consider some of the most widely used statistical tools both conceptually and as implemented in Python modules. This chapter provides many numerical examples painstakingly detailed in math and in code.

© The Author(s), under exclusive license to Springer Nature Switzerland AG 2022
J. Unpingco, *Python for Probability, Statistics, and Machine Learning*,
https://doi.org/10.1007/978-3-031-04648-3_3

3.2   Python Modules for Statistics

Python has powerful third-party modules for both numerical and symbolic statistical analysis.

3.2.1   Scipy Statistics Module

Although there are some basic statistical functions in Numpy (e.g., mean, std, median), the real repository for statistical functions is in scipy.stats. There are over 80 continuous probability distributions implemented in scipy.stats and an additional set of more than 10 discrete distributions, along with many other supplementary statistical functions.

To get started with scipy.stats, you have to load the module and create an object that has the distribution you're interested in. For example,

```
>>> import scipy.stats # might take awhile
>>> n = scipy.stats.norm(0,10) # create normal distrib
```

The n variable is an object that represents a normally distributed random variable with mean zero and standard deviation, $\sigma = 10$. Note that the more general term for these two parameters is *location* and *scale*, respectively. Now that we have this defined, we can compute mean, as in the following:

```
>>> n.mean() # we already know this from its definition!
0.0
```

We can also compute higher-order moments as

```
>>> n.moment(4)
30000.0
```

The main public methods for continuous random variables are:

- rvs: random variates
- pdf: probability density function
- cdf: cumulative distribution function
- sf: survival Function (1-CDF)
- ppf: percent point function (Inverse of CDF)
- isf: inverse survival function (Inverse of SF)
- stats: mean, variance, (Fisher's) skew, or (Fisher's) kurtosis
- moment: non-central moments of the distribution

For example, we can compute the value of the pdf at a specific point

```
>>> n.pdf(0)
0.03989422804014327
```

or the `cdf` for the same random variable.

```
>>> n.cdf(0)
0.5
```

You can also create samples from this distribution as in the following:

```
>>> n.rvs(10)
array([15.3244518 , -9.4087413 ,  6.94760096,  0.61627683,
       -3.92073633,
        6.9753351 ,  7.95314387, -3.18127815,  5.69087949,
        0.84197674])
```

Many common statistical tests are already built-in. For example, Shapiro-Wilks tests the null hypothesis that the data were drawn from a normal distribution,[1] as in the following:

```
>>> scipy.stats.shapiro(n.rvs(100))
ShapiroResult(statistic=0.9749656915664673,
  pvalue=0.05362436920404434)
```

The second value in the tuple is the *p*-value (discussed below).

3.2.2 Sympy Statistics Module

Sympy has its own much smaller, but still extremely useful, statistics module that enables symbolic manipulation of statistical quantities. For example,

```
>>> from sympy import stats, sqrt, exp, pi
>>> X = stats.Normal('x',0,10) # create normal random variable
```

We can obtain the probability density function as

```
>>> from sympy.abc import x
>>> stats.density(X)(x)
sqrt(2)*exp(-x**2/200)/(20*sqrt(pi))
>>> sqrt(2)*exp(-x**2/200)/(20*sqrt(pi))
sqrt(2)*exp(-x**2/200)/(20*sqrt(pi))
```

and we can evaluate the cumulative density function as the following:

```
>>> stats.cdf(X)(0)
1/2
```

Note that you can evaluate this numerically by using the `evalf()` method on the output. Sympy provides intuitive ways to consider standard probability questions by using the `stats.P` function, as in the following:

---

[1] We will explain null hypothesis and the rest of it later.

```
>>> stats.P(X>0)  # prob X >0?
1/2
```

There is also a corresponding expectation function `stats.E` you can use to compute complicated expectations using all of Sympy's powerful built-in integration machinery. For example, we can compute $\mathbb{E}(\sqrt{|X|})$ in the following:

```
>>> stats.E(abs(X)**(1/2)).evalf()
2.59995815363879
```

Unfortunately, there is very limited support for multivariate distributions at the time of this writing.

3.2.3 Other Python Modules for Statistics

There are many other important Python modules for statistical work. Two important modules are Seaborn and Statsmodels. As we discussed earlier, Seaborn is library built on top of Matplotlib for very detailed and expressive statistical visualizations, ideally suited for exploratory data analysis. Statsmodels is designed to complement Scipy with descriptive statistics, estimation, and inference for a large variety of statistical models. Statsmodels includes (among many others) generalized linear models, robust linear models, and methods for timeseries analysis, with an emphasis on econometric data and problems. Both these modules are well supported and very well documented and designed to integrate tightly into Matplotlib, Numpy, Scipy, and the rest of the scientific Python stack. Because the focus of this text is more conceptual as opposed to domain-specific, I have chosen not to emphasize either of these, notwithstanding how powerful each is.

3.3 Types of Convergence

The absence of the probability density for the raw data means that we have to argue about sequences of random variables in a structured way. From basic calculus, recall the following convergence notation:

$$x_n \rightarrow x_o$$

for the real number sequence x_n. This means that for any given $\epsilon > 0$, no matter how small, we can exhibit a m such that for any $n > m$, we have

$$|x_n - x_o| < \epsilon$$

Intuitively, this means that once we get past m in the sequence, we get to within ϵ of x_o. This means that nothing surprising happens in the sequence on the long march to infinity, which gives a sense of uniformity to the convergence process. When we argue about convergence for statistics, we want to have the same look-and-feel as we have here, but because we are now talking about random variables, we need other concepts. There are two moving parts for random variables. Recall from our probability chapter that random variables are really functions that map sets into the real line: $X : \Omega \mapsto \mathbb{R}$. Thus, one part is the behavior of the subsets of Ω in terms of convergence. The other part is how the sequences of real values of the random variable behave in convergence.

3.3.1 Almost Sure Convergence

The most straightforward extension into statistics of this convergence concept is *almost sure convergence*, which is also known as *convergence with probability one*

$$\mathbb{P}\{\text{for each } \epsilon > 0 \text{ there is } n_\epsilon > 0 \text{ such that for all } n > n_\epsilon, |X_n - X| < \epsilon\} = 1$$
(3.1)

Note the similarity to the prior notion of convergence for real numbers. When this happens, we write this as $X_n \xrightarrow{as} X$. In this context, almost sure convergence means that if we take any particular $\omega \in \Omega$ and then look at the sequence of real numbers that are produced by each of the random variables,

$$(X_1(\omega), X_2(\omega), X_3(\omega), \ldots, X_n(\omega))$$

then this sequence is just a real-valued sequence in the sense of our convergence on the real line and converges in the same way. If we collect all of the ω for which this is true and the measure of that collection equals one, then we have almost sure convergence of the random variable. Notice how the convergence idea applies to both sides of the random variable: the (domain) Ω side and the (co-domain) real-valued side.

An equivalent and more compact way of writing this is the following:

$$\mathbb{P}\left(\omega \in \Omega : \lim_{n \to \infty} X_n(\omega) = X(\omega)\right) = 1$$

Example To get some feel for the mechanics of this kind of convergence, consider the following sequence of uniformly distributed random variables on the unit interval, $X_n \sim \mathcal{U}[0, 1]$. Now, consider taking the maximum of the set of n such variables as the following:

$$X_{(n)} = \max\{X_1, \ldots, X_n\}$$

In other words, we scan through a list of n uniformly distributed random variables and pick out the maximum over the set. Intuitively, we should expect that $X_{(n)}$ should somehow converge to one. Let us see if we can make this happen almost surely. We want to exhibit m so that the following is true:

$$\mathbb{P}(|1 - X_{(n)}|) < \epsilon \text{ when } n > m$$

Because $X_{(n)} < 1$, we can simplify this as the following:

$$1 - \mathbb{P}(X_{(n)} < 1 - \epsilon) = 1 - (1 - \epsilon)^m \xrightarrow[m \to \infty]{} 1$$

Thus, this sequence converges almost surely. We can work this example out in Python using Scipy to make it concrete with the following code:

```
>>> from scipy import stats
>>> u = stats.uniform()
>>> xn = lambda i: u.rvs(i).max()
>>> xn(5)
0.966717838482003
```

Thus, the xn variable is the same as the $X_{(n)}$ random variable in our example. Figure 3.1 shows a plot of these random variables for different values of n and multiple realizations of each random variable (multiple gray lines). The dark horizontal line is at the 0.95 level. For this example, suppose we are interested in the convergence of the random variable to within 0.05 of one so we are interested in the region between one and 0.95. Thus, in our Eq. 3.1, $\epsilon = 0.05$. Now, we have to find n_ϵ to get the almost sure convergence. From Fig. 3.1, as soon as we get past $n > 60$, we can see that all the realizations start to fit in the region above the 0.95 horizontal line. However, there are still some cases where a particular realization

Fig. 3.1 Almost sure convergence example for multiple realizations of the limiting sequence

will skip below this line. To get the probability guarantee of the definition satisfied, we have to make sure that for whatever n_ϵ we settle on, the probability of this kind of noncompliant behavior should be extremely small, say, less than 1%. Now, we can compute the following to estimate this probability for $n = 60$ over 1000 realizations:

```
>>> import numpy as np
>>> np.mean([xn(60) > 0.95 for i in range(1000)])
0.961
```

So, the probability of having a noncompliant case beyond $n > 60$ is pretty good, but not still what we are after (0.99). We can solve for the m in our analytic proof of convergence by plugging in our factors for ϵ and our desired probability constraint

```
>>> np.log(1-.99)/np.log(.95)
89.78113496070968
```

Now, rounding this up and re-visiting the same estimate as above,

```
>>> import numpy as np
>>> np.mean([xn(90) > 0.95 for i in range(1000)])
0.995
```

which is the result we were looking for. The important thing to understand from this example is that we had to choose convergence criteria for *both* the values of the random variable (0.95) and for the probability of achieving that level (0.99) in order to compute the m. Informally speaking, almost sure convergence means that not only will any particular X_n be close to X for large n, but whole sequence of values will remain close to X with high probability.

3.3.2 Convergence in Probability

A weaker kind of convergence is *convergence in probability* which means the following:

$$\mathbb{P}(|X_n - X| > \epsilon) \to 0$$

as $n \to \infty$ for each $\epsilon > 0$.

This is notationally shown as $X_n \overset{P}{\to} X$. For example, let us consider the following sequence of random variables where $X_n = 1/2^n$ with probability p_n and where $X_n = c$ with probability $1 - p_n$. Then, we have $X_n \overset{P}{\to} 0$ as $p_n \to 1$. This is allowable under this notion of convergence because a diminishing amount of *non-converging* behavior (namely, when $X_n = c$) is possible. Note that we have said nothing about *how $p_n \to 1$*.

Example To get some sense of the mechanics of this kind of convergence, let $\{X_1, X_2, X_3, \ldots\}$ be the indicators of the corresponding intervals:

$$(0, 1], (0, \tfrac{1}{2}], (\tfrac{1}{2}, 1], (0, \tfrac{1}{3}], (\tfrac{1}{3}, \tfrac{2}{3}], (\tfrac{2}{3}, 1]$$

In other words, just keep splitting the unit interval into equal chunks and enumerate those chunks with X_i. Because each X_i is an indicator function, it takes only two values: zero and one. For example, for $X_2 = 1$ if $0 < x \le 1/2$ and zero otherwise. Note that $x \sim \mathcal{U}(0, 1)$. This means that $P(X_2 = 1) = 1/2$. Now, we want to compute the sequence of $P(X_n > \epsilon)$ for each n for some $\epsilon \in (0, 1)$. For X_1, we have $P(X_1 > \epsilon) = 1$ because we already chose ϵ in the interval covered by X_1. For X_2, we have $P(X_2 > \epsilon) = 1/2$; for X_3, we have $P(X_3 > \epsilon) = 1/3$; and so on. This produces the following sequence: $(1, \tfrac{1}{2}, \tfrac{1}{2}, \tfrac{1}{3}, \tfrac{1}{3}, \dots)$. The limit of the sequence is zero so that $X_n \xrightarrow{P} 0$. However, for every $x \in (0, 1)$, the sequence of function values of $X_n(x)$ consists of infinitely many zeros and ones (remember that indicator functions can evaluate to either zero or one). Thus, the set of x for which the sequence $X_n(x)$ converges is empty because the sequence bounces between zero and one. This means that almost sure convergence fails here even though we have convergence in probability. The key distinction is that convergence in probability considers the convergence of a sequence of probabilities, whereas almost sure convergence is concerned about the sequence of values of the random variables over sets of events that *fill out* the underlying probability space entirely (i.e., with probability one).

This is a good example so let us see if we can make it concrete with some Python. The following is a function to compute the different subintervals:

```
>>> make_interval= lambda n: np.array(list(zip(range(n+1),
...                                          range(1,n+1))))/n
```

Now, we can use this function to create a Numpy array of intervals, as in the example:

```
>>> intervals= np.vstack([make_interval(i) for i in range(1,5)])
>>> print(intervals)
[[0.          1.        ]
 [0.          0.5       ]
 [0.5         1.        ]
 [0.          0.33333333]
 [0.33333333  0.66666667]
 [0.66666667  1.        ]
 [0.          0.25      ]
 [0.25        0.5       ]
 [0.5         0.75      ]
 [0.75        1.        ]]
```

The following function computes the bit string in our example: $\{X_1, X_2, \dots, X_n\}$,

```
>>> bits= lambda u:((intervals[:,0] < u) &
         (u<=intervals[:,1])).astype(int)
>>> bits(u.rvs())
array([1, 0, 1, 0, 0, 1, 0, 0, 0, 1])
```

Now that we have the individual bit strings, to show convergence we want to show that the probability of each entry goes to a limit. For example, using ten realizations,

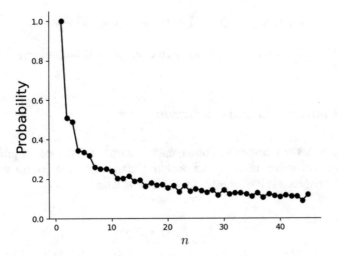

Fig. 3.2 Convergence in probability for the random variable sequence

```
>>> print(np.vstack([bits(u.rvs()) for i in range(10)]))
[[1 1 0 1 0 0 0 1 0 0]
 [1 1 0 1 0 0 0 1 0 0]
 [1 1 0 0 1 0 0 1 0 0]
 [1 0 1 0 0 1 0 0 1 0]
 [1 0 1 0 0 1 0 0 1 0]
 [1 1 0 0 1 0 0 1 0 0]
 [1 1 0 1 0 0 1 0 0 0]
 [1 1 0 0 1 0 0 1 0 0]
 [1 1 0 0 1 0 0 1 0 0]
 [1 1 0 1 0 0 1 0 0 0]]
```

We want the limiting probability of a one in each column to convert to a limit. We can estimate this over 1000 realizations using the following code:

```
>>> np.vstack([bits(u.rvs()) for i in range(1000)]).mean(axis=0)
array([1.    , 0.493, 0.507, 0.325, 0.34 , 0.335, 0.253, 0.24 , 0.248,
       0.259])
```

Note that these entries should approach the $(1, \frac{1}{2}, \frac{1}{2}, \frac{1}{3}, \frac{1}{3}, \ldots)$ sequence we found earlier. Figure 3.2 shows the convergence of these probabilities for a large number of intervals. Eventually, the probability shown on this graph will decrease to zero with large enough n. Again, note that the individual sequences of zeros and ones do not converge, but the probabilities of these sequences converge. This is the key difference between almost sure convergence and convergence in probability. Thus, convergence in probability does *not* imply almost sure convergence. Conversely, almost sure convergence *does* imply convergence in probability.

The following notation should help emphasize the difference between almost sure convergence and convergence in probability, respectively:

$$P\left(\lim_{n\to\infty} |X_n - X| < \epsilon\right) = 1 \text{(almost sure convergence)}$$

$$\lim_{n\to\infty} P(|X_n - X| < \epsilon) = 1 \text{(convergence in probability)}$$

3.3.3 Convergence in Distribution

So far, we have been discussing convergence in terms of sequences of probabilities or sequences of values taken by the random variable. By contrast, the next major kind of convergence is *convergence in distribution* where

$$\lim_{n\to\infty} F_n(t) = F(t)$$

for all t for which F is continuous and F is the cumulative density function. For this case, convergence is only concerned with the cumulative density function, written as $X_n \overset{d}{\to} X$.

Example To develop some intuition about this kind of convergence, consider a sequence of X_n Bernoulli random variables. Furthermore, suppose these are all really just the same random variable X. Trivially, $X_n \overset{d}{\to} X$. Now, suppose we define $Y = 1 - X$, which means that Y has the same distribution as X. Thus, $X_n \overset{d}{\to} Y$. By contrast, because $|X_n - Y| = 1$ for all n, we can never have almost sure convergence or convergence in probability. Thus, convergence in distribution is the weakest of the three forms of convergence in the sense that it is implied by the other two, but implies neither of the two.

As another striking example, we could have $Y_n \overset{d}{\to} Z$ where $Z \sim \mathcal{N}(0, 1)$, but we could also have $Y_n \overset{d}{\to} -Z$. That is, Y_n could converge in distribution to either Z or $-Z$. This may seem ambiguous, but this kind of convergence is practically very useful because it allows for complicated distributions to be approximated by simpler distributions.

3.3.4 Limit Theorems

Now that we have all of these notions of convergence, we can apply them to different situations and see what kinds of claims we can construct from them.

Weak Law of Large Numbers Let $\{X_1, X_2, \ldots, X_n\}$ be an iid (independent, identically distributed) set of random variables with finite mean $\mathbb{E}(X_k) = \mu$ and finite variance. Let $\overline{X}_n = \frac{1}{n}\sum_k X_k$. Then, we have $\overline{X}_n \overset{P}{\to} \mu$. This result is important because we frequently estimate parameters using an averaging process of some kind. This basically justifies this in terms of convergence in probability. Informally, this means that the distribution of $\overline{X}_n$ becomes concentrated around μ as $n \to \infty$.

Strong Law of Large Numbers Let $\{X_1, X_2, \ldots, \}$ be an iid set of random variables. Suppose that $\mu = \mathbb{E}|X_i| < \infty$, then $\overline{X}_n \overset{as}{\to} \mu$. The reason this is called the strong law is that it implies the weak law because almost sure convergence implies convergence in probability. The so-called Kolmogorov criterion gives the convergence of the following:

$$\sum_k \frac{\sigma_k^2}{k^2}$$

as a sufficient condition for concluding that the strong law applies to the sequence $\{X_k\}$ with corresponding $\{\sigma_k^2\}$.

As an example, consider an infinite sequence of Bernoulli trials with $X_i = 1$ if the ith trial is successful. Then $\overline{X}_n$ is the relative frequency of successes in n trials, and $\mathbb{E}(X_i)$ is the probability p of success on the ith trial. With all that established, the weak law says only that if we consider a sufficiently large and fixed n, the probability that the relative frequency will converge to p is guaranteed. The strong law states that if we regard the observation of all the infinite $\{X_i\}$ as one performance of the experiment, the relative frequency of successes will almost surely converge to p. The difference between the strong law and the weak law of large numbers is subtle and rarely arises in practical applications of probability theory.

Central Limit Theorem Although the weak law of large numbers tells us that the distribution of $\overline{X}_n$ becomes concentrated around μ, it does not tell us what that distribution is. The central limit theorem (CLT) says that $\overline{X}_n$ has a distribution that is approximately normal with mean μ and variance σ^2/n. Amazingly, nothing is assumed about the distribution of X_i, except the existence of the mean and variance. The following is the central limit theorem: Let $\{X_1, X_2, \ldots, X_n\}$ be iid with mean μ and variance σ^2. Then,

$$Z_n = \frac{\sqrt{n}(\overline{X}_n - \mu)}{\sigma} \overset{P}{\to} Z \sim \mathcal{N}(0, 1)$$

The loose interpretation of the central limit theorem is that $\overline{X}_n$ can be legitimately approximated by a normal distribution. Because we are talking about convergence in probability here, claims about probability are legitimized, not claims about the random variable itself. Intuitively, this shows that normality arises from sums of small, independent disturbances of finite variance. Technically, the finite variance assumption is essential for normality. Although the Central Limit Theorem provides a powerful, general approximation, the quality of the approximation for a particular situation still depends on the original (usually unknown) distribution.

3.4 Estimation Using Maximum Likelihood

The estimation problem starts with the desire to infer something meaningful from data. For parametric estimation, the strategy is to postulate a model for the data and then use the data to fit model parameters. This leads to two fundamental questions: where to get the model and how to estimate the parameters. The first question is best answered by the maxim: *all models are wrong; some are useful*. In other words, choosing a model depends as much on the application as on the model itself. Think about models as building different telescopes to view the sky. No one would ever claim that the telescope generates the sky! It is same with data models. Models give us multiple perspectives on the data that themselves are proxies for some deeper underlying phenomenon.

Some categories of data may be more commonly studied using certain types of models, but this is usually very domain-specific and ultimately depends on the aims of the analysis. In some cases, there may be strong physical reasons behind choosing a model. For example, one could postulate that the model is linear with some noise as in the following:

$$Y = aX + \epsilon$$

which basically says that you, as the experimenter, dial in some value for X and then read off something directly proportional to X as the measurement, Y, plus some additive noise that you attribute to jitter in the apparatus. Then, the next step is to estimate the parameter a in the model, given some postulated claim about the nature of ϵ. How to compute the model parameters depends on the particular methodology. The two broad rubrics are parametric and nonparametric estimation. In the former, we assume we know the density function of the data and then try to derive the embedded parameters for it. In the latter, we claim only to know that the density function is a member of a broad class of density functions and then use the data to characterize a member of that class. Broadly speaking, the former consumes less data than the latter, because there are fewer unknowns to compute from the data.

Let us concentrate on parametric estimation for now. The tradition is to denote the unknown parameter to be estimated as θ which is a member of a large space of alternates, Θ. To judge between potential θ values, we need an objective function, known as a *risk* function, $L(\theta, \hat{\theta})$, where $\hat{\theta}(\mathbf{x})$ is an estimate for the unknown θ that is derived from the available data $\mathbf{x}$. The most common and useful risk function is the squared error loss:

$$L(\theta, \hat{\theta}) = (\theta - \hat{\theta})^2$$

Although neat, this is not practical because we need to know the unknown θ to compute it. The other problem is because $\hat{\theta}$ is a function of the observed data, it is also a random variable with its own probability density function. This leads to the notion of the *expected risk* function:

$$R(\theta, \hat{\theta}) = \mathbb{E}_\theta(L(\theta, \hat{\theta})) = \int L(\theta, \hat{\theta}(\mathbf{x})) f(\mathbf{x}; \theta) d\mathbf{x}$$

In other words, given a fixed θ, integrate over the probability density function of the data, $f(\mathbf{x})$, to compute the risk. Plugging in for the squared error loss, we compute the mean squared error:

$$\mathbb{E}_\theta(\theta - \hat{\theta})^2 = \int (\theta - \hat{\theta})^2 f(\mathbf{x}; \theta) d\mathbf{x}$$

This has the important factorization into the *bias*

$$\text{bias} = \mathbb{E}_\theta(\hat{\theta}) - \theta$$

with the corresponding variance, $\mathbb{V}_\theta(\hat{\theta})$, as in the following *mean squared error* (MSE):

$$\mathbb{E}_\theta(\theta - \hat{\theta})^2 = \text{bias}^2 + \mathbb{V}_\theta(\hat{\theta})$$

This is an important trade-off that we will return to repeatedly. The idea is the bias is nonzero when the estimator $\hat{\theta}$, integrated over all possible data, $f(\mathbf{x})$, does not equal the underlying target parameter θ. In some sense, the estimator misses the target, no matter how much data is used. When the bias equals zero, the estimated is *unbiased*. For fixed MSE, low bias implies high variance and vice versa. This trade-off was once not emphasized, and instead much attention was paid to the smallest variance of unbiased estimators (see Cramer-Rao bounds). In practice, understanding and exploiting the trade-off between bias and variance and reducing the MSE is more important.

With all this set up, we can now ask how bad can bad get by examining *minimax risk*

$$R_{\text{mmx}} = \inf_{\hat{\theta}} \sup_{\theta} R(\theta, \hat{\theta})$$

where the inf is take over all estimators. Intuitively, this means if we found the worst possible θ and swept over all possible parameter estimators $\hat{\theta}$, and then took the smallest possible risk we could find, we would have the minimax risk. Thus, an estimator, $\hat{\theta}_{\text{mmx}}$, is a *minimax estimator* if it achieves this feat:

$$\sup_{\theta} R(\theta, \hat{\theta}_{\text{mmx}}) = \inf_{\hat{\theta}} \sup_{\theta} R(\theta, \hat{\theta})$$

In other words, even in the face of the worst θ (i.e., the $\sup_\theta$), $\hat{\theta}_{\text{mmx}}$ still achieves the minimax risk. There is a greater theory that revolves around minimax estimators of various kinds, but this is far beyond our scope here. The main thing is that maximum likelihood estimator is approximately minimax under certain technical, but easily satisfiable, conditions. Maximum likelihood is the subject of the next section. Let us get started with the simplest application: coin flipping.

3.4.1 Setting Up the Coin Flipping Experiment

Suppose we have coin and want to estimate the probability of heads (p) for it. We model the distribution of heads and tails as a Bernoulli distribution with the following probability mass function:

$$\phi(x) = p^x (1 - p)^{(1-x)}$$

where x is the outcome, 1 for heads and 0 for tails. Maximum likelihood is a parametric method that requires the specification of a particular model for which we will compute embedded parameters. For n independent flips, we have the joint density as the product of n of these functions as in

$$\phi(\mathbf{x}) = \prod_{i=1}^{n} p_i^x (1 - p)^{(1-x_i)}$$

The following is the *likelihood function*:

$$\mathcal{L}(p; \mathbf{x}) = \prod_{i=1}^{n} p^{x_i} (1 - p)^{1-x_i}$$

This is basically notation. We have just renamed the previous equation to emphasize the p parameter, which is what we want to estimate.

The principle of *maximum likelihood* is to maximize the likelihood as the function of p after plugging in all of the x_i data. We then call this maximizer $\hat{p}$ which is a function of the observed x_i data and, as such, is a random variable with its own distribution. This method therefore ingests data and a postulated probability density and produces a function that estimates the embedded parameter in the assumed probability density. Thus, maximum likelihood generates the *functions* of data that we need to get the underlying parameters of the model. Note that there is no limit to the ways we can functionally manipulate the data we have collected. The maximum likelihood principle gives us a systematic method for constructing these functions subject to the assumed model. This is a point worth emphasizing: the maximum likelihood principle yields functions as solutions the same way solving differential equations yields functions as solutions. It is very, very much harder to produce a function than to produce a value as a solution, even with the assumption of a convenient probability density. Thus, the power of the principle is that you can construct such functions subject to the model assumptions.

Simulating the Experiment We need the following code to simulate coin flipping:

```
>>> from scipy.stats import bernoulli
>>> p_true=1/2.0          # estimate this!
>>> fp=bernoulli(p_true)  # create bernoulli random variate
>>> xs = fp.rvs(100)      # generate some samples
>>> print(xs[:30])        # see first 30 samples
[0 1 0 1 1 0 0 1 1 1 0 1 1 1 0 1 1 0 1 1 0 1 0 1 0 0 1 1 0 1 0 1]
```

Now, we can write out the likelihood function using Sympy. Note that we give the Sympy variables the `positive=True` attribute upon construction because this eases Sympy's internal simplification algorithms.

```
>>> import numpy as np
>>> import sympy
>>> x,p,z = sympy.symbols('x p z', positive=True)
>>> phi = p**x*(1-p)**(1-x) # distribution function
>>> L = np.prod([phi.subs(x,i) for i in xs]) # likelihood function
>>> print(L) # approx 0.5?
p**57*(1 - p)**43
```

Note that, once we plug in the data, the likelihood function is solely a function of the unknown parameter (p in this case). The following code uses calculus to find the extrema of the likelihood function. Note that taking the `log` of L makes the maximization problem tractable but doesn't change the extrema.

```
>>> logL = sympy.expand_log(sympy.log(L))
>>> sol, = sympy.solve(sympy.diff(logL,p),p)
>>> print(sol)
57/100
```

> **Programming Tip**
> Note that `sol,=sympy.solve` statement includes a comma after the `sol` variable. This is because the `solve` function returns a list containing a single element. Using this assignment unpacks that single element into the `sol` variable directly. This is another one of the many small elegancies of Python.

The following code generates Fig. 3.3:

```
fig,ax=subplots()
x=np.linspace(0,1,100)
ax.plot(x,map(sympy.lambdify(p,logJ,'numpy'),x),'k-',lw=3)
ax.plot(sol,logJ.subs(p,sol),'o',
        color='gray',ms=15,label='Estimated')
ax.plot(p_true,logJ.subs(p,p_true),'s',
        color='k',ms=15,label='Actual')
ax.set_xlabel('$p$',fontsize=18,usetex=True)
ax.set_ylabel('Likelihood',fontsize=18)
ax.set_title('Estimate not equal to true value',fontsize=18)
ax.legend(loc=0)
```

Fig. 3.3 Maximum
likelihood estimate vs. true
parameter. Note that the
estimate is slightly off from
the true value. This is a
consequence of the fact that
the estimator is a function of
the data and lacks knowledge
of the true underlying value

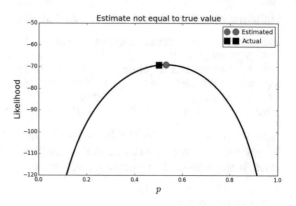

Figure 3.3 shows that our estimator $\hat{p}$ (circle) is not equal to the true value of
p (square), despite being the maximum of the likelihood function. This may sound
disturbing, but keep in mind this estimate is a function of the random data; and since
that data can change, the ultimate estimate can likewise change. Remember that the
estimator is a *function* of the data and is thus also a *random variable*, just like the
data is. This means it has its own probability distribution with corresponding mean
and variance. So, what we are observing is a consequence of that variance.

> **Programming Tip**
>
> In the prior code, we use the `lambdify` function in
> `lambdify(p,logJ,'numpy')` to take a Sympy expression and
> convert it into a Numpy version that is easier to compute. The `lambdify`
> function has an extra argument where you can specify the function space that
> it should use to convert the expression. In the above this is set to Numpy.

Figure 3.4 shows what happens when you run many thousands of coin experiments and compute the maximum likelihood estimate for each experiment, given a
particular number of samples per experiment. This simulation gives us a histogram
of the maximum likelihood estimates, which is an approximation of the probability
distribution of the $\hat{p}$ estimator itself. This figure shows that the sample mean of the
estimator ($\mu = \frac{1}{n}\sum \hat{p}_i$) is pretty close to the true value, but looks can be deceiving.
The only way to know for sure is to check if the estimator is unbiased, namely, if

$$\mathbb{E}(\hat{p}) = p$$

Because this problem is simple, we can solve for this in general noting that the terms
above are either p, if $x_i = 1$, or $1 - p$ if $x_i = 0$. This means that we can write

$$\mathcal{L}(p|\mathbf{x}) = p^{\sum_{i=1}^{n} x_i}(1 - p)^{n - \sum_{i=1}^{n} x_i}$$

with corresponding logarithm as

Fig. 3.4 Histogram of maximum likelihood estimates. The title shows the estimated mean and standard deviation of the samples

$$J = \log(\mathcal{L}(p|\mathbf{x})) = \log(p)\sum_{i=1}^{n} x_i + \log(1-p)\left(n - \sum_{i=1}^{n} x_i\right)$$

Taking the derivative of this gives

$$\frac{dJ}{dp} = \frac{1}{p}\sum_{i=1}^{n} x_i + \frac{(n - \sum_{i=1}^{n} x_i)}{p-1}$$

and solving this for p leads to

$$\hat{p} = \frac{1}{n}\sum_{i=1}^{n} x_i$$

This is our *estimator* for p. Up until now, we have been using Sympy to solve for this based on the data x_i, but now that we have it analytically, we don't have to solve for it each time. To check if this estimator is biased, we compute its expectation

$$\mathbb{E}\left(\hat{p}\right) = \frac{1}{n}\sum_{i}^{n} \mathbb{E}(x_i) = \frac{1}{n}n\mathbb{E}(x_i)$$

by linearity of the expectation and where

$$\mathbb{E}(x_i) = p$$

Therefore,

$$\mathbb{E}\left(\hat{p}\right) = p$$

This means that the estimator is *unbiased*. Similarly,

$$\mathbb{E}\left(\hat{p}^2\right) = \frac{1}{n^2}\mathbb{E}\left[\left(\sum_{i=1}^{n} x_i\right)^2\right]$$

and where

$$\mathbb{E}\left(x_i^2\right) = p$$

and by the independence assumption,

$$\mathbb{E}\left(x_i x_j\right) = \mathbb{E}(x_i)\mathbb{E}(x_j) = p^2$$

Thus,

$$\mathbb{E}\left(\hat{p}^2\right) = \left(\frac{1}{n^2}\right)n\left[p + (n-1)p^2\right]$$

So, the variance of the estimator, $\hat{p}$, is the following:

$$\mathbb{V}(\hat{p}) = \mathbb{E}\left(\hat{p}^2\right) - \mathbb{E}\left(\hat{p}\right)^2 = \frac{p(1-p)}{n}$$

Note that the n in the denominator means that the variance asymptotically goes to zero as n increases (i.e., we consider more and more samples). This is good news because it means that more and more coin flips lead to a better estimate of the underlying p.

Unfortunately, this formula for the variance is practically useless because we need p to compute it and p is the parameter we are trying to estimate in the first place! However, this is where the *plug-in* principle[2] saves the day. It turns out in this situation, you can simply substitute the maximum likelihood estimator, $\hat{p}$, for the p in the above equation to obtain the asymptotic variance for $\mathbb{V}(\hat{p})$. The fact that this works is guaranteed by the asymptotic theory of maximum likelihood estimators.

Nevertheless, looking at $\mathbb{V}(\hat{p})^2$, we can immediately notice that if $p = 0$, then there is no estimator variance because the outcomes are guaranteed to be tails. Also, for any n, the maximum of this variance happens at $p = 1/2$. This is our worst-case scenario, and the only way to compensate is with larger n.

All we have computed is the mean and variance of the estimator. In general, this is insufficient to characterize the underlying probability density of $\hat{p}$, except if we somehow knew that $\hat{p}$ were normally distributed. This is where the powerful *central limit theorem* we discussed in Sect. 3.3.4 comes in. The form of the estimator, which

---

[2] This is also known as the *invariance property* of maximum likelihood estimators. It basically states that the maximum likelihood estimator of any function, say, $h(\theta)$, is the same h with the maximum likelihood estimator for θ substituted in for θ; namely, $h(\theta_{ML})$.

is just a sample mean, implies that we can apply this theorem and conclude that $\hat{p}$ is asymptotically normally distributed. However, it doesn't quantify how many samples n we need. In our simulation this is no problem because we can generate as much data as we like, but in the real world, with a costly experiment, each sample may be precious.[3] In the following, we won't apply the central limit theorem and instead proceed analytically.

Probability Density for the Estimator To write out the full density for $\hat{p}$, we first have to ask what is the probability that the estimator will equal a specific value and the tally up all the ways that could happen with their corresponding probabilities. For example, what is the probability that

$$\hat{p} = \frac{1}{n} \sum_{i=1}^{n} x_i = 0$$

This can only happen one way: when $x_i = 0 \ \forall i$. The probability of this happening can be computed from the density

$$f(\mathbf{x}, p) = \prod_{i=1}^{n} p^{x_i} (1 - p)^{1 - x_i}$$

$$f\left(\sum_{i=1}^{n} x_i = 0, p\right) = (1 - p)^n$$

Likewise, if $\{x_i\}$ has only one nonzero element, then

$$f\left(\sum_{i=1}^{n} x_i = 1, p\right) = np \prod_{i=1}^{n-1} (1 - p)$$

where the n comes from the n ways to pick one element from the n elements x_i. Continuing this way, we can construct the entire density as

$$f\left(\sum_{i=1}^{n} x_i = k, p\right) = \binom{n}{k} p^k (1 - p)^{n-k}$$

where the first term on the right is the binomial coefficient of n things taken k at a time. This is the binomial distribution and it's not the density for $\hat{p}$, but rather for

---

[3] It turns out that the central limit theorem augmented with an Edgeworth expansion tells us that convergence is regulated by the skewness of the distribution [13]. In other words, the more symmetric the distribution, the faster it converges to the normal distribution according to the central limit theorem.

$n\hat{p}$. We'll leave this as-is because it's easier to work with below. We just have to remember to keep track of the n factor.

Confidence Intervals Now that we have the full density for $\hat{p}$, we are ready to ask some meaningful questions. For example, what is the probability the estimator is within ϵ fraction of the true value of p?

$$\mathbb{P}\left(|\hat{p} - p| \leq \epsilon p\right)$$

More concretely, we want to know how often the estimated $\hat{p}$ is trapped within ϵ of the actual value. That is, suppose we ran the experiment 1000 times to generate 1000 different estimates of $\hat{p}$. What percentage of the 1000 so-computed values are trapped within ϵ of the underlying value. Rewriting the above equation as the following:

$$\mathbb{P}\left(p - \epsilon p < \hat{p} < p + \epsilon p\right) = \mathbb{P}\left(np - n\epsilon p < \sum_{i=1}^{n} x_i < np + n\epsilon p\right)$$

Let us plug in some live numbers here for our worst-case scenario (i.e., highest variance scenario) where $p = 1/2$. Then, if $\epsilon = 1/100$, we have

$$\mathbb{P}\left(\frac{99n}{200} < \sum_{i=1}^{n} x_i < \frac{101n}{200}\right)$$

For example for $n = 101$, we have the following:

$$\mathbb{P}\left(\frac{9999}{200} < \sum_{i=1}^{101} x_i < \frac{10201}{200}\right) = f\left(\sum_{i=1}^{101} x_i = 50, p\right)$$

$$= \binom{101}{50}(1/2)^{50}(1 - 1/2)^{101-50} = 0.079$$

This means that in the worst-case scenario for $p = 1/2$, given $n = 101$ trials, we will only get within 1% of the actual $p = 1/2$ about 8% of the time. If you feel disappointed, it is because you've been paying attention. What if the coin was really heavy and it was hard work to repeat this 101 times?

Let us come at this another way: given I could only flip the coin 100 times, how close could I come to the true underlying value with high probability (say, 95%)? In this case, instead of picking a value for ϵ, we are solving for ϵ. Plugging in gives

$$\mathbb{P}\left(50 - 50\epsilon < \sum_{i=1}^{100} x_i < 50 + 50\epsilon\right) = 0.95$$

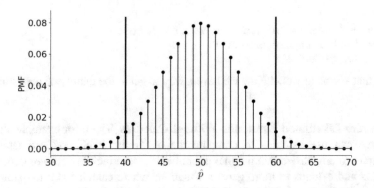

Fig. 3.5 Probability mass function for $\hat{p}$. The two vertical lines form the confidence interval

which we have to solve for ϵ. Fortunately, all the tools we need to solve for this are already in Scipy.

```
>>> from scipy.stats import binom
>>> # n=100, p = 0.5, distribution of the estimator phat
>>> b=binom(100,.5)
>>> # symmetric sum the probability around the mean
>>> g = lambda i:b.pmf(np.arange(-i,i)+50).sum()
>>> print(g(10)) # approx 0.95
0.9539559330706571
```

The two vertical lines in Fig. 3.5 show how far out from the mean we have to go to accumulate 95% of the probability. Now, we can solve this as

$$50 + 50\epsilon = 60$$

which makes $\epsilon = 1/5$ or 20%. So, flipping 100 times means I can only get within 20% of the real p 95% of the time in the worst-case scenario (i.e., $p = 1/2$). The following code verifies the situation:

```
>>> from scipy.stats import bernoulli
>>> b=bernoulli(0.5) # coin distribution
>>> xs = b.rvs(100) # flip it 100 times
>>> phat = np.mean(xs) # estimated p
>>> print(abs(phat-0.5) < 0.5*0.20) # make it w/in interval?
True
```

Let us keep doing this and see if we can get within this interval 95% of the time.

```
>>> out=[]
>>> b=bernoulli(0.5) # coin distribution
>>> for i in range(500):    # number of tries
...     xs = b.rvs(100)     # flip it 100 times
...     phat = np.mean(xs) # estimated p
...     out.append(abs(phat-0.5) < 0.5*0.20 ) # within 20% ?
```

```
...
>>> # percentage of tries w/in 20% interval
>>> print(100*np.mean(out))
97.39999999999999
```

Well, that seems to work! Now we have a way to get at the quality of the estimator, $\hat{p}$.

Maximum Likelihood Estimator Without Calculus The prior example showed how we can use calculus to compute the maximum likelihood estimator. It's important to emphasize that the maximum likelihood principle does *not* depend on calculus and extends to more general situations where calculus is impossible. For example, let X be uniformly distributed in the interval $[0, \theta]$. Given n measurements of X, the likelihood function is the following:

$$L(\theta) = \prod_{i=1}^{n} \frac{1}{\theta} = \frac{1}{\theta^n}$$

where each $x_i \in [0, \theta]$. Note that the slope of this function is not zero anywhere so the usual calculus approach is not going to work here. Because the likelihood is the product of the individual uniform densities, if any of the x_i values were outside of the proposed $[0, \theta]$ interval, then the likelihood would go to zero, because the uniform density is zero outside of the $[0, \theta]$. This is no good for maximization. Thus, observing that the likelihood function is strictly decreasing with increasing θ, we conclude that the value for θ that maximizes the likelihood is the maximum of the x_i values. To summarize, the maximum likelihood estimator is the following:

$$\theta_{ML} = \max_i x_i$$

As always, we want the distribution of this estimator to judge its performance. In this case, this is pretty straightforward. The cumulative density function for the max function is the following:

$$\mathbb{P}\left(\hat{\theta}_{ML} < v\right) = \mathbb{P}(x_0 \leq v \wedge x_1 \leq v \ldots \wedge x_n \leq v)$$

and since all the x_i are uniformly distributed in $[0, \theta]$, we have

$$\mathbb{P}\left(\hat{\theta}_{ML} < v\right) = \left(\frac{v}{\theta}\right)^n$$

So, the probability density function is then

$$f_{\hat{\theta}_{ML}}(\theta_{ML}) = n\theta_{ML}^{n-1}\theta^{-n}$$

Then, we can compute the $\mathbb{E}(\theta_{ML}) = (\theta n)/(n+1)$ with corresponding variance as $\mathbb{V}(\theta_{ML}) = (\theta_{ML}^2 n)/(n+1)^2/(n+2)$.

For a quick sanity check, we can write the following simulation for $\theta = 1$ as in the following:

```
>>> from scipy import stats
>>> rv = stats.uniform(0,1)    # define uniform random variable
>>> mle=rv.rvs((100,500)).max(0) # max along row-dimension
>>> print(np.mean(mle)) # approx n/(n+1) = 100/101 ~= 0.99
0.989942138048
>>> print(np.var(mle)) #approx n/(n+1)**2/(n+2)  ~= 9.61E-5
9.95762009884e-05
```

> **Programming Tip**
>
> The `max(0)` suffix on for the `mle` computation takes the maximum of the so-computed array along the row (`axis=0`) dimension.

You can also plot `hist(mle)` to see the histogram of the simulated maximum likelihood estimates and match it up against the probability density function we derived above.

In this section, we explored the concept of maximum likelihood estimation using a coin flipping experiment both analytically and numerically with the scientific Python stack. We also explored the case when calculus is not workable for maximum likelihood estimation. There are two key points to remember. First, maximum likelihood estimation produces a function of the data that is itself a random variable, with its own probability distribution. We can get at the quality of the so-derived estimators by examining the confidence intervals around the estimated values using the probability distributions associated with the estimators themselves. Second, maximum likelihood estimation applies even in situations where using basic calculus is not applicable [48].

3.4.2 Delta Method

Sometimes we want to characterize the distribution of a *function* of a random variable. In order to extend and generalize the central limit theorem in this way, we need the Taylor series expansion. Recall that the Taylor series expansion is an approximation of a function of the following form:

$$T_r(x) = \sum_{i=0}^{r} \frac{g^{(i)}(a)}{i!}(x-a)^i$$

this basically says that a function g can be adequately approximated about a point a using a polynomial based on its derivatives evaluated at a. Before we state the general theorem, let us examine an example to understand how the mechanics work.

Example Suppose that X is a random variable with $\mathbb{E}(X) = \mu \neq 0$. Furthermore, supposedly we have a suitable function g and we want the distribution of $g(X)$. Applying the Taylor series expansion, we obtain the following:

$$g(X) \approx g(\mu) + g'(\mu)(X - \mu)$$

If we use $g(X)$ as an estimator for $g(\mu)$, then we can say that we approximately have the following:

$$\mathbb{E}(g(X)) = g(\mu)$$

$$\mathbb{V}(g(X)) = (g'(\mu))^2 \mathbb{V}(X)$$

Concretely, suppose we want to estimate the odds, $\frac{p}{1-p}$. For example, if $p = 2/3$, then we say that the odds is $2:1$ meaning that the odds of the one outcome are twice as likely as the odds of the other outcome. Thus, we have $g(p) = \frac{p}{1-p}$, and we want to find $\mathbb{V}(g(\hat{p}))$. In our coin flipping problem, we have the estimator $\hat{p} = \frac{1}{n}\sum X_k$ from the Bernoulli-distributed data X_k individual coin flips. Thus,

$$\mathbb{E}(\hat{p}) = p$$

$$\mathbb{V}(\hat{p}) = \frac{p(1-p)}{n}$$

Now, $g'(p) = 1/(1-p)^2$, so we have

$$\mathbb{V}(g(\hat{p})) = (g'(p))^2 \mathbb{V}(\hat{p})$$

$$= \left(\frac{1}{(1-p)^2}\right)^2 \frac{p(1-p)}{n}$$

$$= \frac{p}{n(1-p)^3}$$

which is an approximation of the variance of the estimator $g(\hat{p})$. Let us simulate this and see how it agrees.

```
>>> from scipy import stats
>>> # compute MLE estimates
>>> d=stats.bernoulli(0.1).rvs((10,5000)).mean(0)
>>> # avoid divide-by-zero
>>> d=d[np.logical_not(np.isclose(d,1))]
>>> # compute odds ratio
>>> odds = d/(1-d)
>>> print('odds ratio=',np.mean(odds),'var=',np.var(odds))
odds ratio= 0.12289206349206351 var= 0.01797950092214664
```

The first number above is the mean of the simulated odds Ratio, and the second is the variance of the estimate. According to the variance estimate equation above, we have $\mathbb{V}(g(1/10)) \approx 0.0137$, which is not too bad for this numerical approximation. Recall we want to estimate the odds from $\hat{p}$. The code above takes 5000 estimates of the $\hat{p}$ to estimate $\mathbb{V}(g)$. The odds ratio for $p = 1/10$ is $1/9 \approx 0.111$.

Programming Tip

The code above uses the np.isclose function to identify the ones from the simulation, and the np.logical_not removes these elements from the data because the odds ratio has a zero in the denominator for these values.

Let us try this again with a probability of heads of 0.5 instead of 0.3.

```
>>> from scipy import stats
>>> d = stats.bernoulli(.5).rvs((10,5000)).mean(0)
>>> d = d[np.logical_not(np.isclose(d,1))]
>>> print('odds ratio=',np.mean(d),'var=',np.var(d))
odds ratio= 0.499379627776666 var= 0.024512322762879256
```

The odds ratio in this case is equal to one, which is not close to what was reported. According to our approximation, we should have $\mathbb{V}(g) = 0.4$, which does not look like what our simulation just reported. This is because the approximation is best when the odds ratio is nearly linear and worse otherwise (see Fig. 3.6).

Delta Method for Variance Stabilization To summarize the delta method more formally, we have n samples for the T_n statistic that estimates θ. If we have

$$\sqrt{n}\,(T_n - \theta) \xrightarrow{P} \mathcal{N}(0, \tau(\theta)^2)$$

then

$$\sqrt{n}\,(g(T_n) - g(\theta)) \xrightarrow{P} \mathcal{N}(0, \tau(\theta)^2(g'(\theta))^2)$$

Fig. 3.6 The odds ratio is close to linear for small values but becomes unbounded as p approaches one. The delta method is more effective for small underlying values of p, where the linear approximation is better

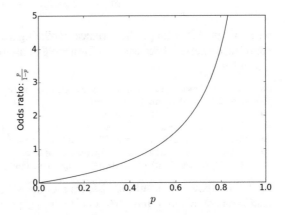

where convergence comes from the central limit theorem. Loosely speaking, this means that if the statistic T_n has a limiting normal distribution, then $g(T_n)$ also has a limiting normal distribution, with an explicit formula for the asymptotic variance of $g(T_n)$! This is a powerful result because it is not generally the case that a nonlinear function of a normal random variable has a normal distribution.

The unfortunate part of the previous example is that the variance depended on the estimated parameter, p. To fix this, we want to find a *variance stabilizing transformation* that makes the asymptotic variance independent of the estimated parameter. That is, we want

$$\sqrt{n}\,(g(T_n) - g(\theta)) \xrightarrow{P} \mathcal{N}(0, c^2)$$

We are almost there using the delta method because we already have

$$\sqrt{n}\,(g(T_n) - g(\theta)) \xrightarrow{P} \mathcal{N}(0, (g'(\theta))^2 \tau(\theta)^2)$$

so by judicious choice of g, we can force

$$g'(\theta)\tau(\theta) = c$$

Returning to our previous Bernoulli example, we had

$$\tau(p) = \sqrt{\frac{p(1-p)}{n}}$$

This time, let us choose

$$g'(p) = 1/\tau(p)$$

which means that

$$g(p) = 2\sqrt{n}\arcsin(\sqrt{p})$$

where we plug in $\hat{p}$ for p. The variance stabilizing transformation is used to compute confidence intervals because it avoids having to re-use the plug-in estimate value for the variance.

```
>>> from scipy.stats import bernoulli, norm
>>> n = 100 # num samples
>>> p = 0.1 # target parameter
>>> # simulate estimating p using MLE
>>> mns = bernoulli(p).rvs((n,5_000)).mean(axis=0)
>>> # variance stabilizing transformation
>>> v = 2*np.sqrt(n)*(np.arcsin(np.sqrt(mns))-np.arcsin(np.sqrt(p)))
```

Figure 3.7 shows the resulting histogram of the MLE estimates using this transformation as compared to the $\mathcal{N}(0, 1)$ density. This shows that the transformed MLE

Fig. 3.7 The variance stabilizing transformation versus the limiting normal distribution

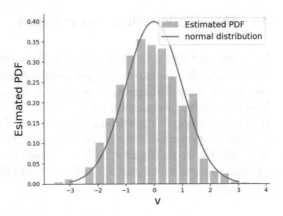

estimates following a normal distribution. To summarize, using this transformation, we have

$$2\sqrt{n}\left(\arcsin(\sqrt{\hat{p}}) - \arcsin(\sqrt{p})\right) \to \mathcal{N}(0, 1)$$

converging in distribution. This means that

$$\mathbb{P}\left\{\left|\arcsin\left(\sqrt{\hat{p}}\right) - \arcsin(\sqrt{p})\right| \le \frac{z_{\alpha/2}}{2\sqrt{n}}\right\} \to 1 - \alpha \quad \text{as } n \to \infty$$

so the $(1 - \alpha)$ confidence interval for $\sqrt{\hat{p}}$ is the following:

$$\left[\sin\left(\arcsin\left(\sqrt{\hat{p}}\right) - \frac{z_{\alpha/2}}{2\sqrt{n}}\right), \sin\left(\arcsin\left(\sqrt{\hat{p}}\right) + \frac{z_{\alpha/2}}{2\sqrt{n}}\right)\right]$$

Beware that the left side might become negative so it will have to be thresholded at zero. To get the confidence interval for $\hat{p}$, we just square the ends of the interval

$$\left[\sin^2\left(\arcsin\left(\sqrt{\hat{p}}\right) - \frac{z_{\alpha/2}}{2\sqrt{n}}\right), \sin^2\left(\arcsin\left(\sqrt{\hat{p}}\right) + \frac{z_{\alpha/2}}{2\sqrt{n}}\right)\right]$$

The advantage of using this variance stabilizing transformation is that otherwise we would have

$$\sqrt{n}\left(\hat{p} - p\right) \xrightarrow{d} N(0, p(1 - p))$$

where the variance on the right depends on the unknown p we are trying to estimate with $\hat{p}$.

3.5 Hypothesis Testing and P-Values

It is sometimes very difficult to unequivocally attribute outcomes to causal factors. For example, did your experiment generate the outcome you were hoping for or not? Maybe something did happen, but the effect is not pronounced enough to separate it from inescapable measurement errors or other factors in the ambient environment? Hypothesis testing is a powerful statistical method to address these questions. Let us begin by again considering our coin-tossing experiment with unknown parameter p. Recall that the individual coin flips are Bernoulli distributed. The first step is to establish separate hypotheses. First, H_0 is the so-called null hypothesis. In our case this can be

$$H_0 : \theta < \frac{1}{2}$$

and the alternative hypothesis is then

$$H_1 : \theta \geq \frac{1}{2}$$

With this setup, the question now boils down to figuring out which hypothesis the data is most consistent with. To choose between these, we need a statistical test that is a function, G, of the sample set $\mathbf{X}_n = \{X_i\}_n$ into the real line, where X_i is the heads or tails outcome ($X_i \in \{0, 1\}$). In other words, we compute $G(\mathbf{X}_n)$ and check if it exceeds a threshold c. If not, then we declare H_0 (otherwise, declare H_1). Notationally, this is the following:

$$G(\mathbf{X}_n) < c \Rightarrow H_0$$

$$G(\mathbf{X}_n) \geq c \Rightarrow H_1$$

In summary, we have the observed data $\mathbf{X}_n$ and a function G that maps that data onto the real line. Then, using the constant c as a threshold, the inequality effectively divides the real line into two parts, one corresponding to each of the hypotheses.

Whatever this test G is, it will make mistakes of two types—false negatives and false positives. The false positives arise from the case where we declare H_0 when the test says we should declare H_1. This is summarized in Table 3.1.

Table 3.1 Truth table for hypotheses testing

	Declare H_0	Declare H_1
H_0 true	Correct	False positive (type I error)
H_1 true	False negative (type II error)	Correct (true-detect)

For this example, here are the false positives (a.k.a. false alarms):

$$P_{FA} = \mathbb{P}\left(G(\mathbf{X}_n) > c \mid \theta \leq \frac{1}{2}\right)$$

Or, equivalently,

$$P_{FA} = \mathbb{P}\left(G(\mathbf{X}_n) > c \mid H_0\right)$$

Likewise, the other error is a false negative, which we can write analogously as

$$P_{FN} = \mathbb{P}\left(G(\mathbf{X}_n) < c \mid H_1\right)$$

By choosing some acceptable values for either of these errors, we can solve for the other one. The practice is usually to pick a value of P_{FA} and then find the corresponding value of P_{FN}. Note that it is traditional in engineering to speak about *detection probability*, which is defined as

$$P_D = 1 - P_{FN} = \mathbb{P}\left(G(\mathbf{X}_n) > c \mid H_1\right)$$

In other words, this is the probability of declaring H_1 when the test exceeds the threshold. This is otherwise known as the *probability of a true detection* or *true-detect*.

3.5.1 Back to the Coin Flipping Example

In our previous maximum likelihood discussion, we wanted to derive an estimator for the *value* of the probability of heads for the coin flipping experiment. For hypthesis testing, we want to ask a softer question: Is the probability of heads greater or less than $1/2$? As we just established, this leads to the two hypotheses:

$$H_0: \theta < \frac{1}{2}$$

versus

$$H_1: \theta > \frac{1}{2}$$

Let us assume we have five observations. Now we need the G function and a threshold c to help pick between the two hypotheses. Let us count the number of heads observed in five observations as our criterion. Thus, we have

$$G(\mathbf{X}_5) := \sum_{i=1}^{5} X_i$$

Fig. 3.8 Power function for the all-heads test. The dark circle indicates the value of the function indicating α

and suppose further that we pick H_1 only if exactly five out of five observations are heads. We'll call this the *all-heads* test.

Now, because all of the X_i are random variables, so is G, and we must find the corresponding probability mass function for G. Assuming the individual coin tosses are independent, the probability of five heads is θ^5. This means that the probability of rejecting the H_0 hypothesis (and choosing H_1, because there are only two choices here) based on the unknown underlying probability is θ^5. In the parlance, this is known as the *power function* as in denoted by β as in

$$\beta(\theta) = \theta^5$$

Let us get a quick plot of this in Fig. 3.8.

Now, we have the following false alarm probability:

$$P_{FA} = \mathbb{P}(G(\mathbf{X}_n) = 5 | H_0) = \mathbb{P}(\theta^5 | H_0)$$

Notice that this is a function of θ, which means there are many false alarm probability values that correspond to this test. To be on the conservative side, we'll pick the supremum (i.e., maximum) of this function, which is known as the *size* of the test, traditionally denoted by α,

$$\alpha = \sup_{\theta \in \Theta_0} \beta(\theta)$$

with domain $\Theta_0 = \{\theta < 1/2\}$ which in our case is

$$\alpha = \sup_{\theta < \frac{1}{2}} \theta^5 = \left(\frac{1}{2}\right)^5 = 0.03125$$

Likewise, for the detection probability,

$$\mathbb{P}_D(\theta) = \mathbb{P}(\theta^5 | H_1)$$

which is again a function of the parameter θ. The problem with this test is that the P_D is pretty low for most of the domain of θ. For instance, values in the nineties for P_D only happen when $\theta > 0.98$. In other words, if the coin produces heads 98 times out of 100, then we can detect H_1 reliably. Ideally, we want a test that is zero for the domain corresponding to H_0 (i.e., Θ_0) and equal to one otherwise. Unfortunately, even if we increase the length of the observed sequence, we cannot escape this effect with this test. You can try plotting θ^n for larger and larger values of n to see this.

Majority Vote Test Due to the problems with the detection probability in the all-heads test, maybe we can think of another test that will have the performance we want. Suppose we reject H_0 if the majority of the observations are heads. Then, using the same reasoning as above, we have

$$\beta(\theta) = \sum_{k=3}^{5} \binom{5}{k} \theta^k (1 - \theta)^{5-k}$$

Figure 3.9 shows the power function for both the majority vote and the all-heads tests.

In this case, the new test has *size*

$$\alpha = \sup_{\theta < \frac{1}{2}} \theta^5 + 5\theta^4 (-\theta + 1) + 10\theta^3 (-\theta + 1)^2 = \frac{1}{2}$$

As before we only get to upward of 90% for detection probability only when the underlying parameter $\theta > 0.75$. Let us see what happens when we consider more than five samples. For example, let us suppose that we have $n = 100$ samples and we want to vary the threshold for the majority vote test. For example, let us have a new test where we declare H_1 when $k = 60$ out of the 100 trials turns out to be

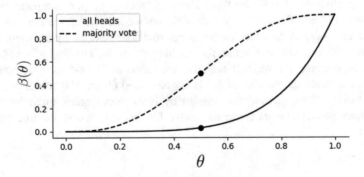

Fig. 3.9 Compares the power function for the all-heads test with that of the majority vote test

heads. What is the β function in this case?

$$\beta(\theta) = \sum_{k=60}^{100} \binom{100}{k} \theta^k (1 - \theta)^{100-k}$$

This is too complicated to write by hand, but the statistics module in Sympy has all the tools we need to compute this.

```
>>> from sympy.stats import P, Binomial
>>> theta = S.symbols('theta',real=True)
>>> X = Binomial('x',100,theta)
>>> beta_function = P(X>60)
>>> print(beta_function.subs(theta,0.5)) # alpha
0.0176001001088524
>>> print(beta_function.subs(theta,0.70))
0.979011423996075
```

These results are much better than before because the β function is much steeper. If we declare H_1 when we observe 60 out of 100 trials are heads, then we wrongly declare heads approximately 1.8% of the time. Otherwise, if it happens that the true value for $p > 0.7$, we will conclude correctly approximately 97% of the time. A quick simulation can sanity check these results as shown below:

```
>>> from scipy import stats
>>> rv=stats.bernoulli(0.5) # true p = 0.5
>>> # number of false alarms ~ 0.018
>>> print(sum(rv.rvs((1000,100)).sum(axis=1)>60)/1000.)
0.015
```

The above code is pretty dense, so let us unpack it. In the first line, we use the `scipy.stats` module to define the Bernoulli random variable for the coin flip. Then, we use the `rvs` method of the variable to generate 1000 trials of the experiment where each trial consists of 100 coin flips. This generates a 1000×100 matrix where the rows are the individual trials and the columns are the outcomes of each respective set of 100 coin flips. The `sum(axis=1)` part computes the sum across the columns. Because the values of the embedded matrix are only 1 or 0, this gives us the count of flips that are heads per row. The next `>60` part computes the boolean 1000-long vector of values that are bigger than 60. The final `sum` adds these up. Again, because the entries in the array are `True` or `False`, the `sum` computes the count of times the number of heads has exceeded 60 per 100 coin flips in each of 1000 trials. Then, dividing this number by 1000 gives a quick approximation of false alarm probability we computed above for this case where the true value of $p = 0.5$.

3.5.2 Receiver Operating Characteristic

Because the majority vote test is a binary test, we can compute the *receiver operating characteristic* (ROC) which is the graph of the (P_{FA}, P_D). The term comes from radar systems but is a very general method for consolidating all of these issues into a single graph. Let us consider a typical signal processing example with two hypotheses. In H_0, there is noise but no signal present at the receiver

$$H_0: X = \epsilon$$

where $\epsilon \sim \mathcal{N}(0, \sigma^2)$ represents additive noise. In the alternative hypothesis, there is a deterministic signal at the receiver:

$$H_1: X = \mu + \epsilon$$

Again, the problem is to choose between these two hypotheses. For H_0, we have $X \sim \mathcal{N}(0, \sigma^2)$, and for H_1, we have $X \sim \mathcal{N}(\mu, \sigma^2)$. Recall that we only observe values for x and must pick either H_0 or H_1 from these observations. Thus, we need a threshold, c, to compare x against in order to distinguish the two hypotheses. Figure 3.10 shows the probability density functions under each of the hypotheses. The dark vertical line is the threshold c. The gray shaded area is the probability of detection, P_D, and the shaded area is the probability of false alarm, P_{FA}. The test evaluates every observation of x and concludes H_0 if $x < c$ and H_1 otherwise.

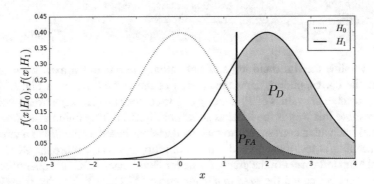

Fig. 3.10 The two density functions for the H_0 and H_1 hypotheses. The shaded gray area is the detection probability, and the shaded dark gray area is the probability of false alarm. The vertical line is the decision threshold

Fig. 3.11 The receiver operating characteristic (ROC) corresponding to Fig. 3.10

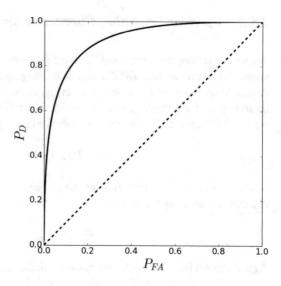

As we slide the threshold left and right along the horizontal axis, we naturally change the corresponding areas under each of the curves shown in Fig. 3.10 and thereby change the values of P_D and P_{FA}. The contour that emerges from sweeping the threshold this way is the ROC as shown in Fig. 3.11. This figure also shows the diagonal line which corresponds to making decisions based on the flip of a fair coin. Any meaningful test must do better than coin flipping, so the more the ROC bows up to the top left corner of the graph, the better. Sometimes ROCs are quantified into a single number called the *area under the curve* (AUC), which varies from 0.5 to 1.0 as shown. In our example, what separates the two probability density functions is the value of μ. In a real situation, this would be determined by signal processing methods that include many complicated trade-offs. The key idea is that whatever those trade-offs are, the test itself boils down to the separation between these two density functions—good tests separate the two density functions and bad tests do not. Indeed, when there is no separation, we arrive at the diagonal-line coin flipping situation we just discussed.

What values for P_D and P_{FA} are considered *acceptable* depends on the application. For example, suppose you are testing for a fatal disease. It could be that you are willing to except a relatively high P_{FA} value if that corresponds to a good P_D because the test is relatively cheap to administer compared to the alternative of missing a detection. On the other hand, maybe a false alarm triggers an expensive response, so that minimizing these alarms is more important than potentially missing a detection. These trade-offs can only be determined by the application and design factors.

3.5.3 P-Values

There are a lot of moving parts in hypothesis testing. What we need is a way to consolidate the findings. The idea is that we want to find the minimum level at which the test rejects H_0. Thus, the *p*-value is the probability, under H_0, that the test-statistic is at least as extreme as what was actually observed. Informally, this means that smaller values imply that H_0 should be rejected, although this doesn't mean that large values imply that H_0 should be retained. This is because a large *p*-value can arise from either H_0 being true or the test having low statistical power.

If H_0 is true, the *p*-value is uniformly distributed in the interval $(0, 1)$. If H_1 is true, the distribution of the *p*-value will concentrate closer to zero. For continuous distributions, this can be proven rigorously and implies that if we reject H_0 when the corresponding *p*-value is less than α, then the probability of a false alarm is α. Perhaps it helps to formalize this a bit before computing it. Suppose $\tau(X)$ is a test statistic that rejects H_0 as it gets bigger. Then, for each sample x, corresponding to the data we actually have on-hand, we define

$$p(x) = \sup_{\theta \in \Theta_0} \mathbb{P}_\theta(\tau(X) > \tau(x))$$

This equation states that the supremum (i.e., maximum) probability that the test statistic, $\tau(X)$, exceeds the value for the test statistic on this particular data $(\tau(x))$ over the domain Θ_0 is defined as the *p*-value. Thus, this embodies a worst-case scenario over all values of θ.

Here's one way to think about this. Suppose you rejected H_0, and someone says that you just got *lucky* and somehow just drew data that happened to correspond to a rejection of H_0. What *p*-values provide is a way to address this by capturing the probability of just such a favorable data-draw. Thus, suppose that your *p*-value is 0.05. Then, what you are showing is that the odds of just drawing that data sample, given H_0 is in force, is just 5%. This means that there's a 5% chance that you somehow lucked out and got a favorable draw of data.

Let us make this concrete with an example. Given the majority vote rule above, suppose we actually do observe three of five heads. Given the H_0, the probability of observing this event is the following:

$$p(x) = \sup_{\theta \in \Theta_0} \sum_{k=3}^{5} \binom{5}{k} \theta^k (1-\theta)^{5-k} = \frac{1}{2}$$

For the all-heads test, the corresponding computation is the following:

$$p(x) = \sup_{\theta \in \Theta_0} \theta^5 = \frac{1}{2^5} = 0.03125$$

From just looking at these p-values, you might get the feeling that the second test is better, but we still have the same detection probability issues we discussed above; so, p-values help in summarizing some aspects of our hypothesis testing, but they do *not* summarize all the salient aspects of the *entire* situation.

3.5.4 Test Statistics

As we have seen, it is difficult to derive good test statistics for hypothesis testing without a systematic process. The Neyman-Pearson test is derived from fixing a false-alarm value (α) and then maximizing the detection probability

$$L(\mathbf{x}) = \frac{f_{X|H_1}(\mathbf{x})}{f_{X|H_0}(\mathbf{x})} \mathop{\gtrless}_{H_0}^{H_1} \gamma$$

where L is the likelihood ratio and where the threshold γ is chosen such that

$$\int_{x:L(\mathbf{x})>\gamma} f_{X|H_0}(\mathbf{x})d\mathbf{x} = \alpha$$

The Neyman-Pearson test is one of a family of tests that use the likelihood ratio.

Example Suppose we have a receiver and we want to distinguish whether just noise (H_0) or signal plus noise (H_1) is received. For the noise-only case, we have $x \sim \mathcal{N}(0, 1)$, and for the signal plus noise case, we have $x \sim \mathcal{N}(1, 1)$. In other words, the mean of the distribution shifts in the presence of the signal. This is a very common problem in signal processing and communications. The Neyman-Pearson test then boils down to the following:

$$L(x) = e^{-\frac{1}{2}+x} \mathop{\gtrless}_{H_0}^{H_1} \gamma$$

Now we have to find the threshold γ that solves the maximization problem that characterizes the Neyman-Pearson test. Taking the natural logarithm and rearranging gives,

$$x \underset{H_0}{\overset{H_1}{\gtrless}} \frac{1}{2} + \log \gamma$$

The next step is find γ corresponding to the desired α by computing it from the following:

$$\int_{1/2+\log \gamma}^{\infty} f_{X|H_0}(x)dx = \alpha$$

For example, taking $\alpha = 1/100$ gives $\gamma \approx 6.21$. To summarize the test in this case, we have

$$x \underset{H_0}{\overset{H_1}{\gtrless}} 2.32$$

Thus, if we measure X and see that its value exceeds the threshold above, we declare H_1 and otherwise declare H_0. The following code shows how to solve this example using Sympy and Scipy. First, we set up the likelihood ratio:

```
>>> import sympy as S
>>> from sympy import stats
>>> s = stats.Normal('s',1,1) # signal+noise
>>> n = stats.Normal('n',0,1) # noise
>>> x = S.symbols('x',real=True)
>>> L = stats.density(s)(x)/stats.density(n)(x)
```

Next, to find the γ value

```
>>> g = S.symbols('g',positive=True) # define gamma
>>> v=S.integrate(stats.density(n)(x),
...                 (x,S.Rational(1,2)+S.log(g),S.oo))
```

Programming Tip

Providing additional information regarding the Sympy variable by using the keyword argument `positive=True` helps the internal simplification algorithms work faster and better. This is especially useful when dealing with complicated integrals that involve special functions. Furthermore, note that we used the `Rational` function to define the $1/2$ fraction, which is another way of providing hints to Sympy. Otherwise, it's possible that the floating-point representation of the fraction could disguise the simple fraction and thereby miss internal simplification opportunities.

We want to solve for g in the above expression. Sympy has some built-in numerical solvers as in the following:

```
>>> print(S.nsolve(v-0.01,3.0)) # approx 6.21
6.21116124253284
```

Note that in this situation, it is better to use the numerical solvers because Sympy `solve` may grind along for a long time to resolve this.

Generalized Likelihood Ratio Test The likelihood ratio test can be generalized using the following statistic:

$$\Lambda(\mathbf{x}) = \frac{\sup_{\theta \in \Theta_0} L(\theta)}{\sup_{\theta \in \Theta} L(\theta)} = \frac{L(\hat{\theta}_0)}{L(\hat{\theta})}$$

where $\hat{\theta}_0$ maximizes $L(\theta)$ subject to $\theta \in \Theta_0$ and $\hat{\theta}$ is the maximum likelihood estimator. The intuition behind this generalization of the likelihood ratio test is that the denominator is the usual maximum likelihood estimator and the numerator is the maximum likelihood estimator, but over a restricted domain (Θ_0). This means that the ratio is always less than unity because the maximum likelihood estimator over the entire space will always be at least as maximal as that over the more restricted space. When this Λ ratio gets small enough, it means that the maximum likelihood estimator over the entire domain (Θ) is larger which means that it is safe to reject the null hypothesis H_0. The tricky part is that the statistical distribution of Λ is usually eye-wateringly difficult. Fortunately, Wilks theorem says that with sufficiently large n, the distribution of $-2 \log \Lambda$ is approximately chi-square (see Sect. 2.6.3) with $r - r_0$ degrees of freedom, where r is the number of free parameters for Θ and r_0 is the number of free parameters in Θ_0. With this result, if we want an approximate test at level α, we can reject H_0 when $-2 \log \Lambda \geq \chi^2_{r-r_0}(\alpha)$ where $\chi^2_{r-r_0}(\alpha)$ denotes the $1 - \alpha$ quantile of the $\chi^2_{r-r_0}$ chi-square distribution. However, the problem with this result is that there is no definite way of knowing how big n should be. The advantage of this generalized likelihood ratio test is that it can test multiple hypotheses simultaneously, as illustrated in the following example.

Example Let us return to our coin flipping example, except now we have three different coins. The likelihood function is then

$$L(p_1, p_2, p_3) = \text{binom}(k_1; n_1, p_1)\text{binom}(k_2; n_2, p_2)\text{binom}(k_3; n_3, p_3)$$

where `binom` is the binomial distribution with the given parameters. For example,

$$\text{binom}(k; n, p) = \sum_{k=0}^{n} \binom{n}{k} p^k (1 - p)^{n-k}$$

The null hypothesis is that all three coins have the same probability of heads, $H_0 : p = p_1 = p_2 = p_3$. The alternative hypothesis is that at least one of these probabilities is different. Let us consider the numerator of the Λ first, which will give us the maximum likelihood estimator of p. Because the null hypothesis is that all the p values are equal, we can just treat this as one big binomial distribution with $n = n_1 + n_2 + n_3$, and $k = k_1 + k_2 + k_3$ is the total number of heads observed for any coin. Thus, under the null hypothesis, the distribution of k is binomial with parameters n and p. Now, what is the maximum likelihood estimator for this distribution? We have worked this problem before and have the following:

$$\hat{p}_0 = \frac{k}{n}$$

In other words, the maximum likelihood estimator under the null hypothesis is the proportion of ones observed in the sequence of n trials total. Now, we have to substitute this in for the likelihood under the null hypothesis to finish the numerator of Λ

$$L(\hat{p}_0, \hat{p}_0, \hat{p}_0) = \texttt{binom}(k_1; n_1, \hat{p}_0)\texttt{binom}(k_2; n_2, \hat{p}_0)\texttt{binom}(k_3; n_3, \hat{p}_0)$$

For the denominator of Λ, which represents the case of maximizing over the entire space, the maximum likelihood estimator for each separate binomial distribution is likewise

$$\hat{p}_i = \frac{k_i}{n_i}$$

which makes the likelihood in the denominator the following:

$$L(\hat{p}_1, \hat{p}_2, \hat{p}_3) = \texttt{binom}(k_1; n_1, \hat{p}_1)\texttt{binom}(k_2; n_2, \hat{p}_2)\texttt{binom}(k_3; n_3, \hat{p}_3)$$

for each of the $i \in \{1, 2, 3\}$ binomial distributions. Then, the Λ statistic is then the following:

$$\Lambda(k_1, k_2, k_3) = \frac{L(\hat{p}_0, \hat{p}_0, \hat{p}_0)}{L(\hat{p}_1, \hat{p}_2, \hat{p}_3)}$$

Wilks theorems states that $-2 \log \Lambda$ is chi-square distributed. We can compute this example with the statistics tools in Sympy and Scipy.

```
>>> from scipy.stats import binom, chi2
>>> import numpy as np
>>> # some sample parameters
>>> p0,p1,p2 = 0.3,0.4,0.5
>>> n0,n1,n2 = 50,180,200
>>> brvs= [ binom(i,j) for i,j in zip((n0,n1,n2),(p0,p1,p2))]
>>> def gen_sample(n=1):
...         'generate samples from separate binomial distributions'
...         if n==1:
...             return [i.rvs() for i in brvs]
...         else:
...             return [gen_sample() for k in range(n)]
...
```

> **Programming Tip**
> Note the recursion in the definition of the `gen_sample` function where a
> conditional clause of the function calls itself. This is a quick way to reusing
> code and generating vectorized output. Using `np.vectorize` is another
> way, but the code is simple enough in this case to use the conditional clause. In
> Python, it is generally bad for performance to have code with nested recursion
> because of how the stack frames are managed. However, here we are only
> recursing once so this is not an issue.

Next, we compute the logarithm of the numerator of the Λ statistic,

```
>>> k0,k1,k2 = gen_sample()
>>> print(k0,k1,k2)
12 68 103
>>> pH0 = sum((k0,k1,k2))/sum((n0,n1,n2))
>>> numer = np.sum([np.log(binom(ni,pH0).pmf(ki))
...                 for ni,ki in
...                     zip((n0,n1,n2),(k0,k1,k2))])
>>> print(numer)
-15.545863836567923
```

Note that we used the null hypothesis estimate for the $\hat{p}_0$. Likewise, for the
logarithm of the denominator, we have the following:

```
>>> denom = np.sum([np.log(binom(ni,pi).pmf(ki))
...                 for ni,ki,pi in
...                     zip((n0,n1,n2),(k0,k1,k2),(p0,p1,p2))])
>>> print(denom)
-8.424106480792464
```

Now, we can compute the logarithm of the Λ statistic as follows and see what the
corresponding value is according to Wilks theorem:

```
>>> chsq=chi2(2)
>>> logLambda =-2*(numer-denom)
>>> print(logLambda)
14.243514711550919
>>> print(1-chsq.cdf(logLambda))
0.0008073467083287156
```

Because the value reported above is less than the 5% significance level, we reject the
null hypothesis that all the coins have the same probability of heads. Note that there
are two degrees of freedom because the difference in the number of parameters
between the null hypothesis (p) and the alternative (p_1, p_2, p_3) is two. We can
build a quick Monte Carlo simulation to check the probability of detection for this
example using the following code, which is just a combination of the last few code
blocks:

```
>>> c= chsq.isf(.05) # 5% significance level
>>> out = []
>>> for k0,k1,k2 in gen_sample(100):
```

```
...      pH0 = sum((k0,k1,k2))/sum((n0,n1,n2))
...      numer = np.sum([np.log(binom(ni,pH0).pmf(ki))
...                       for ni,ki in
...                           zip((n0,n1,n2),(k0,k1,k2))])
...      denom = np.sum([np.log(binom(ni,pi).pmf(ki))
...                       for ni,ki,pi in
...                           zip((n0,n1,n2),(k0,k1,k2),(p0,p1,p2))])
...      out.append(-2*(numer-denom)>c)
...
>>> print(np.mean(out)) # estimated probability of detection
0.59
```

The above simulation shows the estimated probability of detection, for this set of example parameters. This relative low probability of detection means that while the test is unlikely (i.e., at the 5% significance level) to mistakenly pick the null hypothesis, it is likewise missing many of the H_1 cases (i.e., low probability of detection). The trade-off between which is more important is up to the particular context of the problem. In some situations, we may prefer additional false alarms in exchange for missing fewer H_1 cases.

Permutation Test The permutation test is good way to test whether or not samples come from the same distribution. For example, suppose that

$$X_1, X_2, \ldots, X_m \sim F$$

and also

$$Y_1, Y_2, \ldots, Y_n \sim G$$

That is, Y_i and X_i come from different distributions. Suppose we have some test statistic, for example

$$T(X_1, \ldots, X_m, Y_1, \ldots, Y_n) = |\overline{X} - \overline{Y}|$$

Under the null hypothesis for which $F = G$, any of the $(n + m)!$ permutations are equally likely. Thus, suppose for each of the $(n + m)!$ permutations, we have the computed statistic:

$$\{T_1, T_2, \ldots, T_{(n+m)!}\}$$

Then, under the null hypothesis, each of these values is equally likely. The distribution of T under the null hypothesis is the *permutation distribution* that puts weight $1/(n + m)!$ on each T-value. Suppose t_o is the observed value of the test statistic and assume that large T rejects the null hypothesis, then the *p*-value for the permutation test is the following:

$$\mathbb{P}(T > t_o) = \frac{1}{(n + m)!} \sum_{j=1}^{(n+m)!} I(T_j > t_o)$$

where $I()$ is the indicator function. For large $(n + m)!$, we can sample randomly from the set of all permutations to estimate this p-value.

Example Let us return to our coin flipping example from last time, but now we have only two coins. The hypothesis is that both coins have the same probability of heads. We can use the built-in function in Numpy to compute the random permutations.

```
>>> x=binom(10,0.3).rvs(5)  # p=0.3
>>> y=binom(10,0.5).rvs(3)  # p=0.5
>>> z = np.hstack([x,y])  # combine into one array
>>> t_o = abs(x.mean()-y.mean())
>>> out = []  # output container
>>> for k in range(1000):
...     perm = np.random.permutation(z)
...     T=abs(perm[:len(x)].mean()-perm[len(x):].mean())
...     out.append((T>t_o))
...
>>> print('p-value = ', np.mean(out))
p-value =   0.0
```

Note that the size of total permutation space is $8! = 40320$ so we are taking relatively few (i.e., 1000) random permutations from this space.

Wald Test The Wald test is an asymptotic test. Suppose we have $H_0 : \theta = \theta_0$ and otherwise $H_1 : \theta \neq \theta_0$, the corresponding statistic is defined as the following:

$$W = \frac{\hat{\theta}_n - \theta_0}{\text{se}}$$

where $\hat{\theta}$ is the maximum likelihood estimator and se is the standard error:

$$\text{se} = \sqrt{\mathbb{V}(\hat{\theta}_n)}$$

Under general conditions, $W \xrightarrow{d} \mathcal{N}(0, 1)$. Thus, an asymptotic test at level α rejects when $|W| > z_{\alpha/2}$ where $z_{\alpha/2}$ corresponds to $\mathbb{P}(|Z| > z_{\alpha/2}) = \alpha$ with $Z \sim \mathcal{N}(0, 1)$. For our favorite Coin flipping example, if $H_0 : \theta = \theta_0$, then

$$W = \frac{\hat{\theta} - \theta_0}{\sqrt{\hat{\theta}(1 - \hat{\theta})/n}}$$

We can simulate this using the following code at the usual 5% significance level:

```
>>> from scipy import stats
>>> theta0 = 0.5  # H0
>>> k=np.random.binomial(1000,0.3)
>>> theta_hat = k/1000.  # MLE
```

```
>>> W = (theta_hat-theta0)/np.sqrt(theta_hat*(1-theta_hat)/1000)
>>> c = stats.norm().isf(0.05/2) # z_{alpha/2}
>>> print(abs(W)>c) # if true, reject H0
True
```

This rejects H_0 because the true $\theta = 0.3$ and the null hypothesis is that $\theta = 0.5$.
Note that $n = 1000$ in this case which puts us well inside the asymptotic range of
the result. We can re-do this example to estimate the detection probability for this
example as in the following code:

```
>>> theta0 = 0.5 # H0
>>> c = stats.norm().isf(0.05/2.) # z_{alpha/2}
>>> out = []
>>> for i in range(100):
...     k=np.random.binomial(1000,0.3)
...     theta_hat = k/1000. # MLE
...     W = (theta_hat-theta0)/np.sqrt(theta_hat*(1-theta_hat)/1000.)
...     out.append(abs(W)>c) # if true, reject H0
...
>>> print(np.mean(out)) # detection probability
1.0
```

3.5.5 Testing Multiple Hypotheses

Thus far, we have focused primarily on two competing hypotheses. Now, we
consider multiple comparisons. The general situation is the following. We test the
null hypothesis against a sequence of n competing hypotheses H_k. We obtain p-
values for each hypothesis, so now we have multiple p-values to consider $\{p_k\}$. To
boil this sequence down to a single criterion, we can make the following argument.
Given n independent hypotheses that are all untrue, the probability of getting at least
one false alarm is the following:

$$P_{FA} = 1 - (1 - p_0)^n$$

where p_0 is the individual p-value threshold (say, 0.05). The problem here is that
$P_{FA} \rightarrow 1$ as $n \rightarrow \infty$. If we want to make many comparisons at once and control
the overall false alarm rate, the overall p-value should be computed under the
assumption that none of the competing hypotheses is valid. The most common
way to address this is with the Bonferroni correction which says that the individual
significance level should be reduced to p/n. Obviously, this makes it much harder
to declare significance for any particular hypothesis. The natural consequence of
this conservative restriction is to reduce the statistical power of the experiment, thus
making it more likely the true effects will be missed.

In 1995, Benjamini and Hochberg devised a simple method that tells which p-
values are statistically significant. The procedure is to sort the list of p-values in
ascending order, choose a false-discovery rate (say, q), and then find the largest p-
value in the sorted list such that $p_k \leq kq/n$, where k is the p-value's position in the
sorted list. Finally, declare that p_k value and all the others less than it statistically

Table 3.2 Example
contingency table

	Infection	No infection	Total
Male	13	11	24
Female	12	1	13
Total	25	12	37

significant. This procedure guarantees that the proportion of false positives is less than q (on average). The Benjamini-Hochberg procedure (and its derivatives) is fast and effective and is widely used for testing hundreds of primarily false hypotheses when studying genetics or diseases. Additionally, this procedure provides better statistical power than the Bonferroni correction.

3.5.6 Fisher Exact Test

Contingency tables represent the partitioning of a sample population of two categories between two different classifications as shown in Table 3.2. The question is whether or not the observed table corresponds to a random partition of the sample population, constrained by the marginal sums. Note that because this is a two-by-two table, a change in any of the table entries automatically affects all of the other terms because of the row and column sum constraints. This means that equivalent questions like "Under a random partition, what is the probability that a particular table entry is at least as large as a given value?" can be meaningfully posed.

The Fisher exact test addresses this question. The idea is to compute the probability of a particular entry of the table, conditioned upon the marginal row and column sums

$$\mathbb{P}(X_{i,j}|r_1, r_2, c_1, c_2)$$

where $X_{i,j}$ is (i, j) table entry, r_1 represents the sum of the first row, r_2 represents the sum of the second row, c_1 represents the sum of the first column, and c_2 is the sum of the second column. This probability is given by the *hypergeometric distribution*. Recall that the hypergeometric distribution gives the probability of sampling (without replacement) k items from a population of N items consisting of exactly two different kinds of items

$$\mathbb{P}(X = k) = \frac{\binom{K}{k}\binom{N-K}{n-k}}{\binom{N}{n}}$$

where N is the population size, K is the total number of possible favorable draws, n is the number of draws, and k is the number of observed favorable draws. With the corresponding identification of variables, the hypergeometric distribution gives the desired conditional probability: $K = r_1, k = x, n = c_1, N = r_1 + r_2$.

Table 3.2 shows the data for $x = 13$ male infections among a population of $r_1 = 24$ males, in a total population of $c_1 = 25$ infected persons, including $r_2 = 13$ females. The `scipy.stats` module has the Fisher exact test implemented as shown below:

```
>>> from scipy.stats import fisher_exact
>>> table = [[13,11],[12,1]]
>>> odds_ratio, p_value=fisher_exact(table)
>>> p_value
0.02718387758955712
```

The default for `fisher_exact` is the two-sided test. The following result is for the `less` option:

```
>>> odds_ratio, p_value=fisher_exact(table,alternative='less')
>>> p_value
0.018976707519532877
```

This means that the *p*-value is computed by summing over the probabilities of contingency tables that are *less* extreme than the given table. To understand what this means, we can use the `hypergeom` function to compute the probabilities of these with the number of infected men less than or equal to 13.

```
>>> from scipy.stats import hypergeom
>>> hg = hypergeom(37, 24, 25)
>>> probs = [(hg.pmf(i)) for i in range(14)]
>>> print(probs)
[0.0, 0.0, 0.0, 0.0, 0.0, 0.0, 0.0, 0.0, 0.0, 0.0, 0.0, 0.0,
 0.0014597467322717626, 0.017516960787261115]
>>> print(sum(probs))
0.018976707519532877
```

This is the same as the prior *p*-value result we obtained from `fisher_exact`. Another option is `greater` which derives from the following analogous summation:

```
>>> odds_ratio, p_value=fisher_exact(table,alternative='greater')
>>> probs = [hg.pmf(i) for i in range(13,25)]
>>> print(probs)
[0.017516960787261115, 0.08257995799708828, 0.2018621195484381,
 0.28386860561499044, 0.24045340710916852, 0.12467954442697629,
 0.039372487713781906, 0.00738234144633414, 0.0007812001530512284,
 4.261091743915799e-05, 1.0105355914424832e-06,
 7.017608273906114e-09]
>>> print(p_value)
0.9985402532677288
>>> print(sum(probs))
0.9985402532677288
```

Finally, the two-sided version excludes those individual table probabilities that are less than that of the given table:

```
>>> _,p_value=fisher_exact(table)
>>> probs = [ hg.pmf(i) for i in range(25) ]
>>> print(sum(i for i in probs if i<= hg.pmf(13)))
0.027183877589557117
>>> print(p_value)
0.0271838775895712
```

Thus, for this particular contingency table, we could reasonably conclude that 13 infected males in this total population is statistically significant with a p-value less than five percent.

Performing this kind of analysis for tables larger than 2x2 easily becomes computationally challenging due to the nature of the underlying combinatorics and usually requires specialized approximations.

Contingency Table Odds Ratio Using a two-by-two contingency table, we can write the odds ratio:

$$\texttt{odds\_ratio} = \frac{\frac{\pi_{0,0}}{\pi_{0,1}}}{\frac{\pi_{1,0}}{\pi_{1,1}}}$$

For the multinomial distribution, we have the maximum likelihood estimators for the individual probabilities:

$$\hat{\pi}_{i,j} = X_{i,j}/N$$

Via the plug-in principle, the MLE of the odds ratio is therefore

$$\texttt{odds\_ratioMLE} = \frac{X_{0,0}X_{1,1}}{X_{0,1}X_{1,0}}$$

Note that under independence, $\pi_{i,j} = \pi_{i,\bullet}\pi_{\bullet,j}$ so that $\texttt{odds\_ratio} = 1$ Whereas tests like the Fisher exact test inform the validity of the association in the contingency table, the odds ratio quantifies the strength of that association with the value of one indicating independence or absence of association. Consider the following example from the $\texttt{fisher\_exact}$ documentation:

	Atlantic	Indian
Whales	8	2
Sharks	1	5

```
>>> table = [[8,2],[1,5]]
>>> odds_ratio, p_value = fisher_exact(table)
>>> print(f'{odds_ratio=}, {p_value=}')
odds_ratio=20.0, p_value=0.03496503496503495
```

The interpretation for the odds ratio above indicates that the association for whales in the Atlantic versus the Indian ocean is 20 times stronger than for sharks in the Atlantic versus the Indian Ocean. If the threshold for significance in the *p*-value is the usual five percent, then the *p*-value above indicates that there is a valid association in the contingency table.

3.5.7 Contingency Table Protocols

Contingency tables summarize the joint relationships between categorical variables where each cell denotes the count of the cross-tabulated data for that column and row. As shown in the two-by-two table below, the rows usually indicate the independent (exogenous or explanatory) variables, and the columns indicate the dependent (endogenous or response) variables. The • subscript denotes summation of the entries along that positional index. Contingency tables can have $I \times J$ rows by columns and can extend into multiple dimensions, but for our discussion we consider at most three dimensions $I \times J \times K$, simply because the notation becomes unwieldy, even though the same principles extend to higher dimensional contingency tables. For certain situations, it may be unrealistic to designate the variables independent versus dependent variables so the contingency table can also be considered a two dimensional probability density histogram.

In this section, we start out with the simplest two-by-two contingency tables and examine common methods to understand and analyze them. Then, we move onto larger two-dimensional and three-dimensional tables and consider methods of analysis for these.

X/Y	Y_0	Y_1	Total
X_0	$n_{0,0}$	$n_{0,1}$	$n_{0,\bullet}$
X_1	$n_{1,0}$	$n_{1,1}$	$n_{1,\bullet}$
Total	$n_{\bullet,0}$	$n_{\bullet,1}$	$n_{\bullet,\bullet}$

Note that the rows are called *factors* and the columns *cases*. The rectangular grids of the contingency table impose structural relationships among the constituent variables that are reflected in the probability models used to represent the data. For example, the simplest relationship between the columns and rows is no relationship at all, which can be characterized by the Poisson distribution with the count in each cell as a separate Poisson random sample process with its own Poisson mean parameter, say $\mu_{i,j}$. This means that the marginal sums of the contingency table are likewise Poisson distributed. At the other extreme, we can consider the total of sum of all cells and the marginals as fixed, which is modeled by the hypergeometric distribution. In between these two cases are the multinomial distributions, or their binary special cases, the binomial distributions.

Example: Voting Partisanship Consistency Let us consider the following two-by-two contingency table for the *longitudinal* study characterized by repeated observations of the same individuals. A total of 433 males were observed voting in 2004 and then again in 2008. The table summarizes whether voting behavior with respect to political party was consistent between these two elections. For example, of 191 men who voted Democrat in 2004, 175 of them voted Democrat in 2008. For this study, the 191 Democrats and 242 Republicans were selected in 2004, and the same group was surveyed again in 2008. Because the future voting behavior was unknown when the 191 Democrats and 242 Republicans were chosen for the study, the column proportions are unknown. The row marginals are fixed (no one dropped out in the interim), but the column marginals are not. The null hypothesis is that there is no difference in the consistency of the voting pattern between Democrats and Republicans.

X/Y	Democrat-2008	Republican-2008	Total
Democrat-2004	175	16	191
Republican-2004	54	188	242
Total	229	204	433

We can model each row as a separate binomial distribution with parameter p. If the null hypothesis is valid, then there is no statistical difference between the two rows, so they can be collapsed into a single row with overall binomial probability $p = (175 + 188)/433$ for voting for the same party. Then, the expected value ($\hat{\mu}_{i,j}$) for each of the cells using this value is the following.

X/Expected Y	Democrat-2008	Republican-2008	Total
Democrat-2004	160.1	30.9	191
Republican-2004	39.1	202.9	242
Total	199.2	233.8	433

Taking the squared difference between what we observed and what the model expects under the null hypothesis gives us the χ_1^2 statistic, $X^2 = 15.3$, which is larger than the 0.05 threshold (i.e., 3.84), so we can confidently reject the null hypothesis and conclude that there is a difference in partisanship voting between the two political parties.

As interesting as this rejection may be, we want to know why the null hypothesis has failed. To investigate, we can compute the Pearson standardized residuals

$$e_{i,j} = \frac{n_{i,j} - \hat{\mu}_{i,j}}{\sqrt{\hat{\mu}_{i,j}}}$$

which are asymptotically normal with zero mean, but, unfortunately, their asymptotic variances are less than unity. A residual that is easier to interpret is the *standardized* residual that results from dividing the numerator by its standard error $\sqrt{p(1-p)}$.

X/Standardized residuals	Democrat-2008	Republican-2008
Democrat-2004	3.19	−7.27
Republican-2004	6.46	−2.83

This shows that more Democrats than expected continue to vote Democrat in 2008 (3.19) and also that more Republicans from 2004 switched to Democrat in 2008 (6.46). There were fewer than expected Democrats voting Republican in 2008 (-7.27) and fewer than expected Republicans voting Republican in 2008 (-2.83). Thus, voting behavior shifted toward the Democrats from 2004 to 2008.

Example: Cigarette Smoking and Lung Cancer The previous longitudinal study had fixed *row* marginals, but now we consider a *case-control* or *retrospective* study with the *column* marginals fixed. People once thought that lung cancer was caused by air pollution generated by burning coal and not by cigarette smoking. The table below shows some relevant data from a 1950 study by Doll and Hill taken from 20 hospitals in London, England. For each of the 709 patients admitted, they recorded the smoking behavior of a noncancer patient at the same hospital of the same gender and within the same 5-year age bracket. A smoker was defined as a person who had smoked at least one cigarette a day for at least a year.

Smoker/Case	Cancer Cases	Controls	Total
Yes	688	650	1338
No	21	59	80
Total	709	709	1418

The column-sums are deliberately matched between the two cases. Case-control studies provide the conditional distribution of smoking behavior given cancer status. Thus, the two following conditional probabilities

$$\mathbb{P}(\text{smoker}|\text{cancer}) = 688/709$$

and

$$\mathbb{P}(\text{smoker}|\text{control}) = 650/709$$

should be statistically equal under the null hypothesis. Obviously, we would prefer to analyze $\mathbb{P}(\text{cancer}|\text{smoker})$ but the data was not collected to determine this (i.e., by randomly partitioning participants and then forcing one group to smoke!). Under the null hypothesis, cancer status has nothing to do with being a smoker, so the distribution of controls and cases should be statistically equivalent. Because the column-sums are fixed, we need to know only one entry in each column to define the table. For the 0th column the count of the $n_{0,0}$ is binomially distributed

$$n_{0,0} \sim \binom{n_{\bullet,0}}{n_{0,0}} p^{n_{0,0}} (1-p)^{n_{\bullet,0}-n_{0,0}}$$

and similarly for the $n_{0,1}$ column. The null hypothesis is that each $n_{0,0}$ and $n_{0,1}$ of the table are independent. For the smoker data, we can estimate $\hat{p} = 1338/1418$, and using the Pearson χ^2 test discussed below, we can compute the expected values for each entry in the table

Smoker/Case	Cancer Cases	Controls	Total
Yes	669	669	1338
No	40	40	80
Total	709	709	1418

and compute the χ_1^2 statistic as 19.1292, which is above the 3.84 value so we reject the null hypothesis. Scipy has the tools to compute these parameters

```
>>> from scipy.stats import fisher_exact, chi2_contingency
>>> from scipy.stats.contingency import expected_freq
>>> expected_freq([[688,650],[21,59]])
array([[669., 669.],
       [ 40.,  40.]])
```

and to compute the χ_1^2 test

```
>>> (xistat, pvalue,
...   degfreedom, expectation) = chi2_contingency([[688,650],
...                                                [21,59]],
...                                           correction=False)
>>> print(f'{xistat=},{pvalue=},{degfreedom=}')
xistat=19.129222720478325,pvalue=1.2216007893346798e-05,
degfreedom=1
>>> print(f'{expectation=}')
expectation=array([[669., 669.],
       [ 40.,  40.]])
```

The expected value array is the same one we computed earlier with `expected_freq`. The reported `pvalue` is far below the usual 0.05 value, so we reject the null hypothesis.

Example: Cross-Sectional Study For this study, the population is sampled at a specific point in time, and the resulting contingency table has *no fixed* marginals but does maintain a fixed total of samples, $n_{\bullet,\bullet}$. This means that the entries of the table are determined by the multinomial distribution so that the marginals M_1 and N_1 are random variables. The following code illustrates a contingency table created from the multinomial distribution:

```
>>> from scipy.stats import multinomial
>>> table=multinomial(100,[.1,.2,.3,.4]).rvs().reshape((2,2))
>>> print(table) # contingency table
[[12 26]
 [24 38]]
>>> print(table.sum(axis=0)) # column sums
[36 64]
>>> print(table.sum(axis=1)) # row sums
[38 62]
```

Let us consider the following contingency table where only N is fixed. The marginal sums N_1, M_1 are binomially distributed with parameters p_n, p_m, respectively, as a consequence of the multinomial distribution of the table entries.

X/Y	Y_1	Y_2	Total
X_1	$n_{0,0}$	-	M_1
X_2	-	-	M_2
Total	N_1	N_2	N

Under the null hypothesis, $n_{0,0} \sim \text{Binom}(N, p_m p_n)$, which means that the probability for that entry is independent of the row or column (i.e., $p_{0,0} = p_m p_n$). The test statistic (sometimes called the *association test*) is the following:

$$Y = \frac{n_{0,0}}{N} - \frac{M_1}{N}\frac{N_1}{N}$$

which says that $\hat{p}_{0,0} = \hat{p}_n \hat{p}_m$, which is the claim of the null hypothesis. We next determine the distribution of this statistic under the null hypothesis. The expectation is zero, $\mathbb{E}(Y) = 0$. Furthermore, we have

$$\mathbb{E}(Y^2) = \frac{p_m p_n \left(p_m \left((1 - 3N)p_n + N - 1 \right) + (N - 1)p_n + N + 1 \right)}{N^2}$$

With these two facts, using $\hat{p}_m = \frac{M_1}{N}$ and $\hat{p}_n = \frac{N_1}{N}$, we can compute the normalized score as

$$Z = \frac{N \left(N n_{0,0} - M_1 N_1 \right)}{\sqrt{M_1}\sqrt{N_1}\sqrt{M_1 \left((N-1)N + (1 - 3N)N_1 \right) + N \left(N(N+1) + (N-1)N_1 \right)}}$$

Let us use a Monte Carlo simulation to check the distribution of the statistic Y against the asymptotic normal distribution with the given variance (see Fig. 3.12).

```
>>> from scipy.stats import binom, norm
>>> p_m, p_n = 0.1, 0.2 # simulation parameters
>>> N, nsamples = 500, 1000
>>> M1, N1, n00 = binom(N,p_m), binom(N,p_n), binom(N,p_n*p_m)
>>> # plotted below in Figure
>>> association = n00.rvs(nsamples)/N-M1.rvs(nsamples)*N1.rvs(nsamples)/N/N
```

Fig. 3.12 Association test density vs. histogram

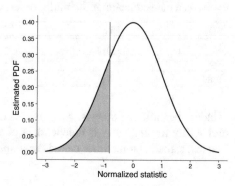

Fig. 3.13 Shaded area correspondence to p-value ≈ 0.2

Let us consider a sample contingency table under this protocol using the values in the previous code:

X/Y	Y_1	Y_2	Total
X_1	7	41	48
X_2	95	357	452
Total	102	398	500

The code below computes the normalized statistic for this table gives Fig. 3.13 which shows that rejecting the null hypothesis (i.e., that the aforementioned entries are all mutually independent) is not reasonable (i.e., shaded area for p-value ≈ 0.2).

```
>>> val = 7/N - 48/N*102/N
>>> p_m, p_n = 48/N, 102/N
>>> std = np.sqrt((p_m*p_n*(1+N+(-1+N)*p_n
...        +p_m*(-1+N+(1-3*N)*p_n)))/N**2)
>>> Y = val/std
>>> nrv = norm(0,1) # standard normal distribution
>>> nrv.cdf(Y) # p-value
0.21174395725924777
```

The chi-square statistic is proportional to the square of Y above, so we can also test against the chi-square statistic with one degree of freedom. Here is the Monte

Fig. 3.14 Asymptotic
distribution for Y^2 vs.
histogram

Carlo simulation for Y^2. Figure 3.14 shows the corresponding comparison with the
asymptotic χ^2 distribution.

```
>>> from scipy.stats import chi2
>>> std = np.sqrt((p_m*p_n*(1+N+(N-1)*p_n+
...          p_m*(-1+N+(1-3*N)*p_n)))/N**2)
>>> d = (n00.rvs(nsamples)/N-M1.rvs(nsamples)*\
...          N1.rvs(nsamples)/N/N)**2/(std**2)
```

Continuing the contingency table example above using the chi-square statistic shows
that the null hypothesis is rejected again. Note that the Pearson chi-square statistic
is $\frac{N}{N-1}Y^2$ so it shares the same asymptotic behavior for large N. The following code
computes the *p*-value for the ξ^2 statistic:

```
>>> p_m, p_n = 48/N, 102/N
>>> val = 7/N - p_m*p_n
>>> std = np.sqrt((p_m*p_n*(1+N+(N-1)*p_n+
...               p_m*(N-1+(1-3*N)*p_n)))/N**2)
>>> Y2 = (val/std)**2
>>> 1-chirv.cdf(Y2) # p-value
0.42348791451849566
```

Case-Control Contingency Table Analysis We can use the same pattern of
analysis for case-control contingency tables. Recall that in this case, the column-
sums are *fixed*. Here is the corresponding contingency table:

X/Y	Y_1	Y_2	Total
X_1	$n_{0,0}$	-	M_1
X_2	$n_{1,0}$	-	M_2
Total	N_1	N_2	N

In this case N_1, N_2, and N are fixed with $M_1 \sim \text{Binom}(N, p_m)$. The null
hypothesis is that there is no relationship between the counts in the rows. That
is, the two factors (rows) are irrelevant to the cases (columns). This means that

$n_{0,0} \sim \texttt{Binom}(M_1, p_1)$ and $n_{1,0} \sim \texttt{Binom}(M_2, p_2)$ with the null hypothesis that $H_0 : p_1 = p_2$. Notice that once M_1 is drawn, M_2 is fixed because the row marginals sum to N. We form the following statistic:

$$Y = \hat{p}_1 - \hat{p}_2$$

with $\hat{p}_1 = \frac{n_{0,0}}{M_1}$ and $\hat{p}_2 = \frac{n_{1,0}}{M_2}$ Under the null hypothesis, we have $\mathbb{E}(Y) = 0$ and

$$\mathbb{E}(Y^2) = \mathbb{V}(Y) = \frac{1}{M_1^2}\mathbb{V}(n_{0,0}) + \frac{1}{M_2^2}\mathbb{V}(n_{1,0})$$

and because we have $\mathbb{V}(n_{0,0}) = M_1 p_1(1 - p_1)$ with corresponding $\mathbb{V}(n_{1,0}) = M_2 p_2(1 - p_2)$ so that we have

$$\mathbb{V}(Y) = \frac{1}{M_1}p_1(1 - p_1) + \frac{1}{M_2}p_2(1 - p_2)$$

Note that if we scale the Y by $\sqrt{N}$ to form the new statistic,

$$Y_1 = \sqrt{N}Y = \sqrt{N}(\hat{p}_1 - \hat{p}_2)$$

We then obtain,

$$\mathbb{V}(Y_1) = \frac{N}{M_1}p_1(1 - p_1) + \frac{N}{M_2}p_2(1 - p_2) = \left(\frac{N}{M_1} + \frac{N}{M_2}\right)p_1(1 - p_1)$$

under the null hypothesis that $p_1 = p_2$. Then, using the estimate $\hat{p}_m = \frac{M_1}{N}$, we obtain the following:

$$\mathbb{V}(Y_1) = \frac{p_1(1 - p_1)}{p_m(1 - p_m)}$$

Thus, we have the parameters for the asymptotically normal distribution of the statistic, Y_1, to obtain

$$Z = \sqrt{N}\left(\frac{n_{0,0}}{M_1} - \frac{n_{1,0}}{N - M_1}\right)\sqrt{\frac{\frac{M_1}{N}(1 - \frac{M_1}{N})}{\frac{n_{0,0}}{M_1}(1 - \frac{n_{0,0}}{M_1})}}$$

The following code simulates the data under these conditions. Figure 3.15 shows the sample histogram and corresponding asymptotic density.

```
>>> nsamples, N = 1000, 500
>>> p_1, p_m = 0.3, 0.8
>>> M1 = binom(N,p_m)
```

Fig. 3.15 Histogram and
asymptotic density for
case-control analysis

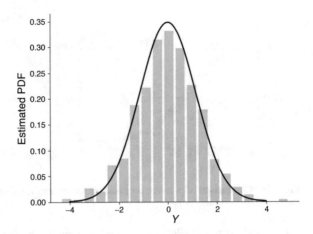

```
>>> def gen_samples(n=1):
...     M = M1.rvs()
...     n00 = binom(M,p_1).rvs(n)
...     n10 = binom(N-M,p_1).rvs(n)
...     return (n00/M-n10/(N-M))*np.sqrt(N)
...
>>> y = gen_samples(nsamples)
```

Using the parameters above, let us consider a sample contingency table for the
null hypothesis under this protocol with $p_m = 0.8$ and $p_1 = 0.3$,

X/Y	Y_1	Y_2	Total
X_1	116	290	406
X_2	27	67	94
Total	143	357	500

The Y_1 statistic is then $Z \approx -0.029$. This is very close to the mean so the p-value
is approximately 0.5, which means that we cannot reject the null hypothesis that the
rows are independent, which is reassuring because we generated this contingency
table under the null hypothesis anyway. Let us consider a contingency table that is
not generated under the null hypothesis with $p_1 = 0.3$, $p_2 = 0.8$:

X/Y	Y_1	Y_2	Total
X_1	116	290	406
X_2	67	27	94
Total	183	317	500

This means that $n_{1,0}$ was generated under $\text{Binom}(M_2, p_2)$. For this case we have
$Z \approx -8.25$ with corresponding p-value ≈ 0. The Fisher exact test is frequently
used for this situation. Figure 3.16 (on the left) shows the comparison of the p-
values obtained using the Fisher exact test and the Z statistic. Note that the p-value

Fig. 3.16 Comparison of p-values for Fisher exact test and Z-statistic for case-control analysis

corresponding to the Fisher exact test is bounded below by the corresponding p-value of our Z statistic. This gap (rightmost figure) means that there are cases where the Fisher exact test fails to reject the null hypothesis that would otherwise have been rejected by the Z statistic. The difference between the two is significant near where $n_{2,1} = 27$, indicating that this is a particularly sensitive value for the contingency table in terms of rejecting the null hypothesis.

Pearson Chi-Square Test For $I \times J$ tables larger than two by two, the Pearson chi-square statistic is commonly used to analyze contingency tables when N is large compared to the number of columns in the table. We define $\pi_{i,j} = \mathbb{P}(X = i, Y = j)$ with $\pi_{\bullet,\bullet} = 1$. The parameters for the corresponding multinomial distributions are estimated using the usual proportions $\hat{\pi}_{i,j} = n_{i,j}/N$. For independence, the joint outcome factors into a product of the marginals, $\pi_{i,j} = \pi_{i,\bullet}\pi_{\bullet,j}$. Without independence, there are $IJ - 1$ free parameters in the table because the constraint $\pi_{\bullet,\bullet} = 1$; however, *with* independence we have $(I - 1) + (J - 1)$ free parameters. This is because the original constraint breaks into $\pi_{\bullet,\bullet} = \sum_j \pi_{\bullet,j} \sum_i \pi_{i,\bullet} = 1$ and the marginal probabilities $\sum_j \pi_{\bullet,j} = 1$ and $\sum_i \pi_{i,\bullet} = 1$. Thus, there are one fewer free parameters each by row and column and a $(IJ-1)-(I+J-2) = (I-1)(J-1)$ reduction in free parameters for the same table without independence.

The Pearson chi-square test is the squared difference of the observed count minus the estimated expected count divided by the estimated expected count. The expected count under the independence assumption is

$$N\hat{\pi}_{i,\bullet}\hat{\pi}_{\bullet,j} = n_{i,\bullet}n_{\bullet,j}/N$$

So, the Pearson chi-square test statistic is the following:

$$X^2 = \sum_{i=1}^{I}\sum_{j=1}^{J} \frac{(n_{i,j} - n_{i,\bullet}n_{\bullet,j}/N)^2}{n_{i,\bullet}n_{\bullet,j}/N}$$

When the null hypothesis is false, we expect large values for the squared difference in X^2 and therefore reject the null hypothesis (i.e., not independent) when $X > \chi^2_{(I-1)(J-1),1-\alpha}$.

Odds Ratio for Association. The odds ratio provides a consistent and interpretable measure of association and is fundamental to contingency table analysis. It is incorrect, but common, to use the value of the *p*-value as a way to measure such association. Consider the following two-by-two contingency table:

X/Y	Y_1	Y_2	Total
X_1	$n_{0,0}$	$n_{0,1}$	M_1
X_2	$n_{1,0}$	$n_{1,1}$	M_2
Total	N_1	N_2	N

If the marginal sums are fixed, then choosing any one of the $n_{i,j}$ values determines the other three. We can divide all the values by N to obtain the corresponding probabilities. Note that these would be the estimators for the parameters for the multinomial distribution.

X/Y	Y_1	Y_2	Total
X_1	$p_{0,0}$	$p_{0,1}$	$p_{0,\bullet}$
X_2	$p_{1,0}$	$p_{1,1}$	$p_{1,\bullet}$
Total	$p_{\bullet,0}$	$p_{\bullet,1}$	1

The odds across the first row is the following:

$$\text{odds}_0 = \frac{p_{0,0}}{p_{0,1}}$$

and similarly along the next row. Thus, the odds ratio for the table is then

$$\text{oddsRatio} = \frac{p_{0,0} p_{1,1}}{p_{0,1} p_{1,0}}$$

To get a feel for the odds ratio in this situation, consider the following table from the `scipy.stats.fisher_exact` Documentation:

Animal/Ocean	Atlantic	Indian
whales	8	2
sharks	1	5

Using the `fisher_exact` test, we compute the following:

```
>>> from scipy.stats import fisher_exact
>>> odds_ratio, p_value = fisher_exact([[8,2],[1,5]])
>>> print(f'{odds_ratio=}, {p_value=}')
odds_ratio=20.0, p_value=0.03496503496503495
```

The interpretation is that the odds of finding a whale in the Atlantic Ocean as opposed to the Indian Ocean is about 20 times that than of finding a shark in the Atlantic Ocean versus the Indian Ocean.

Interestingly, the odds ratio and the two marginals ($p_{0,\bullet}$, $p_{\bullet,0}$) completely characterize the contingency table in that each of them can be set independently of the others and produce a valid table. In this sense, these three numbers are like the *orthogonal coordinates* of the table that summarize the information without mutually interfering with each other. This is not the *only* way to parameterize the table, as one could use the marginals and the $p_{0,0}$ value to derive the required parameters, but in this case, changing the value of $p_{0,0}$ affects the marginals. The technical term for this non-mutual-interfering behavior of a parameterization is *variation independent*.

Consider the following code that solves for the table probabilities where the marginals are set to 0.1 and 0.2, respectively, with the odds ratio set to 5:

```
>>> import sympy as S
>>> S.var('p:2(:2)',positive=True)
(p00, p01, p10, p11)
>>> eqs = [p00+p01- .1,          # marginal
...        p00+p10-.2,           # marginal
...        p00*p11/p01/p10 - 5,  # odds ratio
...        p00+p11+p01+p10-1,    # probabilities sum
...        ]
>>> sol, = S.solve(eqs)
>>> print((p00+p01).subs(sol)) # marginal
0.100000000000000
>>> print((p00+p10).subs(sol)) # marginal
0.200000000000000
>>> print((p00*p11/p01/p10).subs(sol)) # odds ratio
5.00000000000000
>>> print(sol)
{p00: 0.0500000000000000, p01: 0.0500000000000000,
  p10: 0.150000000000000, p11: 0.750000000000000}
```

Notice the solution provides all the terms for the contingency table. Now, if we leave everything the same and double the odds ratio, we get the following:

```
>>> eqs[2] = p00*p11/p01/p10 - 10 # increase odds ratio
>>> sol, = S.solve(eqs)
>>> print((p00+p01).subs(sol)) # marginal unchanged
0.100000000000000
>>> print((p00+p10).subs(sol)) # marginal unchanged
0.200000000000000
>>> print((p00*p11/p01/p10).subs(sol)) # new odds ratio
10.0000000000000
```

Notice that the marginals have stayed the same even though the odds ratio has doubled. This is what *variation independent* means.

The parameterization using the marginals and the odds ratio also provides a way to assess the degree of independence among the rows and columns because if the rows and columns are independent, then the odds ratio is one. This follows easily by plugging in the independence condition $p_{i,j} = p_{\bullet,j} p_{i,\bullet}$. The MLE estimate of the odds ratio is composed of the MLE estimates of each entry in the table

$$\widehat{OR} = \frac{\hat{p}_{0,0} \hat{p}_{1,1}}{\hat{p}_{0,1} \hat{p}_{1,0}}$$

where $\hat{p}_{i,j} = n_{i,j}/n_{\bullet,\bullet}$. Under the multinomial distribution, the asymptotic variance of the MLE estimate of the log odds ratio is the following:

$$\mathbb{V}\left(\widehat{OR}\right) = \left(\frac{1}{n_{0,0}} + \frac{1}{n_{0,1}} + \frac{1}{n_{1,0}} + \frac{1}{n_{1,1}}\right)$$

Importantly, this does *not* depend on the value of the odds ratio itself. Let us look at a quick example using this estimate for the following contingency table:

Smoker/Case	Cancer Cases	Controls
Yes	483	477
No	1101	1121

The estimated odds ratio is the following:

$$\frac{483 \times 1121}{477 \times 1101} \approx 1.03$$

which is close to one, hinting at independence. Because we have the variance of the estimator above, we can compute the asymptotic standard error,

$$\sqrt{\frac{1}{438} + \frac{1}{447} + \frac{1}{1101} + \frac{1}{1121}} \approx 0.0786$$

Then, with the 95% confidence interval, our odds ratio estimate becomes

$$1.03 \pm 0.154$$

Wide Contingency Table So far, we have been only using two-by-two contingency table, but the same pattern of analysis follows for wider tables. For example, consider the following data of 52 males who had knee surgeries with results classified as excellent (E), good (G), and fair/poor (F) according to type of injury (twist, direct, both).

```
>>> data = np.array([[21,11,4],
...                   [3,2,2],
...                   [7,1,1]]).reshape((3,3))
>>> xa = xr.DataArray(data,
```

```
...            coords = {'i':['twist','direct','both'],
...                      'j':['excellent', 'good', 'poor']})
>>> xa
<xarray.DataArray (i: 3, j: 3)>
array([[21, 11,  4],
       [ 3,  2,  2],
       [ 7,  1,  1]])
Coordinates:
  * i            (i) <U6 'twist' 'direct' 'both'
  * j            (j) <U9 'excellent' 'good' 'poor'
```

We can model this as product-multinomial with each type of injury representing a separate population with its own multinomial distribution. The null hypothesis is that the surgical results have nothing to do with the kind of injury.

$$H_0: \; p_{i,j} = p_{i,\bullet} p_{\bullet,j}$$

We can estimate the expected table of each of the outcomes as the following:

```
>>> expected = xa.sum('j')*xa.sum('i')/xa.sum() # expected
        table under H0
>>> expected
<xarray.DataArray (i: 3, j: 3)>
array([[21.46153846,  9.69230769,  4.84615385],
       [ 4.17307692,  1.88461538,  0.94230769],
       [ 5.36538462,  2.42307692,  1.21153846]])
Coordinates:
  * i            (i) <U6 'twist' 'direct' 'both'
  * j            (j) <U9 'excellent' 'good' 'poor'
```

As before, we can use the χ^2 statistic to check the null hypothesis.

```
>>> float((((xa - expected)**2/expected).sum())) # X^2 chi-squared
        statistic
3.2288420744641946
```

To compute the degrees of freedom, we have to tally the number of unknowns in under the product-multinomial hypothesis (three injury-types times two free p_i parameters per type) minus the number of parameters we estimated under the null hypothesis. We can compute the threshold for the 5% p-value using the `isf` function in `stats.chi2` as follows:

```
>>> degrees_freedom = 2*3 - 2
>>> stats.chi2(degrees_freedom).isf(.05) # chi^2 threshold
9.487729036781158
```

Note that we set the Yates continuity `correction` flag to `False` to get the values to match exactly even though it doesn't matter in this case. The main idea is that we can use the same pattern of analysis for contingency tables bigger than two-by-two.

```
>>> chi2_stat, pvalue, dof, expected_array =
        stats.chi2_contingency(xa.values,correction=False)
```

where `expected_array` is the same as `expected` we computed above and `chi2_stat` is the χ^2 statistic as above.

Simpson's Paradox Simpson's paradox is important to understand as we move onward to study high-dimensional contingency tables. Let us get started with an example regarding treatments and outcomes across gender.

```
>>> data = np.array([[60, 20,40,80],
...                  [100,50,10,30]]).reshape((2,2,2))
>>> xa = xr.DataArray(data,
...         coords = {'treatment':[1,2],
...                   'sex': ['male','female'],
...                   'outcome':['success','failure']})
>>> xa
<xarray.DataArray (treatment: 2, sex: 2, outcome: 2)>
array([[[ 60,  20],
        [ 40,  80]],

       [[100,  50],
        [ 10,  30]]])
Coordinates:
  * treatment  (treatment) int64 1 2
  * sex        (sex) <U6 'male' 'female'
  * outcome    (outcome) <U7 'success' 'failure'
```

Let us consider the males separately,

```
>>> xa.sel(sex='male') # treatment 1 works for males
<xarray.DataArray (treatment: 2, outcome: 2)>
array([[ 60,  20],
       [100,  50]])
Coordinates:
  * treatment  (treatment) int64 1 2
    sex        <U6 'male'
  * outcome    (outcome) <U7 'success' 'failure'
```

This result indicates that `Treatment 1` has a better success rate than `Treatment 2` (i.e., 60/80 vs. 100/150). Likewise considering the females separately,

```
>>> xa.sel(sex='female') # treatment 1 works for females
<xarray.DataArray (treatment: 2, outcome: 2)>
array([[40, 80],
       [10, 30]])
Coordinates:
  * treatment  (treatment) int64 1 2
    sex        <U6 'female'
  * outcome    (outcome) <U7 'success' 'failure'
```

shows that the success rate for `Treatment 1` is better also. Now, let us combine the genders and obtain the following table:

```
>>> # treatment 2 is better than treatment 1
>>> xa.sum('sex')
<xarray.DataArray (treatment: 2, outcome: 2)>
array([[100, 100],
       [110,  80]])
Coordinates:
  * treatment   (treatment) int64 1 2
  * outcome     (outcome) <U7 'success' 'failure'
```

Surprisingly, the success rate for Treatment 2 is better here (110/190 vs. 100/200). This is the paradox—why is the superior treatment (i.e., Treatment 1) for each separate group different for the pooled group? To unpack this paradox, we need to understand models of independence for three-dimensional tables.

Model of Complete Independence Complete independence means that all three dimensions are independent of each other. Consider the following table regarding personality type (*A* vs. *B*), cholesterol (CHL) level, and blood pressure (BP):

```
>>> data = np.array([[716,79],
...                  [207,25],
...                  [819,67],
...                  [186,22]]).reshape((2,2,2))
>>> xa = xr.DataArray(data,
...              coords ={'personality': ['A','B'],
...                       'cholesterol': ['normal','high'],
...                       'blood_pressure': ['normal','high'],
...                       }
...              ).astype(np.int64)
>>> xa
<xarray.DataArray (personality:2, cholesterol:2, blood_pressure:2)>
array([[[716,  79],
        [207,  25]],

       [[819,  67],
        [186,  22]]])
Coordinates:
  * personality      (personality) <U1 'A' 'B'
  * cholesterol      (cholesterol) <U6 'normal' 'high'
  * blood_pressure   (blood_pressure) <U6 'normal' 'high'
```

Under the null hypothesis, we compute the expected table as shown

```
>>> expected =(xa.sum(['personality','cholesterol'])*
...            xa.sum(['cholesterol','blood_pressure'])*
...            xa.sum(['personality','blood_pressure'])/xa.sum()**2)
>>> expected.transpose('personality','cholesterol', 'blood_pressure')
<xarray.DataArray (personality: 2, cholesterol: 2, blood_
pressure: 2)>
array([[[739.88436419,  74.06518791],
        [193.66396207,  19.38648583]],

       [[788.15335387,  78.89709403],
        [206.29831987,  20.65123223]]])
Coordinates:
```

```
* blood_pressure  (blood_pressure) <U6 'normal' 'high'
* personality     (personality) <U1 'A' 'B'
* cholesterol     (cholesterol) <U6 'normal' 'high'
```

where the `transpose` at the end is just to orient the data for comparison with `xa`. Now, we compute the usual Pearson χ^2 test

```
>>> degrees_freedom = 7 - 3
>>> print(((xa-expected)**2/expected).sum()) # chi^2 statistic
<xarray.DataArray ()>
array(8.73016317)
>>> stats.chi2(degrees_freedom).isf(.05) # chi^2 threshold
9.487729036781158
```

where the degrees of freedom tally the number of parameters in the original table $(8 - 1 = 7)$ minus the number of parameters we estimated under this hypothesis (one for each of the three dimensions). Thus, we cannot reject the null hypothesis here, and this indicates there may be a more complicated relationship among the three dimensions.

Models of One Factor Independent of the Others As opposed to the hypothesis of complete independence for all three dimensions, another possible hypothesis is that one of the dimensions is independent of the other two joint dimensions. Let us consider the following three-dimensional data regarding 97 10-year-old schoolchildren classified according to classroom behavior, risk of home conditions, and adversity of school conditions. Classroom behavior was judged by teachers to be *deviant* or *non-deviant*. Risk of home conditions identify as not-at-risk (N) or at risk (R). Adversity of school condition was judged either *low, medium,* or *high*. Here is the corresponding dataframe where the index names have the index variable names (e.g., (i), (j), (k)) included in the coordinate names for clarity.

```
>>> data = np.array([[16,7,15,34,5,3],[1,1,3,8,1,3]]).reshape((2,3,2))
>>> xa = xr.DataArray(data, coords = {
...                       'behavior(i)':['deviant','non-deviant'],
...                       'adversity(j)':['low','medium','high'],
...                       'risk(k)': ['N','R']
...                       }
...                   ).astype(np.int64)
>>> xa
<xarray.DataArray (behavior(i): 2, adversity(j): 3, risk(k): 2)>
array([[[16,  7],
        [15, 34],
        [ 5,  3]],

       [[ 1,  1],
        [ 3,  8],
        [ 1,  3]]])
Coordinates:
  * behavior(i)   (behavior(i)) <U11 'deviant' 'non-deviant'
  * adversity(j)  (adversity(j)) <U6 'low' 'medium' 'high'
  * risk(k)       (risk(k)) <U1 'N' 'R'
```

Consider the hypothesis that the rows are independent of *joint* columns and layers (third dimension),

$$H_1 : p_{i,j,k} = p_{i,\bullet,\bullet} p_{\bullet,j,k}$$

This means that behavior is independent of *joint* home-risk and adversity and likewise

$$H_2 : p_{i,j,k} = p_{\bullet,\bullet,k} p_{i,j,\bullet}$$

means that home-risk is independent of joint behavior and adversity and finally,

$$H_3 : p_{i,j,k} = p_{\bullet,j,\bullet} p_{i,\bullet,k}$$

means that adversity is independent of joint behavior and home-risk.

Let us test the H_1 hypothesis that behavior is independent of joint risk and adversity. The following is the expected table under this hypothesis:

```
>>> expected =
          xa.sum(['adversity(j)','risk(k)'])*xa.sum('behavior(i)')/xa.sum(
>>> expected
<xarray.DataArray (behavior(i):2, adversity(j):3, risk(k):2)>
array([[[14.02061856,  6.59793814],
        [14.84536082, 34.63917526],
        [ 4.94845361,  4.94845361]],

       [[ 2.97938144,  1.40206186],
        [ 3.15463918,  7.36082474],
        [ 1.05154639,  1.05154639]]])
Coordinates:
  * behavior(i)    (behavior(i)) <U11 'deviant' 'non-deviant'
  * adversity(j)   (adversity(j)) <U6 'low' 'medium' 'high'
  * risk(k)        (risk(k)) <U1 'N' 'R'
```

The first xa.sum isolates the behavior dimension by summing over the other dimensions. The second xa.sum isolates the join risk and adversity dimensions by summing over the behavior dimension. Next, we compute the degrees of freedom and the χ^2 statistic:

```
>>> degrees_freedom = 12-1-6
>>> print('X^2 = ',((xa-expected)**2/expected).sum()) # chi^2
          statistic
X^2 =  <xarray.DataArray ()>
array(6.19138681)
>>> stats.chi2(degrees_freedom).isf(.05) # chi^2 threshold
11.070497693516355
```

Thus, we cannot reject H_1 hypothesis that behavior is independent of joint home-risk and adversity. Next, let us test the H_2 Hypothesis that home-risk is independent of joint behavior and adversity:

$$H_2 : p_{i,j,k} = p_{\bullet,\bullet,k} p_{i,j,\bullet}$$

The following is the expected table under this hypothesis:

```
>>> expected = xa.sum(['adversity(j)','behavior(i)'])*xa.
        sum('risk(k)')/xa.sum()
>>> expected
<xarray.DataArray (risk(k):2, behavior(i):2, adversity(j):3)>
array([[[ 9.72164948, 20.71134021,  3.3814433 ],
        [ 0.84536082,  4.64948454,  1.69072165]],

       [[13.27835052, 28.28865979,  4.6185567 ],
        [ 1.15463918,  6.35051546,  2.30927835]]])
Coordinates:
  * risk(k)        (risk(k)) <U1 'N' 'R'
  * behavior(i)    (behavior(i)) <U11 'deviant' 'non-deviant'
  * adversity(j)   (adversity(j)) <U6 'low' 'medium' 'high'
```

Next, we compute the degrees of freedom and the χ^2 statistic:

```
>>> degrees_freedom = (12-1) -5 -1
>>> print('X^2 = ',((xa-expected)**2/expected).sum()) # chi^2
        statistic
X^2 =  <xarray.DataArray ()>
array(12.64459949)
>>> stats.chi2(degrees_freedom).isf(.05) # chi^2 threshold
11.070497693516355
```

In this case, we reject the null hypothesis that home-risk is independent of joint behavior and adversity. Finally, we examine the H_3 hypothesis that adversity is independent of joint behavior and home-risk:

$$H_3 : p_{i,j,k} = p_{\bullet,j,\bullet} p_{i,\bullet,k}$$

The following is the expected table under this hypothesis:

```
>>> expected = xa.sum('adversity(j)')*xa.sum(['behavior(i)',
        'risk(k)'])/xa.sum()
>>> expected
<xarray.DataArray (behavior(i):2, risk(k):2, adversity(j):3)>
array([[[ 9.27835052, 22.26804124,  4.45360825],
        [11.34020619, 27.21649485,  5.44329897]],

       [[ 1.28865979,  3.09278351,  0.6185567 ],
        [ 3.09278351,  7.42268041,  1.48453608]]])
Coordinates:
  * behavior(i)    (behavior(i)) <U11 'deviant' 'non-deviant'
  * risk(k)        (risk(k)) <U1 'N' 'R'
  * adversity(j)   (adversity(j)) <U6 'low' 'medium' 'high'
```

Next, we compute the degrees of freedom and the χ^2 statistic:

```
>>> degrees_freedom = (2*2-1)*(3-1)
>>> print('X^2 = ',((xa-expected)**2/expected).sum()) # chi^2
        statistic
X^2 =  <xarray.DataArray ()>
array(15.06798653)
>>> stats.chi2(degrees_freedom).isf(.05) # chi^2 threshold
12.59158724374398
```

In this case, we can reject the null hypothesis that adversity is independent of joint behavior and home-risk.

Returning to Simpson's Paradox Data Now that we have considered several options for applying Pearson's χ^2 test to three-dimensional contingency tables, we can revisit our earlier data for Simpson's paradox in Sect. 3.5.7 and investigate further. Recall the dataframe from Simpson's paradox:

```
>>> data=np.array([[60,20,40,80],
...                [100,50,10,30]]).reshape((2,2,2))
>>> xa = xr.DataArray(data,
...         coords = {'treatment': [1,2],
...                   'sex': ['male','female'],
...                   'outcome': ['success','failure']}
...         ).astype(np.int64)
>>> xa
<xarray.DataArray (treatment: 2, sex: 2, outcome: 2)>
array([[[ 60,  20],
        [ 40,  80]],

       [[100,  50],
        [ 10,  30]]])
Coordinates:
  * treatment  (treatment) int64 1 2
  * sex        (sex) <U6 'male' 'female'
  * outcome    (outcome) <U7 'success' 'failure'
```

Recall that the paradox was that although treatment 1 was best for each gender separately, treatment 2 was best when the genders were pooled. Consider the hypothesis that gender is independent of joint treatment and outcome (i.e., $p_{\bullet,j,\bullet}$).

```
>>> expected = xa.sum('sex')*xa.sum(['treatment','outcome'])/
        xa.sum()
>>> expected
<xarray.DataArray (treatment: 2, outcome: 2, sex: 2)>
array([[[58.97435897, 41.02564103],
        [58.97435897, 41.02564103]],

       [[64.87179487, 45.12820513],
        [47.17948718, 32.82051282]]])
Coordinates:
  * treatment  (treatment) int64 1 2
  * outcome    (outcome) <U7 'success' 'failure'
  * sex        (sex) <U6 'male' 'female'
```

As before, we compute the degrees of freedom and the χ^2 statistic as shown below:

```
>>> degrees_freedom = (8-1)-3-1
>>> print('X^2 = ',((xa-expected)**2/expected).sum()) # chi^2
        statistic
X^2 = <xarray.DataArray ()>
array(109.60319911)
>>> stats.chi2(degrees_freedom).isf(.05) # chi^2 threshold
7.814727903251178
```

The result below suggests that we can reject the hypothesis of independence between gender and joint treatment and outcome. If the result had been otherwise (i.e., *not* reject the independence hypothesis), then we may feel safe about marginalizing over gender to examine treatment and outcome, but this result for this data shows that we *cannot* do this and thus illustrates that Simpson's paradox is a symptom of this result. In other words, because the model was rejected means that there is a relationship between gender and treatment-outcome so we cannot pool the genders without addressing this relationship. The rejection of the independence hypothesis here just highlights this fact, but it does not explain what that relationship is, which would require further analysis to uncover.

Models of Conditional Independence We have examined multiple ways of examining the idea of independence for the three-dimensional contingency table. Now, let us consider conditional independence starting with the following contingency table regarding personality type (*A* vs. *B*), cholesterol (CHL) level, and blood pressure (BP):

```
>>> data = np.array([[716,79],
...                   [207,25],
...                   [819,67],
...                   [186,22]]).reshape((2,2,2))
>>> xa = xr.DataArray(data,
...          coords = {'personality': ['A','B'],
...                    'cholesterol': ['normal','high'],
...                    'blood_pressure': ['normal','high']}
...              ).astype(np.int64)
>>> xa
<xarray.DataArray (personality: 2, cholesterol: 2,
      blood_pressure: 2)>
array([[[716,  79],
        [207,  25]],

       [[819,  67],
        [186,  22]]])
Coordinates:
  * personality     (personality) <U1 'A' 'B'
  * cholesterol     (cholesterol) <U6 'normal' 'high'
  * blood_pressure  (blood_pressure) <U6 'normal' 'high'
```

Consider the following model of conditional independence between cholesterol and blood pressure given personality type:

$$p_{i,j,k} = p_{i,j,\bullet} p_{i,\bullet,k} / p_{i,\bullet,\bullet}$$

The following is the expected table under this hypothesis:

```
>>> expected =(xa.sum('blood_pressure')*
...             xa.sum('cholesterol')/xa.sum(['cholesterol',
...                                     'blood_pressure']))
>>> expected
<xarray.DataArray (personality: 2, cholesterol: 2,
blood_pressure: 2)>
array([[[714.49367089,  80.50632911],
        [208.50632911,  23.49367089]],

       [[813.9213894 ,  72.0786106 ],
        [191.0786106 ,  16.9213894 ]]])
Coordinates:
  * personality     (personality) <U1 'A' 'B'
  * cholesterol     (cholesterol) <U6 'normal' 'high'
  * blood_pressure  (blood_pressure) <U6 'normal' 'high'
```

As before, we compute the degrees of freedom and the χ^2 statistic:

```
>>> degrees_freedom = 2
>>> print('X^2 =',((xa-expected)**2/expected).sum()) # chi^2
        statistic
X^2 = <xarray.DataArray ()>
array(2.18757144)
>>> stats.chi2(degrees_freedom).isf(.05) # chi^2 threshold
5.991464547107983
```

Given these values, we *cannot* reject the null hypothesis that given personality type, cholesterol and blood pressure are statistically independent, and perhaps we should investigate the relationship between cholesterol and blood pressure for each personality type separately.

In this section, we discussed the structure of statistical hypothesis testing and defined the various terms that are commonly used for this process, along with the illustrations of what they mean in our running coin flipping example. From an engineering standpoint, hypothesis testing is not as common as confidence intervals and point estimates. On the other hand, hypothesis testing is very common in social and medical science, where one must deal with practical constraints that may limit the sample size or other aspects of the hypothesis testing rubric. In engineering, we can usually have much more control over the samples and models we employ because they are typically inanimate objects that can be measured repeatedly and consistently. This is obviously not so with human studies, which generally have other ethical and legal considerations. We also detailed using contingency tables for specific study protocols, which is a very common use case for healthcare and medical applications, along with Fisher exact test, which is a powerful nonparametric test for such tables.

3.6 Confidence Intervals

In a previous coin flipping discussion, we discussed estimation of the underlying probability of getting a heads. There, we derived the estimator as

$$\hat{p}_n = \frac{1}{n} \sum_{i=1}^{n} X_i$$

where $X_i \in \{0, 1\}$. Confidence intervals allow us to estimate how close we can get to the true value that we are estimating. Logically, that seems strange, doesn't it? We really don't know the exact value of what we are estimating (otherwise, why estimate it?), and yet, somehow we know how close we can get to something we admit we don't know. Ultimately, we want to make statements like the *probability of the value in a certain interval is 90%*. Unfortunately, that is something we will not be able to say using our methods. Note that Bayesian estimation gets closer to this statement by using *credible intervals*, but that is a story for another day. In our situation, the best we can do is say roughly the following: *if we ran the experiment multiple times, then the confidence interval would trap the true parameter 90% of the time*.

Let us return to our coin flipping example and see this in action. One way to get at a confidence interval is to use Hoeffding's inequality from Sect. 2.11 specialized to our Bernoulli variables as

$$\mathbb{P}(|\hat{p}_n - p| > \epsilon) \le 2 \exp(-2n\epsilon^2)$$

Now, we can form the interval $\mathbb{I} = [\hat{p}_n - \epsilon_n, \hat{p}_n + \epsilon_n]$, where ϵ_n is carefully constructed as

$$\epsilon_n = \sqrt{\frac{1}{2n} \log \frac{2}{\alpha}}$$

which makes the right side of the Hoeffding's inequality equal to α. Thus, we finally have

$$\mathbb{P}(p \notin \mathbb{I}) = \mathbb{P}(|\hat{p}_n - p| > \epsilon_n) \le \alpha$$

Thus, $\mathbb{P}(p \in \mathbb{I}) \ge 1 - \alpha$. As a numerical example, let us take $n = 100$, $\alpha = 0.05$; then plugging into everything we have gives $\epsilon_n = 0.136$. So, the 95% confidence interval here is therefore

$$\mathbb{I} = [\hat{p}_n - \epsilon_n, \hat{p}_n + \epsilon_n] = [\hat{p}_n - 0.136, \hat{p}_n + 0.136]$$

The following code sample is a simulation to see if we can really trap the underlying parameter in our confidence interval.

```
>>> from scipy import stats
>>> import numpy as np

>>> b= stats.bernoulli(.5) # fair coin distribution
>>> nsamples = 100
>>> # flip it nsamples times for 200 estimates
>>> xs = b.rvs(nsamples*200).reshape(nsamples,-1)
>>> phat = np.mean(xs,axis=0) # estimated p
>>> # edge of 95% confidence interval
>>> epsilon_n=np.sqrt(np.log(2/0.05)/2/nsamples)
>>> pct=np.logical_and(phat-epsilon_n<=0.5,
...                     0.5 <= (epsilon_n +phat)
...                    ).mean()*100
>>> print('Interval trapped correct value ', pct,'% of the
        time')
Interval trapped correct value  99.0 % of the time
```

The result shows that the estimator and the corresponding interval was able to trap the true value at least 95% of the time. This is how to interpret the action of confidence intervals.

However, the usual practice is to not use Hoeffding's inequality and instead use arguments around asymptotic normality. The definition of the standard error is the following:

$$se = \sqrt{\mathbb{V}(\hat{\theta}_n)}$$

where $\hat{\theta}_n$ is the point-estimator for the parameter θ, given n samples of data X_n, and $\mathbb{V}(\hat{\theta}_n)$ is the variance of $\hat{\theta}_n$. Likewise, the estimated standard error is $\widehat{se}$. For example, in our coin flipping example, the estimator was $\hat{p} = \sum X_i/n$ with corresponding variance $\mathbb{V}(\hat{p}_n) = p(1-p)/n$. Plugging in the point estimate gives us the estimated standard error: $\widehat{se} = \sqrt{\hat{p}(1-\hat{p})/n}$. Because maximum likelihood estimators are asymptotically normal,[4] we know that $\hat{p}_n \sim \mathcal{N}(p, \widehat{se}^2)$. Thus, if we want a $1 - \alpha$ confidence interval, we can compute

$$\mathbb{P}(|\hat{p}_n - p| < \xi) > 1 - \alpha$$

but since we know that $(\hat{p}_n - p)$ is asymptotically normal, $\mathcal{N}(0, \widehat{se}^2)$, we can instead compute

$$\int_{-\xi}^{\xi} \mathcal{N}(0, \widehat{se}^2)dx > 1 - \alpha$$

---

[4] Certain technical regularity conditions must hold for this property of maximum likelihood estimator to work. See [48] for more details.

Fig. 3.17 The gray circles are the point estimates that are bounded above and below by both asymptotic confidence intervals and Hoeffding intervals. The asymptotic intervals are tighter because the underpinning asymptotic assumptions are valid for these estimates

This looks ugly to compute because we need to find ξ, but Scipy has everything we need for this.

```
>>> # compute estimated se for all trials
>>> se=np.sqrt(phat*(1-phat)/xs.shape[0])
>>> # generate random variable for trial 0
>>> rv=stats.norm(0, se[0])
>>> # compute 95% confidence interval for that trial 0
>>> np.array(rv.interval(0.95))+phat[0]
array([0.3623159, 0.5576841])
>>> def compute_CI(i):
...     return stats.norm.interval(0.95,loc=i,
...                     scale=np.sqrt(i*(1-i)/xs.shape[0]))
...
>>> lower,upper = compute_CI(phat)
```

Figure 3.17 shows the asymptotic confidence intervals and the Hoeffding-derived confidence intervals. As shown, the Hoeffding intervals are a bit more generous than the asymptotic estimates. However, this is only true so long as the asymptotic approximation is valid. In other words, there exists some number of n samples for which the asymptotic intervals may not work. So, even though they may be a bit more generous, the Hoeffding intervals do not require arguments about asymptotic convergence. In practice, nonetheless, asymptotic convergence is always in play (even if not explicitly stated).

Confidence Intervals and Hypothesis Testing It turns out that there is a close dual relationship between hypothesis testing and confidence intervals. Consider the following hypothesis test for a normal distribution, $H_0 : \mu = \mu_0$ versus $H_1 : \mu \neq \mu_0$. A reasonable test has the following rejection region:

$$\left\{ x :\mid \bar{x} - \mu_0 \mid > z_{\alpha/2} \frac{\sigma}{\sqrt{n}} \right\}$$

where $\mathbb{P}(Z > z_{\alpha/2}) = \alpha/2$ and $\mathbb{P}(-z_{\alpha/2} < Z < z_{\alpha/2}) = 1 - \alpha$ and where $Z \sim \mathcal{N}(0, 1)$. This is the same thing as saying that the region corresponding to acceptance of H_0 is then

$$\bar{x} - z_{\alpha/2}\frac{\sigma}{\sqrt{n}} \leq \mu_0 \leq \bar{x} + z_{\alpha/2}\frac{\sigma}{\sqrt{n}} \tag{3.2}$$

Because the test has size α, the false alarm probability, $\mathbb{P}(H_0\text{rejected} \mid \mu = \mu_0) = \alpha$. Likewise, the $\mathbb{P}(H_0\text{accepted} \mid \mu = \mu_0) = 1 - \alpha$. Putting this all together with interval defined above means that

$$\mathbb{P}\left(\bar{x} - z_{\alpha/2}\frac{\sigma}{\sqrt{n}} \leq \mu_0 \leq \bar{x} + z_{\alpha/2}\frac{\sigma}{\sqrt{n}} \middle| H_0\right) = 1 - \alpha$$

Because this is valid for any μ_0, we can drop the H_0 condition and say the following:

$$\mathbb{P}\left(\bar{x} - z_{\alpha/2}\frac{\sigma}{\sqrt{n}} \leq \mu_0 \leq \bar{x} + z_{\alpha/2}\frac{\sigma}{\sqrt{n}}\right) = 1 - \alpha$$

As may be obvious by now, the interval in Eq. 3.6 above *is* the $1 - \alpha$ confidence interval! Thus, we have just obtained the confidence interval by inverting the acceptance region of the level α test. The hypothesis test fixes the *parameter* and then asks what sample values (i.e., the acceptance region) are consistent with that fixed value. Alternatively, the confidence interval fixes the sample value and then asks what parameter values (i.e., the confidence interval) make this sample value most plausible. Note that sometimes this inversion method results in disjoint intervals (known as *confidence sets*).

3.7   Sufficient Statistics

The concept of sufficient statistics provides a powerful way to think about and organize statistical data [32]. Recall that a statistic is a random variable in the sense that it is a mapping between events and real numbers. For example, the mean is a statistic that computes the following:

$$\overline{X} = \frac{1}{n}\sum_{i=1}^{n} X_n$$

Thus, for every set of $\{X_i\}$ random variables, we compute the mean. Thus, because the mean is a function of random variables, then the statistic is also a random variable. In the general case, the statistic condenses the set of random variables to a single value, which is useful for reducing the dimension of the problem. However, in this reduction, we need to ensure that we do not somehow lose any relevant information. Concretely, suppose we have triples from a Bernoulli distribution. These consist of three-dimensional vectors of zeros and ones. Thus, there are 2^3 possible three-dimensional vectors. Given the statistic

$$t(x_1 + x_2 + x_3) = x_1 + x_2 + x_3$$

there are exactly four possible values $(0, 1, 2, 3)$. Each of these values induces a partition on the original eight possibilities. For example, $0 \mapsto (0, 0, 0)$ and $1 \mapsto \{(0, 0, 1), (0, 1, 0), (0, 0, 1)\}$, and so on. Thus, we have four possible values to keep track of versus the original eight. The sufficient statistic has condensed the original eight values into four.

Importantly, many statistics can induce the same partition. For example, $t = (x_1 + x_2 + x_3)^2$ induces the same partition as above. From the condensation standpoint, statistics that induce the same partition are equally useful. Now, we are in a position to define a *sufficient statistic*.

Sufficient Statistic Let $\{X_1, \cdots, X_n\}$ be a a random sample from $f_\theta(X)$ where θ is the parameter for f. A statistic (s) is sufficient if and only if the conditional distribution of $X_1, \cdots, X_n$ given s does not depend on θ for any value of s. Intuitively, this means that the sufficient statistic does not *lose* any information about θ.

Returning to our Bernoulli distribution example. For $s = X_1 + X_2 + X_3$, we have the conditional probability

$$\mathbb{P}((0, 0, 0)|s = 0) = \frac{(1-p)^3}{(1-p)^3} = 1$$

Note that this does not depend on the parameter p. Let us consider another statistic $t = X_1 X_2 + X_3$

$$\mathbb{P}((0, 0, 0)|t = 0) = \frac{(1-p)^3}{(1-p)^3 + 2p(1-p)^2} = \frac{1-p}{1+p}$$

In this case, the conditional probability *is* a function of the parameter, which means that t is *not* a sufficient statistic. The condensation idea of sufficient statistics is useful for constructing estimators, as embodied by the Rao-Blackwell theorem.

Rao-Blackwell Theorem Let $X_1, \cdots, X_n$ be a random sample from density $f_\theta(X_1, \cdots, X_n)$ with $S_1, \cdots, S_k$ a set of jointly sufficient statistics. Let T be the unbiased estimator of $\tau(\theta)$. Define $T' = \mathbb{E}(T|S_1, \cdots, S_k)$. Then,

$$\mathbb{V}_\theta(T') \leq \mathbb{V}_\theta(T)$$

Loosely speaking, the theorem says that given an unbiased estimator, another unbiased estimator that is a function of sufficient statistics can be derived that will not have a larger variance. To derive this new unbiased estimator requires computing the conditional expectation. The jointly sufficient statistics referred to above pertains when no *single* sufficient statistic exists. Otherwise, the definition for jointly sufficient statistics is the same as for an individual sufficient statistic.

Example Consider our usual Bernoulli distribution samples, $X_1, \cdots, X_n$ with corresponding probability mass function, $f_p(x) = p^x(1-p)^{1-x}$. To keep it simple, suppose our estimator of the parameter p uses only the first sample. Thus, in terms of our notation, we have $T = X_1$. We know that $S = \sum X_i$ is a sufficient statistic. To derive T' as described in the Theorem, we need to compute the conditional expectation

$$\mathbb{E}(X_1 | S = s) = \frac{s}{n}$$

The expectation is equivalent to observing s successes in n samples. In other words, what is the probability that the first sample is a success (i.e., $X_1 = 1$)? Thus, we have the derived estimator

$$T' = \sum_{i=1}^{n} X_i / n$$

with corresponding variance $p(1 - p)/n$, whereas our original estimator has variance $np(1 - p)$. As long as $n > 1$, the derived estimator has a lower variance, as promised by the theorem.

Exponential Family Exponential families have probability density functions with the following Format:

$$f_\theta(x) = \exp\left(K(x)p(\theta) + S(x) + q(\theta)\right)$$

The notion of sufficient statistic is equivalent to the factorization of the density into two factors, where one factor depends on the data only through the sufficient statistic and the other factor does not. This is known as the *Fisher-Neyman Factorization* theorem. With respect to this format for the exponential family, we can conclude that the statistic

$$\sum_{i=1}^{n} K(X_i)$$

is sufficient for θ. This is because the joint PDF of $\{X_1, X_2, \cdots, X_n\}$ can be written as the following two factors:

$$\left\{ \exp\left(p(\theta) \sum_{i=1}^{n} K(x_i) + nq(\theta) \right) \right\} \left\{ \exp\left(\sum_{i=1}^{n} S(x_i) \right) \right\}$$

Note that the first factor only depends on the data via the $\sum_{i=1}^{n} K(x_i)$ which identifies it as the sufficient statistic.

Example Consider $\{X_1, X_2, \cdots, X_n\}$ independently distributed according to the Poisson distribution $(t = 1)$. Then the joint distribution is the following:

$$f_\lambda(X_1, X_2, \cdots, X_n) = \prod_{i=1}^{n} \frac{e^{-\lambda} \lambda^{x_i}}{x_i!}$$

To fit this into the exponential family format, we rewrite the univariate Poisson density

$$f_\lambda(X) = \frac{e^{-\lambda} \lambda^x}{x!} = \exp(x \ln \lambda - \ln(x!) - \lambda)$$

For the joint distribution, by identifying $K(x) = x$, we have the sufficient statistic for the λ parameter

$$\sum_{i=1}^{n} X_i$$

Note that $\frac{1}{n} \sum_{i=1}^{n} X_i$ is also a sufficient statistic (see Rao-Blackwell theorem).

3.8 Linear Regression

Linear regression gets to the heart of statistics: Given a set of data points, what is the relationship of the data in-hand to data yet seen? How should information from one dataset propagate to other data? Linear regression offers the following model to address this question:

$$\mathbb{E}(Y|X = x) \approx ax + b$$

That is, given specific values for X, assume that the conditional expectation is a linear function of those specific values. However, because the observed values are not the expectations themselves, the model accommodates this with an additive noise term. In other words, the observed variable (a.k.a. response, target, dependent variable) is modeled as

$$\mathbb{E}(Y|X = x_i) + \epsilon_i \approx ax + b + \epsilon_i = y$$

where $\mathbb{E}(\epsilon_i) = 0$ and the ϵ_i are iid and where the distribution function of ϵ_i depends on the problem, even though it is often assumed Gaussian. The $X = x$ values are known as independent variables, covariates, or regressors.

Let us see if we can use all of the methods we have developed so far to understand this form of regression. The first task is to determine how to estimate the unknown linear parameters, a and b. To make this concrete, let us assume that $\epsilon \sim \mathcal{N}(0, \sigma^2)$. Bear in mind that $\mathbb{E}(Y|X = x)$ is a deterministic function of x. In other words, the variable x changes with each draw, but after the data have been collected these are no longer random quantities. Thus, for fixed x, y is a random variable generated by ϵ. Perhaps we should denote ϵ as ϵ_x to emphasize this, but because ϵ is an independent, identically distributed (iid) random variable at each fixed x, this would be excessive. Because of Gaussian additive noise, the distribution of y is completely characterized by its mean and variance.

$$\mathbb{E}(y) = ax + b$$

$$\mathbb{V}(y) = \sigma^2$$

Using the maximum likelihood procedure, we write out the log-likelihood function as

$$\mathcal{L}(a, b) = \sum_{i=1}^{n} \log \mathcal{N}(ax_i + b, \sigma^2) \propto \frac{1}{2\sigma^2} \sum_{i=1}^{n} (y_i - ax_i - b)^2$$

Note that we suppressed the terms that are irrelevant to the maximum-finding. Taking the derivative of this with respect to a gives the following equation:

$$\frac{\partial \mathcal{L}(a, b)}{\partial a} = 2 \sum_{i=1}^{n} x_i(b + ax_i - y_i) = 0$$

Likewise, we do the same for the b parameter

$$\frac{\partial \mathcal{L}(a, b)}{\partial b} = 2 \sum_{i=1}^{n} (b + ax_i - y_i) = 0$$

The following code simulates some data and uses Numpy tools to compute the parameters as shown:

```
>>> import numpy as np
>>> a = 6;b = 1 # parameters to estimate
>>> x = np.linspace(0,1,100)
>>> y = a*x + np.random.randn(len(x))+b
```

Fig. 3.18 The panel on the left shows the data and regression line. The panel on the right shows a histogram of the regression errors

```
>>> p,var_=np.polyfit(x,y,1,cov=True) # fit data to line
>>> y_ = np.polyval(p,x) # estimated by linear regression
```

The graph on the left of Fig. 3.18 shows the regression line plotted against the data. The estimated parameters are noted in the title. The histogram on the right of Fig. 3.18 shows the residual errors in the model. It is always a good idea to inspect the residuals of any regression for normality. These are the differences between the fitted line for each x_i value and the corresponding y_i value in the data. Note that the x term does not have to be uniformly monotone.

To decouple the deterministic variation from the random variation, we can fix the index and write separate problems of the form

$$y_i = ax_i + b + \epsilon_i$$

where $\epsilon_i \sim \mathcal{N}(0, \sigma^2)$. What could we do with just this one component of the problem? In other words, suppose we had m-samples of this component as in $\{y_{i,k}\}_{k=1}^m$. Following the usual procedure, we could obtain estimates of the mean of y_i as

$$\hat{y}_i = \frac{1}{m} \sum_{k=1}^m y_{i,k}$$

However, this tells us nothing about the individual parameters a and b because they are not separable in the terms that are computed, namely, we may have

$$\mathbb{E}(y_i) = ax_i + b$$

but we still only have one equation and the two unknowns, a and b. How about we fix another component j as in

$$y_j = ax_j + b + \epsilon_i$$

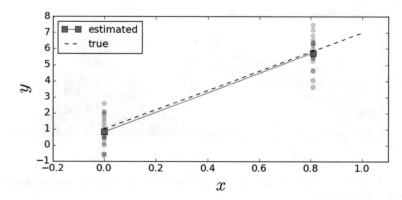

Fig. 3.19 The fitted and true lines are plotted with the data values. The squares at either end of the solid line show the mean value for each of the data groups shown

Then, we have

$$\mathbb{E}(y_j) = ax_j + b$$

so at least now we have two equations and two unknowns and we know how to estimate the left hand sides of these equations from the data using the estimators $\hat{y}_i$ and $\hat{y}_j$. Let us see how this works in the code sample below (Fig. 3.19).

```
>>> x0, xn =x[0],x[80]
>>> # generate synthetic data
>>> y_0 = a*x0 + np.random.randn(20)+b
>>> y_1 = a*xn + np.random.randn(20)+b
>>> # mean along sample dimension
>>> yhat = np.array([y_0,y_1]).mean(axis=1)
>>> a_,b_=np.linalg.solve(np.array([[x0,1],
...                                  [xn,1]]),yhat)
```

Programming Tip

The prior code uses the `solve` function in the Numpy `linalg` module, which contains the core linear algebra codes in Numpy that incorporate the battle-tested LAPACK library.

We can write out the solution for the estimated parameters for this case where $x_0 = 0$

$$\hat{a} = \frac{\hat{y}_i - \hat{y}_0}{x_i}$$

$$\hat{b} = \hat{y}_0$$

The expectations and variances of these estimators are the following:

$$\mathbb{E}(\hat{a}) = \frac{ax_i}{x_i} = a$$

$$\mathbb{E}(\hat{b}) = b$$

$$\mathbb{V}(\hat{a}) = \frac{2\sigma^2}{x_i^2}$$

$$\mathbb{V}(\hat{b}) = \sigma^2$$

The expectations show that the estimators are unbiased. The estimator $\hat{a}$ has a variance that decreases as larger points x_i are selected. That is, it is better to have samples further out along the horizontal axis for fitting the line. This variance quantifies the *leverage* of those distant points.

Regression from Projection Methods Let us see if we can apply our knowledge of projection methods to the general case. In vector notation, we can write the following:

$$\mathbf{y} = a\mathbf{x} + b\mathbf{1} + \epsilon$$

where $\mathbf{1}$ is the vector of all ones. Let us use the inner-product notation,

$$\langle \mathbf{x}, \mathbf{y} \rangle = \mathbb{E}(\mathbf{x}^T \mathbf{y})$$

Then, by taking the inner-product with some $\mathbf{x}_1 \in \mathbf{1}^\perp$ we obtain[5]

$$\langle \mathbf{y}, \mathbf{x}_1 \rangle = a \langle \mathbf{x}, \mathbf{x}_1 \rangle$$

Recall that $\mathbb{E}(\epsilon) = \mathbf{0}$. We can finally solve for a as

$$\hat{a} = \frac{\langle \mathbf{y}, \mathbf{x}_1 \rangle}{\langle \mathbf{x}, \mathbf{x}_1 \rangle} \tag{3.3}$$

That was pretty neat but now we have the mysterious $\mathbf{x}_1$ vector. Where does this come from? If we project $\mathbf{x}$ onto the $\mathbf{1}^\perp$, then we get the MMSE approximation to $\mathbf{x}$ in the $\mathbf{1}^\perp$ space. Thus, we take

$$\mathbf{x}_1 = P_{\mathbf{1}^\perp}(\mathbf{x})$$

Remember that $P_{\mathbf{1}^\perp}$ is a projection matrix so the length of $\mathbf{x}_1$ is at most $\mathbf{x}$. This means that the denominator in the $\hat{a}$ equation above is really just the length of the $\mathbf{x}$ vector

---

[5] The space of all vectors, $\mathbf{a}$ such that $\langle \mathbf{a}, \mathbf{1} \rangle = 0$ is denoted $\mathbf{1}^\perp$.

in the coordinate system of $P_{1\perp}$. Because the projection is orthogonal (namely, of minimum length), the Pythagorean theorem gives this length as the following:

$$\langle \mathbf{x}, \mathbf{x}_1 \rangle^2 = \langle \mathbf{x}, \mathbf{x} \rangle - \langle \mathbf{1}, \mathbf{x} \rangle^2$$

The first term on the right is the length of the $\mathbf{x}$ vector, and last term is the length of $\mathbf{x}$ in the coordinate system orthogonal to $P_{1\perp}$, namely, that of $\mathbf{1}$. We can use this geometric interpretation to understand what is going on in typical linear regression in much more detail. The fact that the denominator is the orthogonal projection of $\mathbf{x}$ tells us that the choice of $\mathbf{x}_1$ has the strongest effect (i.e., largest value) on reducing the variance of $\hat{a}$. That is, the more $\mathbf{x}$ is aligned with $\mathbf{1}$, the worse the variance of $\hat{a}$. This makes intuitive sense because the closer $\mathbf{x}$ is to $\mathbf{1}$, the more constant it is, and we have already seen from our one-dimensional example that distance between the x terms pays off in reduced variance. We already know that $\hat{a}$ is an unbiased estimator, and, because we chose $\mathbf{x}_1$ deliberately as a projection, we know that it is also of minimum variance. Such estimators are known as minimum-variance unbiased estimators (MVUE).

In the same spirit, let us examine the numerator of $\hat{a}$ in Eq. 3.8. We can write $\mathbf{x}_1$ as the following:

$$\mathbf{x}_1 = \mathbf{x} - P_1 \mathbf{x}$$

where P_1 is projection matrix of $\mathbf{x}$ onto the $\mathbf{1}$ vector. Using this, the numerator of $\hat{a}$ becomes

$$\langle \mathbf{y}, \mathbf{x}_1 \rangle = \langle \mathbf{y}, \mathbf{x} \rangle - \langle \mathbf{y}, P_1 \mathbf{x} \rangle$$

Note that,

$$P_1 = \mathbf{1}\mathbf{1}^T \frac{1}{n}$$

so that writing this out explicitly gives

$$\langle \mathbf{y}, P_1 \mathbf{x} \rangle = \left(\mathbf{y}^T \mathbf{1} \right) \left(\mathbf{1}^T \mathbf{x} \right) / n = \left(\sum y_i \right) \left(\sum x_i \right) / n$$

and similarly, we have the following for the denominator:

$$\langle \mathbf{x}, P_1 \mathbf{x} \rangle = \left(\mathbf{x}^T \mathbf{1} \right) \left(\mathbf{1}^T \mathbf{x} \right) / n = \left(\sum x_i \right) \left(\sum x_i \right) / n$$

So, plugging all of this together gives the following:

$$\hat{a} = \frac{\mathbf{x}^T \mathbf{y} - (\sum x_i)(\sum y_i)/n}{\mathbf{x}^T \mathbf{x} - (\sum x_i)^2/n}$$

with corresponding variance

$$\mathbb{V}(\hat{a}) = \sigma^2 \frac{\|\mathbf{x}_1\|^2}{\langle \mathbf{x}, \mathbf{x}_1 \rangle^2}$$

$$= \frac{\sigma^2}{\|\mathbf{x}\|^2 - n(\overline{x^2})}$$

where $\overline{x} = \sum_{i=1}^{n} x_i / n$. Using the same approach with $\hat{b}$ gives

$$\hat{b} = \frac{\langle \mathbf{y}, \mathbf{x}^{\perp} \rangle}{\langle \mathbf{1}, \mathbf{x}^{\perp} \rangle} \tag{3.4}$$

$$= \frac{\langle \mathbf{y}, \mathbf{1} - P_{\mathbf{x}}(\mathbf{1}) \rangle}{\langle \mathbf{1}, \mathbf{1} - P_{\mathbf{x}}(\mathbf{1}) \rangle} \tag{3.5}$$

$$= \frac{\mathbf{x}^T \mathbf{x} (\sum y_i)/n - \mathbf{x}^T \mathbf{y} (\sum x_i)/n}{\mathbf{x}^T \mathbf{x} - (\sum x_i)^2/n} \tag{3.6}$$

where

$$P_{\mathbf{x}} = \frac{\mathbf{x}\mathbf{x}^{\mathbf{T}}}{\|\mathbf{x}\|^2}$$

with variance

$$\mathbb{V}(\hat{b}) = \sigma^2 \frac{\langle \mathbf{1} - P_{\mathbf{x}}(\mathbf{1}), \mathbf{1} - P_{\mathbf{x}}(\mathbf{1}) \rangle}{\langle \mathbf{1}, \mathbf{1} - P_{\mathbf{x}}(\mathbf{1}) \rangle^2}$$

$$= \frac{\sigma^2}{n - \frac{(n\overline{x})^2}{\|\mathbf{x}\|^2}}$$

Qualifying the Estimates Our formulas for the variance above include the unknown σ^2, which we must estimate from the data itself using our plug-in estimates. We can form the residual sum of squares as

$$\text{RSS} = \sum_i (\hat{a}x_i + \hat{b} - y_i)^2$$

Thus, the estimate of σ^2 can be expressed as

$$\hat{\sigma}^2 = \frac{\text{RSS}}{n - 2}$$

where n is the number of samples. This is also known as the *residual mean square*. The $n - 2$ represents the *degrees of freedom* (df). Because we estimated two parameters from the same data, we have $n - 2$ instead of n. Thus, in general, $df = n - p$, where p is the number of estimated parameters. Under the assumption that the noise is Gaussian, the RSS/σ^2 is chi-squared distributed with $n - 2$ degrees of freedom. Another important term is the *sum of squares about the mean* (a.k.a *corrected* sum of squares),

$$SYY = \sum (y_i - \bar{y})^2$$

The SYY captures the idea of not using the x_i data and just using the mean of the y_i data to estimate y. These two terms lead to the R^2 term:

$$R^2 = 1 - \frac{RSS}{SYY}$$

Note that for perfect regression, $R^2 = 1$. That is, if the regression gets each y_i data point exactly right, then $RSS = 0$ this term equals one. Thus, this term is used to measure of goodness of fit. The `stats` module in `scipy` computes many of these terms automatically

```
from scipy import stats
slope,intercept,r_value,p_value,stderr = stats.linregress(x,y)
```

where the square of the `r_value` variable is the R^2 above. The computed p-value is the two-sided hypothesis test with a null hypothesis that the slope of the line is zero. In other words, this tests whether or not the linear regression makes sense for the data for that hypothesis. The Statsmodels module provides a powerful extension to Scipy's stats module by making it easy to do regression and keep track of these parameters. Let us reformulate our problem using the Statsmodels framework by creating a Pandas dataframe for the data:

```
import statsmodels.formula.api as smf
from pandas import DataFrame
import numpy as np
d = DataFrame({'x':np.linspace(0,1,10)}) # create data
d['y'] = a*d.x+ b + np.random.randn(*d.x.shape)
```

Now that we have the input data in the above Pandas dataframe, we can perform the regression as in the following:

```
results = smf.ols('y ~ x', data=d).fit()
```

The $\sim$ symbol is notation for $y = ax + b + \epsilon$, where the constant b is implicit in this usage of Statsmodels. The names in the string are taken from the columns in the dataframe. This makes it very easy to build models with complicated interactions between the named columns in the dataframe. We can examine a report of the model fit by looking at the summary:

```
print(results.summary2())
                    Results: Ordinary least squares
=================================================================
Model:                OLS              Adj. R-squared:      0.808
Dependent Variable:   y                AIC:                 28.1821
Date:                 0000-00-00 00:00 BIC:                 00.0000
No. Observations:     10               Log-Likelihood:      -12.091
Df Model:             1                F-statistic:         38.86
Df Residuals:         8                Prob (F-statistic):  0.000250
R-squared:            0.829            Scale:               0.82158
-----------------------------------------------------------------
               Coef.    Std.Err.    t      P>|t|    [0.025   0.975]
-----------------------------------------------------------------
Intercept      1.5352   0.5327    2.8817   0.0205   0.3067   2.7637
x              5.5990   0.8981    6.2340   0.0003   3.5279   7.6701
```

There is a lot more here than we have discussed so far, but the Statsmodels documentation is the best place to go for complete information about this report. The F-statistic attempts to capture the contrast between including the slope parameter or leaving it off. That is, consider two hypotheses:

$$H_0: \mathbb{E}(Y|X = x) = b$$

$$H_1: \mathbb{E}(Y|X = x) = b + ax$$

In order to quantify how much better adding the slope term is for the regression, we compute the following:

$$F = \frac{\text{SYY} - \text{RSS}}{\hat{\sigma}^2}$$

The numerator computes the difference in the residual squared errors between including the slope in the regression or just using the mean of the y_i values. Once again, if we assume (or can claim asymptotically) that the ϵ noise term is Gaussian, $\epsilon \sim \mathcal{N}(0, \sigma^2)$, then the H_0 hypothesis will follow an F-distribution[6] with degrees of freedom from the numerator and denominator. In this case, $F \sim F(1, n - 2)$. The value of this statistic is reported by Statsmodels above. The corresponding reported probability shows the chance of F exceeding its computed value if H_0 were true. So, the take-home message from all this is that including the slope leads to a much smaller reduction in squared error than could be expected from a favorable draw of n points of this data, under the Gaussian additive noise assumption. This is evidence that including the slope is meaningful for this data.

The Statsmodels report also shows the adjusted R^2 term. This is a correction to the R^2 calculation that accounts for the number of parameters p that the regression is fitting and the sample size n:

---

[6] The $F(m, n)$ F-distribution has two integer degree-of-freedom parameters, m and n.

$$\text{Adjusted } R^2 = 1 - \frac{\text{RSS}/(n-p)}{\text{SYY}/(n-1)}$$

This is always lower than R^2 except when $p = 1$ (i.e., estimating only b). This becomes a better way to compare regressions when one is attempting to fit many parameters with comparatively small n.

Linear Prediction Using linear regression for prediction introduces some other issues. Recall the following expectation:

$$\mathbb{E}(Y|X = x) \approx \hat{a}x + \hat{b}$$

where we have determined $\hat{a}$ and $\hat{b}$ from the data. Given a new point of interest, x_p, we would certainly compute

$$\hat{y}_p = \hat{a}x_p + \hat{b}$$

as the predicted value for $\hat{y}_p$. This is the same as saying that our best prediction for y based on x_p is the above conditional expectation. The variance for this is the following:

$$\mathbb{V}(y_p) = x_p^2 \mathbb{V}(\hat{a}) + \mathbb{V}(\hat{b}) + 2x_p \text{cov}(\hat{a}\hat{b})$$

Note that we have the covariance above because $\hat{a}$ and $\hat{b}$ are derived from the same data. We can work this out below using our previous notation from 3.8:

$$\text{cov}(\hat{a}\hat{b}) = \frac{\mathbf{x}_1^T \mathbb{V}\{\mathbf{y}\mathbf{y}^T\}\mathbf{x}^\perp}{(\mathbf{x}_1^T \mathbf{x})(\mathbf{1}^T \mathbf{x}^\perp)} = \frac{\mathbf{x}_1^T \sigma^2 \mathbf{I} \mathbf{x}^\perp}{(\mathbf{x}_1^T \mathbf{x})(\mathbf{1}^T \mathbf{x}^\perp)}$$

$$= \sigma^2 \frac{\mathbf{x}_1^T \mathbf{x}^\perp}{(\mathbf{x}_1^T \mathbf{x})(\mathbf{1}^T \mathbf{x}^\perp)} = \sigma^2 \frac{(\mathbf{x} - P_1\mathbf{x})^T \mathbf{x}^\perp}{(\mathbf{x}_1^T \mathbf{x})(\mathbf{1}^T \mathbf{x}^\perp)}$$

$$= \sigma^2 \frac{-\mathbf{x}^T P_1^T \mathbf{x}^\perp}{(\mathbf{x}_1^T \mathbf{x})(\mathbf{1}^T \mathbf{x}^\perp)} = \sigma^2 \frac{-\mathbf{x}^T \frac{1}{n}\mathbf{1}\mathbf{1}^T \mathbf{x}^\perp}{(\mathbf{x}_1^T \mathbf{x})(\mathbf{1}^T \mathbf{x}^\perp)}$$

$$= \sigma^2 \frac{-\mathbf{x}^T \frac{1}{n}\mathbf{1}}{(\mathbf{x}_1^T \mathbf{x})} = \frac{-\sigma^2 \bar{x}}{\sum_{i=1}^n (x_i^2 - \bar{x}^2)}$$

After plugging all this in, we obtain the following:

$$\mathbb{V}(y_p) = \sigma^2 \frac{x_p^2 - 2x_p \bar{x} + \|\mathbf{x}\|^2/n}{\|\mathbf{x}\|^2 - n\bar{x}^2}$$

where, in practice, we use the plug-in estimate for the σ^2.

There is an important consequence for the confidence interval for y_p. We cannot simply use the square root of $\mathbb{V}(y_p)$ to form the confidence interval because the model includes the extra ϵ noise term. In particular, the parameters were computed using a set of statistics from the data, but now must include different realizations for the noise term for the prediction part. This means we have to compute

$$\eta^2 = \mathbb{V}(y_p) + \sigma^2$$

Then, the 95% confidence interval $y_p \in (y_p - 2\hat{\eta}, y_p + 2\hat{\eta})$ is the following:

$$\mathbb{P}(y_p - 2\hat{\eta} < y_p < y_p + 2\hat{\eta}) \approx \mathbb{P}(-2 < \mathcal{N}(0, 1) < 2) \approx 0.95$$

where $\hat{\eta}$ comes from substituting the plug-in estimate for σ.

3.8.1 Extensions to Multiple Covariates

With all the machinery we have, it is a short notational hop to consider multiple regressors as in the following:

$$\mathbf{Y} = \mathbf{X}\boldsymbol{\beta} + \boldsymbol{\epsilon}$$

with the usual $\mathbb{E}(\boldsymbol{\epsilon}) = \mathbf{0}$ and $\mathbb{V}(\boldsymbol{\epsilon}) = \sigma^2\mathbf{I}$. Thus, $\mathbf{X}$ is a $n \times p$ full rank matrix of regressors, and $\mathbf{Y}$ is the n-vector of observations. Note that the constant term has been incorporated into $\mathbf{X}$ as a column of ones. The corresponding estimated solution for $\boldsymbol{\beta}$ is the following:

$$\hat{\boldsymbol{\beta}} = (\mathbf{X}^T\mathbf{X})^{-1}\mathbf{X}^T\mathbf{Y}$$

with corresponding variance,

$$\mathbb{V}(\hat{\boldsymbol{\beta}}) = \sigma^2(\mathbf{X}^T\mathbf{X})^{-1}$$

and with the assumption of Gaussian errors, we have

$$\hat{\boldsymbol{\beta}} \sim \mathcal{N}(\boldsymbol{\beta}, \sigma^2(\mathbf{X}^T\mathbf{X})^{-1})$$

The unbiased estimate of σ^2 is the following:

$$\hat{\sigma}^2 = \frac{1}{n - p} \sum \hat{\epsilon}_i^2$$

where $\hat{\boldsymbol{\epsilon}} = \mathbf{X}\hat{\boldsymbol{\beta}} - \mathbf{Y}$ is the vector of residuals. Tukey christened the following matrix as the *hat* matrix (a.k.a. influence matrix),

$$\mathbf{V} = \mathbf{X}(\mathbf{X}^T\mathbf{X})^{-1}\mathbf{X}^T$$

because it maps $\mathbf{Y}$ into $\hat{\mathbf{Y}}$,

$$\hat{\mathbf{Y}} = \mathbf{V}\mathbf{Y}$$

As an exercise you can check that $\mathbf{V}$ is a projection matrix. Note that that matrix is solely a function of $\mathbf{X}$. The diagonal elements of $\mathbf{V}$ are called the *leverage values* and are contained in the closed interval $[1/n, 1]$. These terms measure of distance between the values of x_i and the mean values over the n observations. Thus, the leverage terms depend only on $\mathbf{X}$. This is the generalization of our initial discussion of leverage where we had multiple samples at only two x_i points. Using the hat matrix, we can compute the variance of each residual, $e_i = \hat{y} - y_i$ as

$$\mathbb{V}(e_i) = \sigma^2(1 - v_i)$$

where $v_i = V_{i,i}$. Given the above-mentioned bounds on v_i, these are always less than σ^2.

Degeneracy in the columns of $\mathbf{X}$ can become a problem. This is when two or more of the columns become co-linear. We have already seen this with our single regressor example wherein $\mathbf{x}$ close to $\mathbf{1}$ was bad news. To compensate for this effect we can load the diagonal elements and solve for the unknown parameters as in the following:

$$\hat{\boldsymbol{\beta}} = (\mathbf{X}^T\mathbf{X} + \alpha\mathbf{I})^{-1}\mathbf{X}^T\mathbf{Y}$$

where $\alpha > 0$ is a tunable hyper-parameter. This method is known as *ridge regression* and was proposed in 1970 by Hoerl and Kenndard. It can be shown that this is the equivalent to minimizing the following objective:

$$\|\mathbf{Y} - \mathbf{X}\boldsymbol{\beta}\|^2 + \alpha\|\boldsymbol{\beta}\|^2$$

In other words, the length of the estimated $\boldsymbol{\beta}$ is penalized with larger α. This has the effect of stabilizing the subsequent inverse calculation and also providing a means to trade bias and variance, which we will discuss at length in Sect. 4.7.

Interpreting Residuals Our model assumes an additive Gaussian noise term. We can check the voracity of this assumption by examining the residuals after fitting. The residuals are the difference between the fitted values and the original data:

$$\hat{\epsilon}_i = \hat{a}x_i + \hat{b} - y_i$$

While the p-value and the F-ratio provide some indication of whether or not computing the slope of the regression makes sense, we can get directly at the key assumption of additive Gaussian noise.

For sufficiently small dimensions, the `scipy.stats.probplot` we discussed in the last chapter provides quick visual evidence one way or another by plotting the standardized (Student-ized) residuals

$$r_i = \frac{e_i}{\hat{\sigma}\sqrt{1 - v_i}}$$

where v_i is the *leverage* statistic equal to the diagonal of the influence matrix. The other part of the iid assumption implies homoscedasticity (all r_i have equal variances). Under the additive Gaussian noise assumption, the e_i should also be distributed according to $\mathcal{N}(0, \sigma^2(1 - v_i))$. The normalized residuals r_i should then be distributed according to $\mathcal{N}(0, 1)$. Thus, the presence of any $r_i \notin [-1.96, 1.96]$ should not be common at the 5% significance level and thereby breeds suspicion regarding the homoscedasticity assumption.

The Levene test in `scipy.stats.leven` tests the null hypothesis that all the variances are equal. This basically checks whether or not the standardized residuals vary across x_i more than expected. Under the homoscedasticity assumption, the variance should be independent of x_i. If not, then this is a clue that there is a missing variable in the analysis or that the variables themselves should be transformed (e.g., using the log function) into another format that can reduce this effect. Also, we can use weighted least-squares instead of ordinary least-squares.

Variable Scaling It is tempting to conclude in a multiple regression that small coefficients in any of the β terms implies that those terms are not important. However, simple unit conversions can cause this effect. For example, if one of the regressors is in units of kilometers and the others are in meters, then just the scale factor can give the impression of outsized or under-sized effects. The common way to account for this is to scale the regressors so that

$$x' = \frac{x - \bar{x}}{\sigma_x}$$

This has the side effect of converting the slope parameters into correlation coefficients, which is bounded by ± 1.

Influential Data We have already discussed the idea of leverage. The concept of *influence* combines leverage with outliers. To understand influence, consider Fig. 3.20.

The point on the right in Fig. 3.20 is the only one that contributes to the calculation of the slope for the fitted line. Thus, it is very influential in this sense. Cook's distance is a good way to get at this concept numerically. To compute this, we have to compute the jth component of the estimated target variable with the ith

Fig. 3.20 The point on the right has outsized influence in this data because it is the only one used to determine the slope of the fitted line

Fig. 3.21 The calculated Cook's distance for the data in Fig. 3.20

point deleted. We call this $\hat{y}_{j(i)}$. Then, we compute the following:

$$D_i = \frac{\sum_j (\hat{y}_j - \hat{y}_{j(i)})^2}{p/n \sum_j (\hat{y}_j - y_j)^2}$$

where, as before, p is the number of estimated terms (e.g., $p = 2$ in the bivariate case). This calculation emphasizes the effect of the outlier by predicting the target variable with and without each point. In the case of Fig. 3.20, losing any of the points on the left cannot change the estimated target variable much, but losing the single point on the right surely does. The point on the right does not seem to be an outlier (it *is* on the fitted line), but this is because it is influential enough to rotate the line to align with it. Cook's distance helps capture this effect by leaving each sample out and re-fitting the remainder as shown in the last equation. Figure 3.21 shows the calculated Cook's distance for the data in Fig. 3.20, showing that the data point on the right (sample index 5) has outsized influence on the fitted line. As a rule of thumb, Cook's distance values that are larger than one are suspect.

Fig. 3.22 The upper panel shows data that fit on a line and an outlier point (filled black circle). The lower panel shows the calculated Cook's distance for the data in upper panel and shows that the tenth point (i.e., the outlier) has disproportionate influence

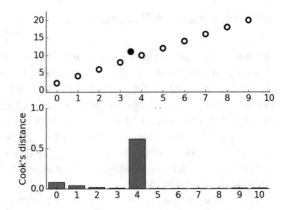

As another illustration of influence, consider Fig. 3.22 which shows some data that nicely line up, but with one outlier (filled black circle) in the upper panel. The lower panel shows so-computed Cook's distance for this data and emphasizes the presence of the outlier. Because the calculation involves leaving a single sample out and re-calculating the rest, it can be a time-consuming operation suitable to relatively small datasets. There is always the temptation to downplay the importance of outliers because they conflict with a favored model, but outliers must be carefully examined to understand why the model is unable to capture them. It could be something as simple as faulty data collection, or it could be an indication of deeper issues that have been overlooked. The following code shows how Cook's distance was computed for Figs. 3.21 and 3.22:

```
>>> fit = lambda i,x,y: np.polyval(np.polyfit(x,y,1),i)
>>> omit = lambda i,x: ([k for j,k in enumerate(x) if j !=i])
>>> def cook_d(k):
...     num = sum((fit(j,omit(k,x),omit(k,y))-fit(j,x,y))**2 for j in x)
...     den = sum((y-np.polyval(np.polyfit(x,y,1),x))**2/len (x)*2)
...     return num/den
...
```

Programming Tip

The function omit sweeps through the data and excludes the ith data element. The embedded enumerate function associates every element in the iterable with its corresponding index.

3.9 Maximum A Posteriori

We saw with maximum likelihood estimation how we could use the principle of maximum likelihood to derive a formula of the data that would estimate the underlying parameters (say, θ). Under that method, the parameter was fixed, but unknown. If we change our perspective slightly and consider the underlying parameter as a random variable in its own right, this leads to additional flexibility in estimation. This method is the simplest of the family of Bayesian statistical methods and is most closely related to maximum likelihood estimation. It is very popular in communications and signal processing and is the backbone of many important algorithms in those areas.

Given that the parameter θ is also a random variable, it has a joint distribution with the other random variables, say, $f(x, \theta)$. Bayes' theorem gives the following:

$$\mathbb{P}(\theta|x) = \frac{\mathbb{P}(x|\theta)\mathbb{P}(\theta)}{\mathbb{P}(x)}$$

The $\mathbb{P}(x|\theta)$ term is the usual likelihood term we have seen before. The term in the denominator is *prior* probability of the data x, and it explicitly makes a very powerful claim: even before collecting or processing any data, we know what the probability of that data is. The $\mathbb{P}(\theta)$ is the prior probability of the parameter. In other words, regardless of the data that is collected, this is the probability of the parameter itself.

In a particular application, whether or not you feel justified making these claims is something that you have to reconcile for yourself and the problem at hand. There are many persuasive philosophical arguments one way or the other, but the main thing to keep in mind when applying any method is whether or not the assumptions are reasonable for the problem at hand.

However, for now, let us just assume that we somehow have $\mathbb{P}(\theta)$, and the next step is the maximizing of this expression over the θ. Whatever results from that maximization is the maximum a posteriori (MAP) estimator for θ. Because the maximization takes place with respect to θ and not x, we can ignore the $\mathbb{P}(x)$ part. To make things concrete, let us return to our original coin flipping problem. From our earlier analysis, we know that the likelihood function for this problem is the following:

$$\ell(\theta) := \theta^k (1 - \theta)^{(n-k)}$$

where the probability of the coin coming up heads is θ. The next step is the prior probability, $\mathbb{P}(\theta)$. For this example, we will choose the $\beta(6, 6)$ distribution (shown in the top left panel of Fig. 3.23). The β family of distributions is a gold mine because it allows for a wide variety of distributions using few input parameters. Now that we have all the ingredients, we turn to maximizing the posterior function,

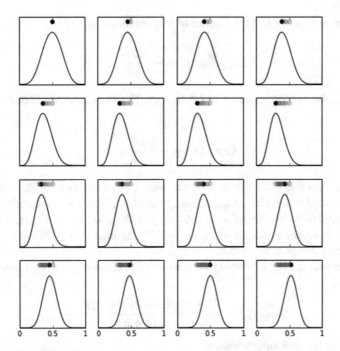

Fig. 3.23 The prior probability is the $\beta(6, 6)$ distribution shown in the top left panel. The dots near the peaks of each of the subgraphs indicate the MAP estimate at that frame

$\mathbb{P}(\theta|x)$. Because the logarithm is convex, we can use it to make the maximization process easier by converting the product to a sum without changing the extrema that we are looking for. Thus, we prefer to work with the logarithm of $\mathbb{P}(\theta|x)$ as in the following:

$$\mathcal{L} := \log \mathbb{P}(\theta|x) = \log \ell(\theta) + \log \mathbb{P}(\theta) - \log \mathbb{P}(x)$$

This is tedious to do by hand and therefore an excellent job for Sympy.

```
>>> from sympy import simplify, log, diff, expand_log, solve
>>> from sympy import stats as st
>>> from sympy.abc import p,k,n
>>> beta_density = st.density(st.Beta('p',6,6))
# setup objective function using sympy.log
>>> obj=expand_log(log(p**k*(1-p)**(n-k)*
                        beta_density(p)))
# use calculus to maximize objective
>>> sol=solve(simplify(diff(obj,p)),p)[0]
>>> sol
(k + 5)/(n + 10)
```

which means that our MAP estimator of θ is the following:

$$\hat{\theta}_{MAP} = \frac{k+5}{n+10}$$

where k is the number of heads in the sample. This is obviously a biased estimator of θ:

$$\mathbb{E}(\hat{\theta}_{MAP}) = \frac{5+n\theta}{10+n} \neq \theta$$

But is this bias *bad*? Why would anyone want a biased estimator? Remember that we constructed this entire estimator using the idea of the prior probability of $\mathbb{P}(\theta)$ which *favors* (biases!) the estimate according to the prior. For example, if $\theta = 1/2$, the MAP estimator evaluates to $\hat{\theta}_{MAP} = 1/2$. No bias there! This is because the peak of the prior probability is at $\theta = 1/2$.

To compute the corresponding variance for this estimator, we need this intermediate result

$$\mathbb{E}(\hat{\theta}_{MAP}^2) = \frac{25 + 10n\theta + n\theta((n-1)p + 1)}{(10+n)^2}$$

which gives the following variance:

$$\mathbb{V}(\hat{\theta}_{MAP}) = \frac{n(1-\theta)\theta}{(n+10)^2}$$

Let us pause and compare this to our previous maximum likelihood (ML) estimator shown below:

$$\hat{\theta}_{ML} = \frac{1}{n}\sum_{i=1}^{n} X_i = \frac{k}{n}$$

As we discussed before, the ML-estimator is unbiased with the following variance:

$$\mathbb{V}(\hat{\theta}_{ML}) = \frac{\theta(1-\theta)}{n}$$

How does this variance compare to that of the MAP? The ratio of the two is the following:

$$\frac{\mathbb{V}(\hat{\theta}_{MAP})}{\mathbb{V}(\hat{\theta}_{ML})} = \frac{n^2}{(n+10)^2}$$

This ratio shows that the variance for the MAP estimator is smaller than that of the ML-estimator. This is payoff for having a biased MAP estimator—it requires fewer samples to estimate if the underlying parameter is consistent with the prior probability. If not, then it will take more samples to pull the estimator away from the bias. In the limit as $n \to \infty$, the ratio goes to one. This means that the benefit of the reduced variance vanishes with enough samples.

The above discussion admits a level of arbitrariness via the prior distribution. We don't have to choose just one prior, however. The following shows how we can use the previous posterior distribution as the prior for the next posterior distribution:

$$\mathbb{P}(\theta|x_{k+1}) = \frac{\mathbb{P}(x_{k+1}|\theta)\mathbb{P}(\theta|x_k)}{\mathbb{P}(x_{k+1})}$$

This is a very different strategy because we are using every data sample x_k as a parameter for the posterior distribution instead of lumping all the samples together in a summation (this is where we got the k term before). This case is much harder to analyze because now every incremental posterior distribution is itself a random function because of the injection of the x random variable. On the other hand, this is more in line with more general Bayesian methods because it is clear that the output of this estimation process is a posterior distribution function, not just a single parameter estimate.

Figure 3.23 illustrates this method. The graph in the top row, far left shows the prior probability ($\beta(6, 6)$), and the dot on the top shows the most recent MAP estimate for θ. Thus, before we obtain any data, the peak of the prior probability is the estimate. The next graph to right shows the effect of $x_0 = 0$ on the incremental prior probability. Note that the estimate has barely moved to the left. This is because the influence of the data has not caused the prior probability to drift away from the original $\beta(6, 6)$-distribution. The first two rows of the figure all have $x_k = 0$ just to illustrate how far left the original prior probability can be moved by those data. The dots on the tops of the sub-graphs show how the MAP estimate changes frame-by-frame as more data is incorporated. The remaining graphs, proceeding top-down, left-to-right show the incremental change in the prior probability for $x_k = 1$. Again, this shows how far to the right the estimate can be pulled from where it started. For this example, there are an equal number of $x_k = 0$ and $x_k = 1$ data, which corresponds to $\theta = 1/2$.

Programming Tip
The following is a quick paraphrase of how Fig. 3.23 was constructed. The first step is to recursively create the posteriors from the data. Note the example data is sorted to make the progression easy to see as a sequence.

```
from sympy.abc import p,x
from scipy.stats import density, Beta, Bernoulli
prior = density(Beta('p',6,6))(p)
likelihood=density(Bernoulli('x',p))(x)
data = (0,0,0,0,0,0,0,1,1,1,1,1,1,1,1)
posteriors = [prior]
```

(continued)

```
for i in data:
    posteriors.append(posteriors[-1]*likelihood.subs(x,i))
```

With the posteriors in hand, the next step is to compute the peak values at each frame using the `fminbound` function from Scipy's `optimize` module.

```
pvals = linspace(0,1,100)
mxvals = []
for i,j in zip(ax.flat,posteriors):
    i.plot(pvals,sympy.lambdify(p,j)(pvals),color='k')
    mxval = fminbound(sympy.lambdify(p,-j),0,1)
    mxvals.append(mxval)
    h = i.axis()[-1]
    i.axis(ymax=h*1.3)
    i.plot(mxvals[-1],h*1.2,'ok')
    i.plot(mxvals[:-1],[h*1.2]*len(mxvals[:-1]),'o')
```

Figure 3.24 is the same as Fig. 3.23 except that the initial prior probability is the $\beta(1.3, 1.3)$-distribution, which has a wider lobe than the $\beta(6, 6)$-distribution. As

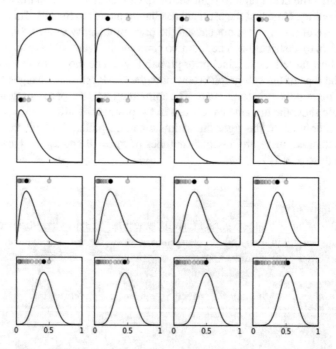

Fig. 3.24 For this example, the prior probability is the $\beta(1.3, 1.3)$ distribution, which has a wider main lobe than the $\beta(6, 6)$ distribution. The dots near the peaks of each of the subgraphs indicate the MAP estimate at that frame

shown in the figure, this prior has the ability to be swayed more violently one way or the other based on the x_k data that is incorporated. This means that it can more quickly adapt to data that is not so consistent with the initial prior and thus does not require a large amount of data in order to *unlearn* the prior probability. Depending on the application, the ability to unlearn the prior probability or stick with it is a design problem for the analyst. In this example, because the data are representative of a $\theta = 1/2$ parameter, both priors eventually settle on an estimated posterior that is about the same. However, if this had not been the case ($\theta \neq 1/2$), then the second prior would have produced a better estimate for the same amount of data.

Because we have the entire posterior density available, we can compute something that is closely related to the confidence interval we discussed earlier; except in this situation, given the Bayesian interpretation, it is called a *credible interval* or *credible set*. The idea is that we want to find a symmetric interval around the peak that accounts for 95% (say) of the posterior density. This means that we can then say the probability that the estimated parameter is within the credible interval is 95%. The computation requires significant numerical processing because even though we have the posterior density in hand, it is hard to integrate analytically and requires numerical quadrature (see Scipy's `integrate` module). Figure 3.25 shows extent of the interval and the shaded region under the posterior density that accounts for 95%.

3.10 Robust Statistics

We considered maximum likelihood estimation (MLE) and maximum a posteriori (MAP) estimation, and in each case we started out with a probability density function of some kind, and we further assumed that the samples were identically distributed and independent (iid). The idea behind robust statistics [29] is to construct estimators that can survive the weakening of either or both of these

Fig. 3.25 The *credible interval* in Bayesian maximum a posteriori is the interval corresponding to the shaded region in the posterior density

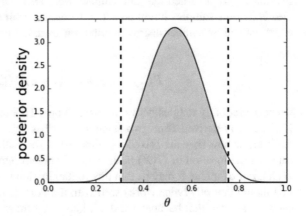

assumptions. More concretely, suppose you have a model that works great except for a few outliers. The temptation is to just ignore the outliers and proceed. Robust estimation methods provide a disciplined way to handle outliers without cherry-picking data that works for your favored model.

The Notion of Location The first notion we need is *location*, which is a general-ization of the idea of *central value*. Typically, we just use an estimate of the mean for this, but we will see later why this could be a bad idea. The general idea of location satisfies the following requirements: Let X be a random variable with distribution F, and let $\theta(X)$ be some descriptive measure of F. Then $\theta(X)$ is said to be a measure of *location* if for any constants a and b, we have the following:

$$\theta(X + b) = \theta(X) + b \tag{3.7}$$

$$\theta(-X) = -\theta(X) \tag{3.8}$$

$$X \geq 0 \Rightarrow \theta(X) \geq 0 \tag{3.9}$$

$$\theta(aX) = a\theta(X) \tag{3.10}$$

The first condition is called *location equivariance* (or *shift-invariance* in signal processing lingo). The fourth condition is called *scale equivariance*, which means that the units that X is measured in should not effect the value of the location estimator. These requirements capture the intuition of *centrality* of a distribution or where most of the probability mass is located.

For example, the sample mean estimator is $\hat{\mu} = \frac{1}{n}\sum X_i$. The first requirement is obviously satisfied as $\hat{\mu} = \frac{1}{n}\sum(X_i+b) = b+\frac{1}{n}\sum X_i = b+\hat{\mu}$. Let us consider the second requirement: $\hat{\mu} = \frac{1}{n}\sum -X_i = -\hat{\mu}$. Finally, the last requirement is satisfied with $\hat{\mu} = \frac{1}{n}\sum aX_i = a\hat{\mu}$.

Robust Estimation and Contamination Now that we have the generalized loca-tion of centrality embodied in the *location* parameter, what can we do with it? Previously, we assumed that our samples were all identically distributed. The key idea is that the samples might be actually coming from a *single* distribution that is contaminated by another nearby distribution, as in the following:

$$F(X) = \epsilon G(X) + (1 - \epsilon)H(X)$$

where ϵ randomly toggles between zero and one. This means that our data samples $\{X_i\}$ actually derived from two separate distributions, $G(X)$ and $H(X)$. We just don't know how they are mixed together. What we really want is an estimator that captures the location of $G(X)$ in the face of random intermittent contamination by $H(X)$. For example, it may be that this contamination is responsible for the outliers in a model that otherwise works well with the dominant F distribution. It can get even worse than that because we don't know that there is only one contaminating

$H(X)$ distribution out there. There may be a whole family of distributions that are contaminating $G(X)$. This means that whatever estimators we construct have to be derived from a more generalized family of distributions instead of from a single distribution, as the maximum-likelihood method assumes. This is what makes robust estimation so difficult—it has to deal with *spaces* of function distributions instead of parameters from a particular probability distribution.

Generalized Maximum Likelihood Estimators M-estimators are generalized maximum likelihood estimators. Recall that for maximum likelihood, we want to maximize the likelihood function as in the following:

$$L_\mu(x_i) = \prod f_0(x_i - \mu)$$

and then to find the estimator $\hat{\mu}$ so that

$$\hat{\mu} = \arg \max_\mu L_\mu(x_i)$$

So far, everything is the same as our usual maximum-likelihood derivation except for the fact that we don't assume a specific f_0 as the distribution of the $\{X_i\}$. Making the definition of

$$\rho = -\log f_0$$

we obtain the more convenient form of the likelihood product and the optimal $\hat{\mu}$ as

$$\hat{\mu} = \arg \min_\mu \sum \rho(x_i - \mu)$$

If ρ is differentiable, then differentiating this with respect to μ gives

$$\sum \psi(x_i - \hat{\mu}) = 0 \tag{3.11}$$

with $\psi = \rho'$, the first derivative of ρ, and for technical reasons we will assume that ψ is increasing. So far, it looks like we just pushed some definitions around, but the key idea is we want to consider general ρ functions that may not be maximum likelihood estimators for *any* distribution. Thus, our focus is now on uncovering the nature of $\hat{\mu}$.

Distribution of M-estimates For a given distribution F, we define $\mu_0 = \mu(F)$ as the solution to the following:

$$\mathbb{E}_F(\psi(x - \mu_0)) = 0$$

It is technical to show, but it turns out that $\hat{\mu} \sim \mathcal{N}(\mu_0, \frac{v}{n})$ with

$$v = \frac{\mathbb{E}_F(\psi(x - \mu_0)^2)}{(\mathbb{E}_F(\psi'(x - \mu_0)))^2}$$

Thus, we can say that $\hat{\mu}$ is asymptotically normal with asymptotic value μ_0 and asymptotic variance v. This leads to the efficiency ratio which is defined as the following:

$$\text{Eff}(\hat{\mu}) = \frac{v_0}{v}$$

where v_0 is the asymptotic variance of the MLE and measures how near $\hat{\mu}$ is to the optimum. In other words, this provides a sense of how much outlier contamination costs in terms of samples. For example, if for two estimates with asymptotic variances v_1 and v_2, we have $v_1 = 3v_2$; then first estimate requires three times as many observations to obtain the same variance as the second. Furthermore, for the sample mean (i.e., $\hat{\mu} = \frac{1}{n} \sum X_i$) with $F = \mathcal{N}$, we have $\rho = x^2/2$ and $\psi = x$ and also $\psi' = 1$. Thus, we have $v = \mathbb{V}(x)$. Alternatively, using the sample median as the estimator for the location, we have $v = 1/(4f(\mu_0)^2)$. Thus, if we have $F = \mathcal{N}(0, 1)$, for the sample median, we obtain $v = 2\pi/4 \approx 1.571$. This means that the sample median takes approximately 1.6 times as many samples to obtain the same variance for the location as the sample mean. The sample median is far more immune to the effects of outliers than the sample mean, so this gives a sense of how much this robustness costs in samples.

M-Estimates as Weighted Means One way to think about M-estimates is a weighted means. Operationally, this means that we want weight functions that can circumscribe the influence of the individual data points, but, when taken as a whole, still provide good estimated parameters. Most of the time, we have $\psi(0) = 0$, and $\psi'(0)$ exists so that ψ is approximately linear at the origin. Using the following definition:

$$W(x) = \begin{cases} \psi(x)/x & \text{if } x \neq 0 \\ \psi'(x) & \text{if } x = 0 \end{cases}$$

We can write our Eq. 3.11 as follows:

$$\sum W(x_i - \hat{\mu})(x_i - \hat{\mu}) = 0 \tag{3.12}$$

Solving this for $\hat{\mu}$ yields the following:

$$\hat{\mu} = \frac{\sum w_i x_i}{\sum w_i}$$

where $w_i = W(x_i - \hat{\mu})$. This is not practically useful because the w_i contains $\hat{\mu}$, which is what we are trying to solve for. The question that remains is how to pick the ψ functions. This is still an open question, but the Huber functions are a well-studied choice.

Huber Functions The family of Huber functions is defined by the following:

$$\rho_k(x) = \begin{cases} x^2 & \text{if } |x| \leq k \\ 2k|x| - k^2 & \text{if } |x| > k \end{cases}$$

with corresponding derivatives $2\psi_k(x)$ with

$$\psi_k(x) = \begin{cases} x & \text{if } |x| \leq k \\ \text{sgn}(x)k & \text{if } |x| > k \end{cases}$$

where the limiting cases $k \to \infty$ and $k \to 0$ correspond to the mean and median, respectively. To see this, take $\psi_\infty = x$ and therefore $W(x) = 1$ and thus the defining Eq. 3.12 results in

$$\sum_{i=1}^{n}(x_i - \hat{\mu}) = 0$$

and then solving this leads to $\hat{\mu} = \frac{1}{n}\sum x_i$. Note that choosing $k = 0$ leads to the sample median, but that is not so straightforward to solve for. Nonetheless, Huber functions provide a way to move between two extremes of estimators for location (namely, the mean vs. the median) with a tunable parameter k. The W function corresponding to Huber's ψ is the following:

$$W_k(x) = \min\left\{1, \frac{k}{|x|}\right\}$$

Figure 3.26 shows the Huber weight function for $k = 2$ with some sample points. The idea is that the location, $\hat{\mu}$, is computed from Eq. 3.12 to lie somewhere in the middle of the weight function so that those terms (i.e., *insiders*) have their values fully reflected in the location estimate. The black circles are the *outliers* that have their values attenuated by the weight function so that only a fraction of their presence is represented in the location estimate.

Breakdown Point So far, our discussion of robustness has been very abstract. A more concrete concept of robustness comes from the breakdown point. In the simplest terms, the breakdown point describes what happens when a single data point in an estimator is changed in the most damaging way possible. For example, suppose we have the sample mean, $\hat{\mu} = \sum x_i/n$, and we take one of the x_i points

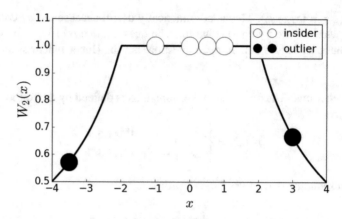

Fig. 3.26 This shows the Huber weight function, $W_2(x)$, and some cartoon data points that are insiders or outsiders as far as the robust location estimate is concerned

to be infinite. What happens to this estimator? It also goes infinite. This means that the breakdown point of the estimator is 0%. On the other hand, the median has a breakdown point of 50%, meaning that half of the data for computing the median could go infinite without affecting the median value. The median is a *rank* statistic that cares more about the relative ranking of the data than the values of the data, which explains its robustness.

The simplest but still formal way to express the breakdown point is to take n data points, $\mathcal{D} = \{(x_i, y_i)\}$. Suppose T is a regression estimator that yields a vector of regression coefficients, $\boldsymbol{\theta}$,

$$T(\mathcal{D}) = \boldsymbol{\theta}$$

Likewise, consider all possible corrupted samples of the data $\mathcal{D}'$. The maximum *bias* caused by this contamination is the following:

$$\text{bias}_m = \sup_{\mathcal{D}'} \|T(\mathcal{D}') - T(\mathcal{D})\|$$

where the sup sweeps over all possible sets of m contaminated samples. Using this, the breakdown point is defined as the following:

$$\epsilon_m = \min\left\{\frac{m}{n} : \text{bias}_m \to \infty\right\}$$

For example, in our least-squares regression, even one point at infinity causes an infinite T. Thus, for least-squares regression, $\epsilon_m = 1/n$. In the limit $n \to \infty$, we have $\epsilon_m \to 0$.

Estimating Scale In robust statistics, the concept of *scale* refers to a measure of the dispersion of the data. Usually, we use the estimated standard deviation for this, but this has a terrible breakdown point. Even more troubling, in order to get a good estimate of location, we have to either somehow know the scale ahead of time or jointly estimate it. None of these methods have easy-to-compute closed form solutions and must be computed numerically.

The most popular method for estimating scale is the *median absolute deviation*

$$MAD = Med(|\mathbf{x} - Med(\mathbf{x})|)$$

In words, take the median of the data $\mathbf{x}$, and then subtract that median from the data itself, and then take the median of the absolute value of the result. Another good dispersion estimate is the *interquartile range*

$$IQR = x_{(n-m+1)} - x_{(n)}$$

where $m = [n/4]$. The $x_{(n)}$ notation means the nth data element after the data have been sorted. Thus, in this notation, $\max(\mathbf{x}) = x_{(n)}$. In the case where $x \sim \mathcal{N}(\mu, \sigma^2)$, then MAD and IQR are constant multiples of σ such that the normalized MAD is the following:

$$MADN(x) = \frac{MAD}{0.675}$$

The number comes from the inverse CDF of the normal distribution corresponding to the 0.75 level. Given the complexity of the calculations, *jointly* estimating both location and scale is a purely numerical matter. Fortunately, the Statsmodels module has many of these ready to use. Let us create some contaminated data in the following code:

```
import statsmodels.api as sm
from scipy import stats
data=np.hstack([stats.norm(10,1).rvs(10),
                stats.norm(0,1).rvs(100)])
```

These data correspond to our model of contamination that we started this section with. As shown in the histogram in Fig. 3.27, there are two normal distributions, one centered neatly at zero, representing the majority of the samples, and another coming less regularly from the normal distribution on the right. Notice that the group of infrequent samples on the right separates the mean and median estimates (vertical dotted and dashed lines). In the absence of the contaminating distribution on the right, the standard deviation for this data should be close to one. However, the usual non-robust estimate for standard deviation (np.std) comes out to approximately three. Using the MADN estimator (sm.robust.scale.mad(data)), we obtain approximately 1.25. Thus, the robust estimate of dispersion is less moved by the presence of the contaminating distribution.

Fig. 3.27 Histogram of sample data. Notice that the group of infrequent samples on the right separates the mean and median estimates indicated by the vertical lines

The generalized maximum likelihood M-estimation extends to joint scale and location estimation using Huber functions. For example,

```
huber = sm.robust.scale.Huber()
loc,scl=huber(data)
```

which implements Huber's *proposal two* method of joint estimation of location and scale. This kind of estimation is the key ingredient to robust regression methods, many of which are implemented in Statsmodels in `statsmodels.formula.api.rlm`. The corresponding documentation has more information.

3.11 Bootstrapping

As we have seen, it can be very difficult or impossible to determine the probability density distribution of the estimator of some quantity. The idea behind the bootstrap is that we can use computation to approximate these functions which would otherwise be impossible to solve for analytically.

Let us start with a simple example. Suppose we have the following set of random variables, $\{X_1, X_2, \ldots, X_n\}$ where each $X_k \sim F$. In other words the samples are all drawn from the same unknown distribution F. Having run the experiment, we thereby obtain the following sample set:

$$\{x_1, x_2, \ldots, x_n\}$$

The sample mean is computed from this set as

$$\bar{x} = \frac{1}{n} \sum_{i=1}^{n} x_i$$

The next question is how close is the sample mean to the true mean, $\theta = \mathbb{E}_F(X)$. Note that the second central moment of X is as follows:

$$\mu_2(F) := \mathbb{E}_F(X^2) - (\mathbb{E}_F(X))^2$$

The standard deviation of the sample mean, $\bar{x}$, given n samples from an underlying distribution F, is the following:

$$\sigma(F) = (\mu_2(F)/n)^{1/2}$$

Unfortunately, because we have only the set of samples $\{x_1, x_2, \ldots, x_n\}$ and not F itself, we cannot compute this and instead must use the estimated standard error

$$\bar{\sigma} = (\bar{\mu}_2/n)^{1/2}$$

where $\bar{\mu}_2 = \sum(x_i - \bar{x})^2/(n-1)$, which is the unbiased estimate of $\mu_2(F)$. However, this is not the only way to proceed. Instead, we could replace F by some estimate, $\hat{F}$ obtained as a piecewise function of $\{x_1, x_2, \ldots, x_n\}$ by placing probability mass $1/n$ on each x_i. With that in place, we can compute the estimated standard error as the following:

$$\hat{\sigma}_B = (\mu_2(\hat{F})/n)^{1/2}$$

which is called the *bootstrap estimate* of the standard error. Unfortunately, the story effectively ends here. In even a slightly more general setting, there is no clean formula $\sigma(F)$ within which F can be swapped for $\hat{F}$.

This is where the computer saves the day. We actually do not need to know the formula $\sigma(F)$ because we can compute it using a resampling method. The key idea is to sample with replacement from $\{x_1, x_2, \ldots, x_n\}$. The new set of n independent draws (with replacement) from this set is the *bootstrap sample*:

$$y^* = \{x_1^*, x_2^*, \ldots, x_n^*\}$$

The Monte Carlo algorithm proceeds by first by selecting a large number of bootstrap samples, $\{y_k^*\}$, then computing the statistic on each of these samples, and then computing the sample standard deviation of the results in the usual way. Thus, the bootstrap estimate of the statistic θ is the following:

$$\hat{\theta}_B^* = \frac{1}{B} \sum_k \hat{\theta}^*(k)$$

with the corresponding square of the sample standard deviation as

$$\hat{\sigma}_B^2 = \frac{1}{B-1} \sum_k (\hat{\theta}^*(k) - \hat{\theta}_B^*)^2$$

The process is much simpler than the notation implies. Let us explore this with a simple example using Python. The next block of code sets up some samples from a $\beta(3, 2)$ distribution:

```
>>> from scipy import stats
>>> rv = stats.beta(3,2)
>>> xsamples = rv.rvs(50)
```

Because this is simulation data, we already know that the mean is $\mu_1 = 3/5$ and the standard deviation of the sample mean for $n = 50$ is $\bar{\sigma} = \sqrt{2}/50$, which we will verify later.

Figure 3.28 shows the $\beta(3, 2)$ distribution and the corresponding histogram of the samples. The histogram represents $\hat{F}$ and is the distribution we sample from to obtain the bootstrap samples. As shown, the $\hat{F}$ is a pretty crude estimate for the F density (smooth solid line), but that's not a serious problem insofar as the following bootstrap estimates are concerned. In fact, the approximation $\hat{F}$ has a naturally tendency to pull toward the bulk of probability mass. This is a feature, not a bug, and is the underlying mechanism that explains bootstrapping, but the formal proofs that exploit this basic idea are far out of our scope here. The next block generates the bootstrap samples

```
>>> yboot = np.random.choice(xsamples,(100,50))
>>> yboot_mn = yboot.mean()
```

and the bootstrap estimate is therefore

```
>>> np.std(yboot.mean(axis=1)) # approx sqrt(1/1250)
0.025598763883825818
```

Figure 3.29 shows the distribution of computed sample means from the bootstrap samples. As promised, the next block shows how to use `sympy.stats` to compute the $\beta(3, 2)$ parameters we quoted earlier. Keep in mind we want to compute the standard deviation of the sample mean *estimator* and not the standard deviation of the samples themselves.

```
>>> import sympy as S
>>> import sympy.stats
>>> for i in range(50): # 50 samples
...     # load sympy.stats Beta random variables
...     # into global namespace using exec
...     execstring = f"x{i} = S.stats.Beta('x{i}',3,2)"
...     exec(execstring)
...
>>> # populate xlist with the sympy.stats random variables
>>> # from above
>>> xlist = [eval(f'x{i}') for i in range(50) ]
>>> # compute sample mean
>>> sample_mean = sum(xlist)/len(xlist)
>>> # compute expectation of sample mean
```

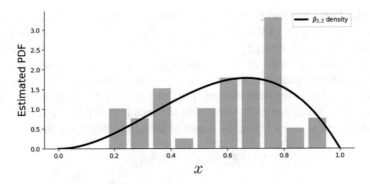

Fig. 3.28 The $\beta(3, 2)$ distribution and the histogram that approximates it

Fig. 3.29 For each bootstrap draw, we compute the sample mean. This is the histogram of those sample means that will be used to compute the bootstrap estimate of the standard deviation

```
>>> sample_mean_1 = S.stats.E(sample_mean).evalf()
>>> # compute 2nd moment of sample mean
>>> sample_mean_2 = S.stats.E(S.expand(sample_mean**2)).evalf()
>>> # standard deviation of sample mean
>>> # use sympy sqrt function
>>> sigma_smn = S.sqrt(sample_mean_2-sample_mean_1**2) # sqrt(2)/50
>>> print(sigma_smn)
0.0282842712474623
```

> **Programming Tip**
> Using the `exec` function enables the creation of a sequence of Sympy random variables. Sympy has the `var` function which can automatically create a sequence of Sympy symbols, but there is no corresponding function in the statistics module to do this for random variables.

Example: Delta Method vs. Bootstrap Recall the delta method from Sect. 3.4.2. Suppose we have a set of Bernoulli coin flips (X_i) with probability of head p. Our maximum likelihood estimator of p is $\hat{p} = \sum X_i/n$ for n flips. We know this estimator is unbiased with $\mathbb{E}(\hat{p}) = p$ and $\mathbb{V}(\hat{p}) = p(1-p)/n$. Suppose we want to use the data to estimate the variance of the Bernoulli trials ($\mathbb{V}(X) = p(1-p)$). By the notation the delta method, $g(x) = x(1-x)$. By the plug-in principle, our maximum likelihood estimator of this variance is then $\hat{p}(1-\hat{p})$. We want the variance of this quantity. Using the results of the delta method, we have

$$\mathbb{V}(g(\hat{p})) = (1 - 2\hat{p})^2 \mathbb{V}(\hat{p})$$

$$\mathbb{V}(g(\hat{p})) = (1 - 2\hat{p})^2 \frac{\hat{p}(1 - \hat{p})}{n}$$

Let us see how useful this is with a short simulation.

```
>>> p = 0.25 # true head-up probability
>>> n = 10
>>> x = stats.bernoulli(p).rvs(n)
>>> print(x)
[0 0 0 0 0 0 1 0 0 0]
```

The maximum likelihood estimator of p is $\hat{p} = \sum X_i/n$:

```
>>> phat = x.mean()
>>> print(phat)
0.1
```

Then, plugging this into the delta method approximant above,

```
>>> print((1-2*phat)**2*(phat)*(1-phat)/n)
0.005760000000000001
```

Now, let us try this using the bootstrap estimate of the variance:

```
>>> phat_b=np.random.choice(x,(50,n)).mean(1)
>>> print(np.var(phat_b*(1-phat_b)))
0.0050490000000000005
```

This shows that the delta method's estimated variance is different from the bootstrap method, but which one is better? For this situation we can solve for this directly using Sympy:

```
>>> import sympy as S
>>> from sympy.stats import E, Bernoulli
>>> xdata =[Bernoulli(i,p) for i in S.symbols('x:10')]
>>> ph = sum(xdata)/len(xdata)
>>> g = ph*(1-ph)
```

> **Programming Tip**
> The argument in the `S.symbols('x:10')` function returns a sequence of Sympy symbols named `x1`, `x2`, and so on. This is shorthand for creating and naming each symbol sequentially.

Note that `g` is the $g(\hat{p}) = \hat{p}(1 - \hat{p})$ whose variance we are trying to estimate. Then, we can plug in for the estimated $\hat{p}$ and get the correct value for the variance:

```
>>> print(E(g**2)-E(g)**2)
0.00442968749999999
```

This case is generally representative—the delta method tends to underestimate the variance, and the bootstrap estimate is better here.

3.11.1 Parametric Bootstrap

In the previous example, we used the $\{x_1, x_2, \ldots, x_n\}$ samples themselves as the basis for $\hat{F}$ by weighting each with $1/n$. An alternative is to *assume* that the samples come from a particular distribution, estimate the parameters of that distribution from the sample set, and then use the bootstrap mechanism to draw samples from the assumed distribution, using the so-derived parameters. For example, the next code block does this for a normal distribution.

```
>>> rv = stats.norm(0,2)
>>> n = 100
>>> xsamples = rv.rvs(n)
>>> # estimate mean and var from xsamples
>>> mn_ = np.mean(xsamples)
>>> std_ = np.std(xsamples)
>>> # bootstrap from assumed normal distribution with
>>> # mn_,std_ as parameters
>>> rvb = stats.norm(mn_,std_) #plug-in distribution
>>> yboot = rvb.rvs((n,500)).var(axis=0)
```

Recall the sample variance estimator is the following:

$$S^2 = \frac{1}{n-1} \sum (X_i - \bar{X})^2$$

Assuming that the samples are normally distributed, this means that $(n-1)S^2/\sigma^2$ has a chi-squared distribution with $n-1$ degrees of freedom. Thus, the variance, $\mathbb{V}(S^2) = 2\sigma^4/(n-1)$. Likewise, the MLE plug-in estimate for this is $\mathbb{V}(S^2) = 2\hat{\sigma}^4/(n-1)$. The following code computes the variance of the sample variance, S^2 using the MLE and bootstrap methods.

```
>>> # MLE-Plugin Variance of the sample mean
>>> print(2*std_**4/(n-1))              # MLE plugin
0.35547420180385103
>>> # Bootstrap variance of the sample mean
>>> print(yboot.var())
0.30032404479182445
>>> # True variance of sample mean
>>> print(2*(2**4)/(n-1))
0.32323232323232326
```

This shows that the bootstrap estimate is better here than the MLE plug-in estimate.

Note that this technique becomes even more powerful with multivariate distributions with many parameters because all the mechanics are the same. Thus, the bootstrap is a great all-purpose method for computing standard errors, but, in the limit, is it converging to the correct value? This is the question of *consistency*. Unfortunately, to answer this question requires more and deeper mathematics than we can get into here. The short answer is that for estimating standard errors, the bootstrap is a consistent estimator in a wide range of cases, and so it definitely belongs in your toolkit.

3.12   Gauss-Markov

The Gauss-Markov model is the fundamental model for noisy parameter estimation because it estimates unobservable parameters given a noisy indirect measurement. Incarnations of the same model appear in all studies of Gaussian models. This case is an excellent opportunity to use everything we have so far learned about projection and conditional expectation.

Following Luenberger [27] let us consider the following problem:

$$\mathbf{y} = \mathbf{W}\boldsymbol{\beta} + \boldsymbol{\epsilon}$$

where $\mathbf{W}$ is a $n \times m$ matrix and $\mathbf{y}$ is a $n \times 1$ vector. Also, $\boldsymbol{\epsilon}$ is a n-dimensional normally distributed random vector with zero-mean and covariance:

$$\mathbb{E}(\boldsymbol{\epsilon}\boldsymbol{\epsilon}^T) = \mathbf{Q}$$

Note that engineering systems usually provide a *calibration mode* where you can estimate $\mathbf{Q}$, so it's not fantastical to assume you have some knowledge of the noise statistics. The problem is to find a matrix $\mathbf{K}$ so that $\hat{\boldsymbol{\beta}} = \mathbf{K}^T\mathbf{y}$ approximates $\boldsymbol{\beta}$. Note that we only have knowledge of $\boldsymbol{\beta}$ via $\mathbf{y}$ so we can't measure it directly. Further, note that $\mathbf{K}$ is a matrix, not a vector, so there are $m \times n$ entries to compute.

We can approach this problem the usual way by trying to solve the MMSE problem:

$$\min_K \mathbb{E}(\|\hat{\beta} - \beta\|^2)$$

which we can write out as

$$\min_K \mathbb{E}(\|\hat{\beta} - \beta\|^2) = \min_K \mathbb{E}(\|\mathbf{K}^T \mathbf{y} - \beta\|^2) = \min_K \mathbb{E}(\|\mathbf{K}^T \mathbf{W} \beta + \mathbf{K}^T \epsilon - \beta\|^2)$$

and since ϵ is the only random variable here, this simplifies to

$$\min_K \|\mathbf{K}^T \mathbf{W} \beta - \beta\|^2 + \mathbb{E}(\|\mathbf{K}^T \epsilon\|^2)$$

The next step is to compute

$$\mathbb{E}(\|\mathbf{K}^T \epsilon\|^2) = \mathrm{Tr}\mathbb{E}(\mathbf{K}^T \epsilon \epsilon^T \mathbf{K}) = \mathrm{Tr}(\mathbf{K}^T \mathbf{Q} \mathbf{K})$$

using the properties of the trace of a matrix. We can assemble everything as

$$\min_K \|\mathbf{K}^T \mathbf{W} \beta - \beta\|^2 + \mathrm{Tr}(\mathbf{K}^T \mathbf{Q} \mathbf{K})$$

Now, if we were to solve this for $\mathbf{K}$, it would be a function of β, which is the same thing as saying that the estimator, $\hat{\beta}$, is a function of what we are trying to estimate, β, which makes no sense. However, writing this out tells us that if we had $\mathbf{K}^T \mathbf{W} = \mathbf{I}$, then the first term vanishes and the problem simplifies to

$$\min_K \mathrm{Tr}(\mathbf{K}^T \mathbf{Q} \mathbf{K})$$

with the constraint

$$\mathbf{K}^T \mathbf{W} = \mathbf{I}$$

This requirement is the same as asserting that the estimator is unbiased,

$$\mathbb{E}(\hat{\beta}) = \mathbf{K}^T \mathbf{W} \beta = \beta$$

To line this problem up with our earlier work, let us consider the ith column of $\mathbf{K}$, $\mathbf{k}_i$. Now, we can rewrite the problem as

$$\min_k (\mathbf{k}_i^T \mathbf{Q} \mathbf{k}_i)$$

with

$$\mathbf{W}^T \mathbf{k}_i = \mathbf{e}_i$$

and we know how to solve this from our previous work on constrained optimization:

$$\mathbf{k}_i = \mathbf{Q}^{-1}\mathbf{W}(\mathbf{W}^T\mathbf{Q}^{-1}\mathbf{W})^{-1}\mathbf{e}_i$$

Now all we have to do is stack these together for the general solution:

$$\mathbf{K} = \mathbf{Q}^{-1}\mathbf{W}(\mathbf{W}^T\mathbf{Q}^{-1}\mathbf{W})^{-1}$$

It's easy when you have all of the concepts lined up! For completeness, the covariance of the error is

$$\mathbb{E}(\hat{\beta} - \beta)(\hat{\beta} - \beta)^T = \mathbf{K}^T\mathbf{Q}\mathbf{K} = (\mathbf{W}^T\mathbf{Q}^{-1}\mathbf{W})^{-1}.$$

Figure 3.30 shows the simulated **y** data as red circles. The black dots show the corresponding estimates, $\hat{\beta}$ for each sample. The black lines show the true value of β versus the average of the estimated β-values, $\widehat{\beta_m}$. The matrix **K** maps the red circles in the corresponding dots. Note there are many possible ways to map the red circles to the plane, but the **K** is the one that minimizes the MSE for β.

Programming Tip

The following snippets provide a quick code walkthrough. To simulate the target data, we define the relevant matrices below:

```
Q = np.eye(3)*0.1 # error covariance matrix
# this is what we are trying estimate
beta = matrix(ones((2,1)))
W = matrix([[1,2],
            [2,3],
            [1,1]])
```

Then, we generate the noise terms and create the simulated data, y,

```
ntrials=50
epsilon=np.random.multivariate_normal((0,0,0),Q,ntrials).T
y=W*beta+epsilon
```

Figure 3.31 shows more detail in the horizontal xy-plane of Fig. 3.30. Figure 3.31 shows the dots, which are individual estimates of $\hat{\beta}$ from the corresponding simulated **y** data. The dashed line is the true value for β, and the filled line ($\widehat{\beta_m}$) is the average of all the dots. The gray ellipse provides an error ellipse for the covariance of the estimated β values.

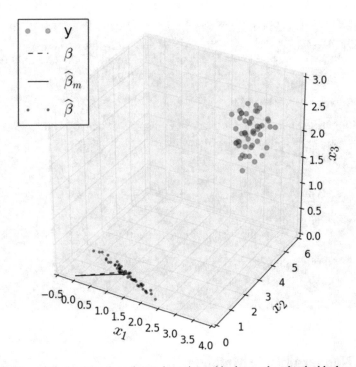

Fig. 3.30 The red circles show the points to be estimated in the *xy*-plane by the black points

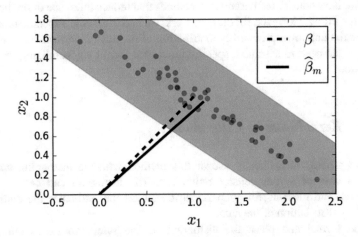

Fig. 3.31 Focusing on the *xy*-plane in Fig. 3.30, the dashed line shows the true value for β versus the mean of the estimated values $\widehat{\beta}_m$

Programming Tip

The following snippets provide a quick walkthrough of the construction of Fig. 3.31. To draw the ellipse, we need to import the patch primitive:

```
from  matplotlib.patches import Ellipse
```

To compute the parameters of the error ellipse based on the covariance matrix of the individual estimates of β in the bm_cov variable below:

```
bm_cov = np.linalg.inv(W.T*Q*W)
U,S,V = np.linalg.svd(bm_cov)
err = np.sqrt((matrix(bm))*(bm_cov)*(matrix(bm).T))
theta = np.arccos(U[0,1])/np.pi*180
```

Then, we draw the add the scaled ellipse in the following:

```
ax.add_patch(Ellipse(bm,err*2/np.sqrt(S[0]),
                         err*2/np.sqrt(S[1]),
                         angle=theta,color='gray'))
```

3.13   Nonparametric Methods

So far, we have considered parametric methods that reduce inference or prediction to parameter-fitting. However, for these to work, we had to assume a specific functional form for the unknown probability distribution of the data. Nonparametric methods eliminate the need to assume a specific functional form by generalizing to classes of functions.

3.13.1   Kernel Density Estimation

We have already made heavy use of this method with the histogram, which is a special case of kernel density estimation. The histogram can be considered the crudest and most useful nonparametric method that estimates the underlying probability distribution of the data.

To be formal and place the histogram on the same footing as our earlier estimations, suppose that $\mathcal{X} = [0, 1]^d$ is the d dimensional unit cube and that h is the *bandwidth* or size of a *bin* or sub-cube. Then, there are $N \approx (1/h)^d$ such bins, each with volume h^d, $\{B_1, B_2, \ldots, B_N\}$. With all this in place, we can write the histogram has a probability density estimator of the form

$$\hat{p}_h(x) = \sum_{k=1}^{N} \frac{\hat{\theta}_k}{h} I(x \in B_k)$$

where

$$\hat{\theta}_k = \frac{1}{n} \sum_{j=1}^{n} I(X_j \in B_k)$$

is the fraction of data points (X_k) in each bin, B_k. We want to bound the bias and variance of $\hat{p}_h(x)$. Keep in mind that we are not trying to estimate a function of x, but the set of all possible probability distribution functions is extremely large and hard to manage. Thus, we need to restrict our attention to the following class of probability distribution of so-called Lipschitz functions:

$$\mathcal{P}(L) = \{p \colon |p(x) - p(y)| \leq L\|x - y\|, \forall x, y\}$$

Roughly speaking, these are the density functions whose slopes (i.e., growth rates) are bounded by L. It turns out that the bias of the histogram estimator is bounded in the following way:

$$\int |p(x) - \mathbb{E}(\hat{p}_h(x))| dx \leq Lh\sqrt{d}$$

Similarly, the variance is bounded by the following:

$$\mathbb{V}(\hat{p}_h(x)) \leq \frac{C}{nh^d}$$

for some constant C. Putting these two facts together means that the risk is bounded by

$$R(p, \hat{p}) = \int \mathbb{E}(p(x) - \hat{p}_h(x))^2 dx \leq L^2 h^2 d + \frac{C}{nh^d}$$

This upper bound is minimized by choosing

$$h = \left(\frac{C}{L^2 nd}\right)^{\frac{1}{d+2}}$$

In particular, this means that

$$\sup_{p \in \mathcal{P}(L)} R(p, \hat{p}) \leq C_0 \left(\frac{1}{n}\right)^{\frac{2}{d+2}}$$

where the constant C_0 is a function of L. There is a theorem [48] that shows this bound in tight, which basically means that the histogram is a really powerful probability density estimator for Lipschitz functions with risk that goes as $\left(\frac{1}{n}\right)^{\frac{2}{d+2}}$. This class of functions is not necessarily smooth because the Lipschitz condition admits non-smooth functions. While this is a reassuring result, we typically do not know which function class (Lipschitz or not) a particular probability belongs to ahead of time. Nonetheless, the rate at which the risk changes with both dimension d and n samples would be hard to understand without this result. Figure 3.32 shows the probability distribution function of the $\beta(2, 2)$ distribution compared to computed histograms for different values of n. The box plots on each of the points show how the variation in each bin of the histogram reduces with increasing n. The risk function $R(p, \hat{p})$ above is based upon integrating the squared difference between the histogram (as a piecewise function of x) and the probability distribution function.

Programming Tip

The following snippet is the main element of the code for Fig. 3.32:

```
def generate_samples(n,ntrials=500):
    phat = np.zeros((nbins,ntrials))
    for k in range(ntrials):
        d = rv.rvs(n)
        phat[:,k],_=histogram(d,bins,density=True)
    return phat
```

The code uses the `histogram` function from Numpy. To be consistent with the risk function $R(p, \hat{p})$, we have to make sure the `bins` keyword argument is formatted correctly using a sequence of bin-edges instead of just a single integer. Also, the `density=True` keyword argument normalizes the histogram appropriately so that the comparison between it and the probability distribution function of the simulated beta distribution is correctly scaled.

3.13.2 Kernel Smoothing

We can extend our methods to other function classes using kernel functions. A one-dimensional smoothing kernel is a smooth function K with the following properties:

$$\int K(x)dx = 1$$

$$\int xK(x)dx = 0$$

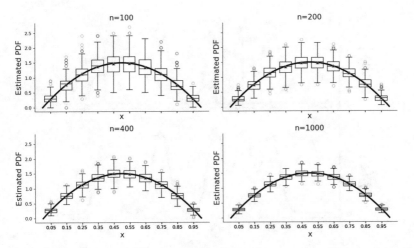

Fig. 3.32 The box plots on each of the points show how the variation in each bin of the histogram reduces with increasing n

$$0 < \int x^2 K(x)dx < \infty$$

For example, $K(x) = I(x)/2$ is the boxcar kernel, where $I(x) = 1$ when $|x| \leq 1$ and zero otherwise. The kernel density estimator is very similar to the histogram, except now we put a kernel function on every point as in the following:

$$\hat{p}(x) = \frac{1}{n} \sum_{i=1}^{n} \frac{1}{h^d} K\left(\frac{\|x - X_i\|}{h}\right)$$

where $X \in \mathbb{R}^d$. Figure 3.33 shows an example of a kernel density estimate using a Gaussian kernel function, $K(x) = e^{-x^2/2}/\sqrt{2\pi}$. There are five data points shown by the vertical lines in the upper panel. The dotted lines show the individual $K(x)$ function at each of the data points. The lower panel shows the overall kernel density estimate, which is the scaled sum of the upper panel.

There is an important technical result in [48] that states that kernel density estimators are minimax in the sense we discussed in the maximum likelihood Sect. 3.4. In broad strokes, this means that the analogous risk for the kernel density estimator is approximately bounded by the following factor:

$$R(p, \hat{p}) \lesssim n^{-\frac{2m}{2m+d}}$$

for some constant C where m is a factor related to bounding the derivatives of the probability density function. For example, if the second derivative of the density

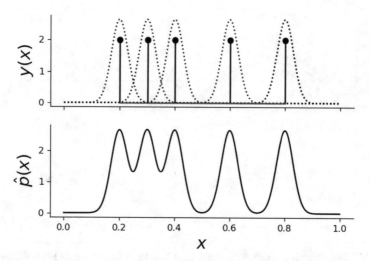

Fig. 3.33 The upper panel shows the individual kernel functions placed at each of the data points. The lower panel shows the composite kernel density estimate which is the sum of the individual functions in the upper panel

function is bounded, then $m = 2$. This means that the convergence rate for this estimator decreases with increasing dimension d.

Cross-Validation As a practical matter, the tricky part of the kernel density estimator (which includes the histogram as a special case) is that we need to somehow compute the bandwidth h term using data. There are several rule-of-thumb methods that for some common kernels, including Silverman's rule and Scott's rule for Gaussian kernels. For example, Scott's factor is to simply compute $h = n^{-1/(d+4)}$, and Silverman's is $h = (n(d + 2)/4)^{(-1/(d+4))}$. Rules of this kind are derived by assuming the underlying probability density function is of a certain family (e.g., Gaussian) and then deriving the best h for a certain type of kernel density estimator, usually equipped with extra functional properties (say, continuous derivatives of a certain order). In practice, these rules seem to work pretty well, especially for unimodal probability density functions. Avoiding these kinds of assumptions means computing the bandwidth from data directly and that is where cross-validation comes in.

Cross-validation is a method to estimate the bandwidth from the data itself. The idea is to write out the following integrated squared error (ISE):

$$ISE(\hat{p}_h, p) = \int (p(x) - \hat{p}_h(x))^2 dx$$

$$= \int \hat{p}_h(x)^2 dx - 2 \int p(x)\hat{p}_h dx + \int p(x)^2 dx$$

The problem with this expression is the middle term,[7]

$$\int p(x)\hat{p}_h dx$$

where $p(x)$ is what we are trying to estimate with $\hat{p}_h$. The form of the last expression looks like an expectation of $\hat{p}_h$ over the density of $p(x)$, $\mathbb{E}(\hat{p}_h)$. The approach is to approximate this with the mean:

$$\mathbb{E}(\hat{p}_h) \approx \frac{1}{n}\sum_{i=1}^{n}\hat{p}_h(X_i)$$

The problem with this approach is that $\hat{p}_h$ is computed using the same data that the approximation utilizes. The way to get around this is to split the data into two equally sized chunks D_1, D_2 and then compute $\hat{p}_h$ for a sequence of different h values over the D_1 set. Then, when we apply the above approximation for the data (Z_i) in the D_2 set,

$$\mathbb{E}(\hat{p}_h) \approx \frac{1}{|D_2|}\sum_{Z_i \in D_2}\hat{p}_h(Z_i)$$

Plugging this approximation back into the integrated squared error provides the objective function:

$$\text{ISE} \approx \int \hat{p}_h(x)^2 dx - \frac{2}{|D_2|}\sum_{Z_i \in D_2}\hat{p}_h(Z_i)$$

Some code will make these steps concrete. We will need some tools from Scikit-learn.

```
>>> from sklearn.model_selection import train_test_split
>>> from sklearn.neighbors import KernelDensity
```

The `train_test_split` function makes it easy to split and keep track of the D_1 and D_2 sets we need for cross-validation. Scikit-learn already has a powerful and flexible implementation of kernel density estimators. To compute the objective function, we need some basic numerical integration tools from Scipy. For this example, we will generate samples from a $\beta(2, 2)$ distribution, which is implemented in the `stats` submodule in Scipy.

```
>>> from scipy.integrate import quad
>>> from scipy import stats
>>> rv= stats.beta(2,2)
```

---

[7] The last term is of no interest because we are only interested in relative changes in the ISE.

```
>>> n=100                        # number of samples to generate
>>> d = rv.rvs(n)[:,None]  # generate samples as column-vector
```

Programming Tip

The use of the [:,None] in the last line formats the Numpy array returned by the rvs function into a Numpy vector with a column dimension of one. This is required by the KernelDensity constructor because the column dimension is used for different features (in general) for Scikit-learn. Thus, even though we only have one feature, we still need to comply with the structured input that Scikit-learn relies upon. There are many ways to inject the additional dimension other than using None. For example, the more cryptic, np.c_, or the less cryptic [:,np.newaxis] can do the same, as can the np.reshape function.

The next step is to split the data into two halves and loop over each of the h_i bandwidths to create a separate kernel density estimator based on the D_1 data:

```
>>> train,test,_,_=train_test_split(d,d,test_size=0.5)
>>> kdes=[KernelDensity(bandwidth=i).fit(train)
...            for i in [.05,0.1,0.2,0.3]]
```

Programming Tip

Note that the single underscore symbol in Python refers to the last evaluated result. The above code unpacks the tuple returned by train_test_split into four elements. Because we are only interested in the first two, we assign the last two to the underscore symbol. This is a stylistic usage to make it clear to the reader that the last two elements of the tuple are unused. Alternatively, we could assign the last two elements to a pair of dummy variables that we do not use later, but then the reader skimming the code may think that those dummy variables are relevant.

The last step is to loop over the so-created kernel density estimators and compute the objective function (Fig. 3.34).

```
>>> for i in kdes:
...        f = lambda x: np.exp(i.score_samples(x))
...        f2 = lambda x: f([[x]])**2
...        h = i.bandwidth
...        if2 = quad(f2,0,1)[0]
...        mn = np.mean(f(test))
...        print('h=%3.2f\t %3.4f'%(h,if2-2*mn))
...
h=0.05              -1.1323
```

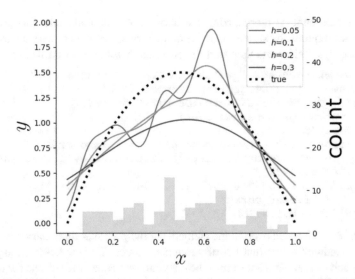

Fig. 3.34 Each line above is a different kernel density estimator for the given bandwidth as an approximation to the true density function. A plain histogram is imprinted on the bottom for reference

```
h=0.10         -1.1336
h=0.20         -1.1330
h=0.30         -1.0810
```

Programming Tip

The lambda functions defined in the last block are necessary because Scikit-learn implements the return value of the kernel density estimator as a logarithm via the `score_samples` function. The numerical quadrature function quad from Scipy computes the $\int \hat{p}_h(x)^2 dx$ part of the objective function.

Scikit-learn has many more advanced tools to automate this kind of hyper-parameter (i.e., kernel density bandwidth) search. To utilize these advanced tools, we need to format the current problem slightly differently by defining the following wrapper class:

```
>>> class KernelDensityWrapper(KernelDensity):
...     def predict(self,x):
...         return np.exp(self.score_samples(x))
...     def score(self,test):
...         f = lambda x: self.predict(x)
...         f2 = lambda x: f([[x]])**2
...         return -(quad(f2,0,1)[0]-2*np.mean(f(test)))
...
```

This is tantamount to reorganizing the above previous code into functions that Scikit-learn requires. Next, we create the dictionary of parameters we want to search over (params) below and then start the grid search with the fit function:

```
>>> from sklearn.model_selection import GridSearchCV
>>> params = {'bandwidth':np.linspace(0.01,0.5,10)}
>>> clf = GridSearchCV(KernelDensityWrapper(),
        param_grid=params,cv=2)
>>> clf.fit(d)
GridSearchCV(cv=2,estimator=KernelDensityWrapper(),
param_grid={'bandwidth':array([0.01,0.06444444,0.11888889,
0.17333333,0.22777778,
0.28222222,0.33666667,0.39111111,0.44555556,0.5])})
>>> print(clf.best_params_)
{'bandwidth': 0.17333333333333334}
```

The grid search iterates over all the elements in the params dictionary and reports the best bandwidth over that list of parameter values. The cv keyword argument above specifies that we want to split the data into two equally sized sets for training and testing. We can also examine the values of the objective function for each point on the grid as follows:

```
>>> clf.cv_results_['mean_test_score']
array([0.60758058,1.06324954,1.11858734,1.13187097,1.12006532,
1.09186225,1.05391076,1.01126161,0.96717292,0.92354959])
```

Keep in mind that the grid search examines multiple folds for cross-validation to compute the above means and standard deviations. Note that there is also a RandomizedSearchCV in case you would rather specify a distribution of parameters instead of a list. This is particularly useful for searching very large parameter spaces where an exhaustive grid search would be too computationally expensive. Although kernel density estimators are easy to understand and have many attractive analytical properties, they become practically prohibitive for large, high-dimensional datasets.

3.13.3 Nonparametric Regression Estimators

Beyond estimating the underlying probability density, we can use nonparametric methods to compute estimators of the underlying function that is generating the data. Nonparametric regression estimators of the following form are known as linear smoothers:

$$\hat{y}(x) = \sum_{i=1}^{n} \ell_i(x) y_i$$

To understand the performance of these smoothers, we can define the risk as the following,

$$R(\hat{y}, y) = \mathbb{E} \left(\frac{1}{n} \sum_{i=1}^{n} (\hat{y}(x_i) - y(x_i))^2 \right)$$

and find the best $\hat{y}$ that minimizes this. The problem with this metric is that we do not know $y(x)$, which is why we are trying to approximate it with $\hat{y}(x)$. We could construct an estimation by using the data at hand as in the following:

$$\hat{R}(\hat{y}, y) = \frac{1}{n} \sum_{i=1}^{n} (\hat{y}(x_i) - Y_i)^2$$

where we have substituted the data Y_i for the unknown function value, $y(x_i)$. The problem with this approach is that we are using the data to estimate the function and then using the same data to evaluate the risk of doing so. This kind of double-dipping leads to overly optimistic estimators. One way out of this conundrum is to use leave-one-out cross-validation, wherein the $\hat{y}$ function is estimated using all but one of the data pairs, (X_i, Y_i). Then, this missing data element is used to estimate the above risk. Notationally, this is written as the following:

$$\hat{R}(\hat{y}, y) = \frac{1}{n} \sum_{i=1}^{n} (\hat{y}_{(-i)}(x_i) - Y_i)^2$$

where $\hat{y}_{(-i)}$ denotes computing the estimator without using the ith data pair. Unfortunately, for anything other than relatively small datasets, it quickly becomes computationally prohibitive to use leave-one-out cross-validation in practice. We'll get back to this issue shortly, but let us consider a concrete example of such a nonparametric smoother.

3.13.4 Nearest Neighbors Regression

The simplest possible nonparametric regression method is the k-nearest neighbors regression. This is easier to explain in words than to write out in math. Given an input x, find the closest one of the k clusters that contains it, and then return the mean of the data values in that cluster. As a univariate example, let us consider the following *chirp* waveform:

$$y(x) = \cos \left(2\pi \left(f_o x + \frac{BW x^2}{2\tau} \right) \right)$$

This waveform is important in high-resolution radar applications. The f_o is the start frequency, and BW/τ is the frequency slope of the signal. For our example, the fact that it is nonuniform over its domain is important. We can easily create some data by sampling the chirp as in the following:

```
>>> xi = np.linspace(0,1,100)[:,None]
>>> xin = np.linspace(0,1,12)[:,None]
>>> f0 = 1 # init frequency
>>> BW = 5 # bandwidth
>>> y = np.cos(2*np.pi*(f0*xin+(BW/2)*xin**2))
```

We can use this data to construct a simple nearest neighbor estimator using Scikit-learn:

```
>>> from sklearn.neighbors import KNeighborsRegressor
>>> knr=KNeighborsRegressor(2)
>>> knr.fit(xin,y)
KNeighborsRegressor(n_neighbors=2)
```

Programming Tip

Scikit-learn has a fantastically consistent interface. The `fit` function above fits the model parameters to the data. The corresponding `predict` function returns the output of the model given an arbitrary input. We will spend a lot more time on Scikit-learn in the machine learning chapter. The `[:,None]` part at the end is just injecting a column dimension into the array in order to satisfy the dimensional requirements of Scikit-learn.

Figure 3.35 shows the sampled signal (gray circles) against the values generated by the nearest neighbor estimator (solid line). The dotted line is the full unsampled chirp signal, which increases in frequency with x. This is important for our example because it adds a non-stationary aspect to this problem in that the function gets

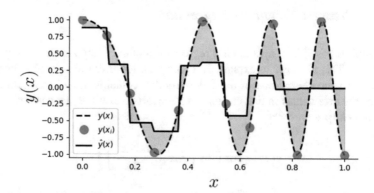

Fig. 3.35 The dotted line shows the chirp signal, and the solid line shows the nearest neighbor estimate. The gray circles are the sample points that we used to fit the nearest neighbor estimator. The shaded area shows the gaps between the estimator and the unsampled chirp

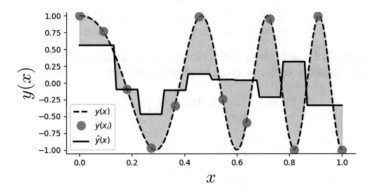

Fig. 3.36 This is the same as Fig. 3.35 except that here there are three nearest neighbors used to build the estimator

progressively wigglier with increasing x. The area between the estimated curve and the signal is shaded in gray. Because the nearest neighbor estimator uses only two nearest neighbors, for each new x, it finds the two adjacent X_i that bracket the x in the training data and then averages the corresponding Y_i values to compute the estimated value. That is, if you take every adjacent pair of sequential gray circles in the Figure, you find that the horizontal solid line splits the pair on the vertical axis. We can adjust the number of nearest neighbors by changing the constructor

```
>>> knr=KNeighborsRegressor(3)
>>> knr.fit(xin,y)
KNeighborsRegressor(n_neighbors=3)
```

which produces Fig. 3.36.

For this example, Fig. 3.36 shows that with more nearest neighbors, the fit performs poorly, especially toward the end of the signal, where there is increasing variation, because the chirp is not uniformly continuous.

Scikit-learn provides many tools for cross-validation. The following code sets up the tools for leave-one-out cross-validation:

```
>>> from sklearn.model_selection import LeaveOneOut
>>> loo=LeaveOneOut()
```

The `LeaveOneOut` object is an iterable that produces a set of disjoint indices of the data—one for fitting the model (training set) and one for evaluating the model (testing set). The next block loops over the disjoint sets of training, and test indices iterates provided by the `loo` variable to evaluate the estimated risk, which is accumulated in the `out` list.

```
>>> out=[]
>>> for train_index, test_index in loo.split(xin):
...     _=knr.fit(xin[train_index],y[train_index])
...     out.append((knr.predict(xi[test_index])-y[test_index])**2)
...
```

```
>>> print('Leave-one-out Estimated Risk: ',np.mean(out),)
Leave-one-out Estimated Risk:   1.0351713662681845
```

The last line in the code above reports leave-one-out's estimated risk.

Linear smoothers of this type can be rewritten in using the following matrix:

$$\mathcal{S} = \left[\ell_i(x_j)\right]_{i,j}$$

so that

$$\hat{\mathbf{y}} = \mathcal{S}\mathbf{y}$$

where $\mathbf{y} = [Y_1, Y_2, \ldots, Y_n] \in \mathbb{R}^n$ and $\hat{\mathbf{y}} = \left[\hat{y}(x_1), \hat{y}(x_2), \ldots, \hat{y}(x_n)\right] \in \mathbb{R}^n$. This leads to a quick way to approximate leave-one-out cross-validation as the following:

$$\hat{R} = \frac{1}{n}\sum_{i=1}^{n}\left(\frac{y_i - \hat{y}(x_i)}{1 - \mathcal{S}_{i,i}}\right)^2$$

However, this does not reproduce the approach in the code above because it assumes that each $\hat{y}_{(-i)}(x_i)$ is consuming one fewer nearest neighbor than $\hat{y}(x)$.

We can get this $\mathcal{S}$ matrix from the knr object as in the following:

```
>>> _= knr.fit(xin,y) # fit on all data
>>> S=(knr.kneighbors_graph(xin)).todense()/knr.n_neighbors
```

The todense part reformats the sparse matrix that is returned into a regular Numpy matrix. The following shows a subsection of this $\mathcal{S}$ matrix:

```
>>> print(S[:5,:5])
[[0.33333333 0.33333333 0.33333333 0.         0.         ]
 [0.33333333 0.33333333 0.33333333 0.         0.         ]
 [0.         0.33333333 0.33333333 0.33333333 0.         ]
 [0.         0.         0.33333333 0.33333333 0.33333333]
 [0.         0.         0.         0.33333333 0.33333333]]
```

The sub-blocks show the windows of the y data that are being processed by the nearest neighbor estimator. For example,

```
>>> print(np.hstack(([knr.predict(xin[:5]),(S*y)[:5]]))) #columns match
[[ 0.55781314  0.55781314]
 [ 0.55781314  0.55781314]
 [-0.09768138 -0.09768138]
 [-0.46686876 -0.46686876]
 [-0.10877633 -0.10877633]]
```

Or, more concisely checking all entries for approximate equality,

```
>>> np.allclose(knr.predict(xin),S*y)
True
```

which shows that the results from the nearest neighbor object and the matrix multiply match.

> **Programming Tip**
>
> Note that because we formatted the returned S as a Numpy matrix, we automatically get the matrix multiplication instead of default element-wise multiplication in the $S*y$ term.

3.13.5 Kernel Regression

For estimating the probability density, we started with the histogram and moved to the more general kernel density estimate. Likewise, we can also extend regression from nearest neighbors to kernel-based regression using the *Nadaraya-Watson* kernel regression estimator. Given a bandwidth $h > 0$, the kernel regression estimator is defined as the following:

$$\hat{y}(x) = \frac{\sum_{i=1}^{n} K\left(\frac{x-x_i}{h}\right) Y_i}{\sum_{i=1}^{n} K\left(\frac{x-x_i}{h}\right)}$$

Unfortunately, Scikit-learn does not implement this regression estimator; however, Jan Hendrik Metzen makes a compatible version available on github.com.

```
>>> from kernel_regression import KernelRegression
```

This code makes it possible to internally optimize over the bandwidth parameter using leave-one-out cross-validation by specifying a grid of potential bandwidth values (gamma), as in the following:

```
>>> kr = KernelRegression(gamma=np.linspace(6e3,7e3,500))
>>> kr.fit(xin,y)
KernelRegression(gamma=6000.0)
```

Figure 3.37 shows the kernel estimator (heavy black line) using the Gaussian kernel compared to the nearest neighbor estimator (solid light black line). As before, the data points are shown as circles. Figure 3.37 shows that the kernel estimator can pick out the sharp peaks that are missed by the nearest neighbor estimator.

Thus, the difference between nearest neighbor and kernel estimation is that the latter provides a smooth moving averaging of points, whereas the former provides a discontinuous averaging. Note that kernel estimates suffer near the boundaries where there is mismatch between the edges and the kernel function. This problem gets worse in higher dimensions because the data naturally drift toward the boundaries (this is a consequence of the *curse of dimensionality*). Indeed, it is not

Fig. 3.37 The heavy blue line is the Gaussian kernel estimator. The light black line is the nearest neighbor estimator. The data points are shown as gray circles. Note that unlike the nearest neighbor estimator, the Gaussian kernel estimator is able to pick out the sharp peaks in the training data

possible to simultaneously maintain local accuracy (i.e., low bias) and a generous neighborhood (i.e., low variance). One way to address this problem is to create a local polynomial regression using the kernel function as a window to localize a region of interest. For example,

$$\hat{y}(x) = \sum_{i=1}^{n} K\left(\frac{x - x_i}{h}\right)(Y_i - \alpha - \beta x_i)^2$$

and now we have to optimize over the two linear parameters α and β. This method is known as *local linear regression* [14, 25]. Naturally, this can be extended to higher-order polynomials. Note that these methods are not yet implemented in Scikit-learn.

3.13.6 Curse of Dimensionality

The so-called curse of dimensionality occurs as we move into higher and higher dimensions. The term was coined by Bellman in 1961 while he was studying adaptive control processes. Nowadays, the term vaguely refers to anything that becomes more complicated as the number of dimensions increases substantially. Nevertheless, the concept is useful for recognizing and characterizing the practical difficulties of high-dimensional analysis and estimation.

Consider the volume of an d-dimensional sphere of radius r:

$$V_s(d, r) = \frac{\pi^{d/2} r^d}{\Gamma\left(\frac{d}{2} + 1\right)}$$

Further, consider the sphere $V_s(d, 1/2)$ enclosed by an d dimensional unit cube. The volume of the cube is always equal to one, but $\lim_{d\to\infty} V_s(d, 1/2) = 0$. What does this mean? It means that the volume of the cube is pushed away from its center, where the embedded hypersphere lives. Specifically, the distance from the center of the cube to its vertices in d dimensions is $\sqrt{d}/2$, whereas the distance from the center of the inscribing sphere is $1/2$. This diagonal distance goes to infinity as d does. For a fixed d, the tiny spherical region at the center of the cube has many long spines attached to it, like a hyper-dimensional sea urchin or porcupine.

Another way to think about this is to consider the $\epsilon > 0$ thick peel of the Hypersphere:

$$\mathcal{P}_\epsilon = V_s(d, r) - V_s(d, r - \epsilon)$$

Then, we consider the following limit:

$$\lim_{d\to\infty} \mathcal{P}_\epsilon = \lim_{d\to\infty} V_s(d, r)\left(1 - \frac{V_s(d, r - \epsilon)}{V_s(d, r)}\right) \tag{3.13}$$

$$= \lim_{d\to\infty} V_s(d, r)\left(1 - \lim_{d\to\infty}\left(\frac{r - \epsilon}{r}\right)^d\right) \tag{3.14}$$

$$= \lim_{d\to\infty} V_s(d, r) \tag{3.15}$$

So, in the limit, the volume of the ϵ-thick peel consumes the volume of the hypersphere.

What are the consequences of this? For methods that rely on nearest neighbors, exploiting locality to lower bias becomes intractable. For example, suppose we have an d dimensional space and a point near the origin we want to localize around. To estimate behavior around this point, we need to average the unknown function about this point, but in a high-dimensional space, the chances of finding neighbors to average are slim. Looked at from the opposing point of view, suppose we have a binary variable, as in the coin flipping problem. If we have 1000 trials, then, based on our earlier work, we can be confident about estimating the probability of heads. Now, suppose we have ten binary variables. Now we have $2^{10} = 1024$ vertices to estimate. If we had the same 1000 points, then at least 24 vertices would not get any data. To keep the same resolution, we would need 1000 samples at each vertex for a grand total of $1000 \times 1024 \approx 10^6$ data points. So, for a tenfold increase in the number of variables, we now have about 1000 times more data points to collect to maintain the same statistical resolution. This is the curse of dimensionality.

Perhaps some code will clarify this. The following code generates samples in two dimensions that are plotted as points in Fig. 3.38 with the inscribed circle in two dimensions. Note that for $d = 2$ dimensions, most of the points are contained in the circle.

```
>>> import numpy as np
>>> v=np.random.rand(1000,2)-1/2.
```

Fig. 3.38 Two-dimensional scatter of points randomly and independently uniformly distributed in the unit square. Note that most of the points are contained in the circle. Counter to intuition, this does not persist as the number of dimensions increases

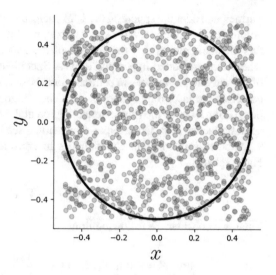

The next code block describes the core computation in Fig. 3.39. For each of the dimensions, we create a set of uniformly distributed random variates along each dimension and then compute how close each d dimensional vector is to the origin. Those that measure one-half are those contained in the hypersphere. The histogram of each measurement is shown in the corresponding panel in Fig. 3.39. The dark vertical line shows the threshold value. Values to the left of this indicate the population that are contained in the hypersphere. Thus, Fig. 3.39 shows that as d increases, fewer points are contained in the inscribed hypersphere. The following code paraphrases the content of Fig. 3.39:

```
fig,ax=subplots()
for d in [2,3,5,10,20,50]:
    v=np.random.rand(5000,d)-1/2.
    ax.hist([np.linalg.norm(i) for i in v])
```

3.13.7 Nonparametric Tests

Determining whether or not two sets of observations derive from the same underlying probability distribution is an important problem. The most popular way to do this is with a standard t-test, but that requires assumptions about normality that may be hard to justify, which leads to nonparametric methods that can get at this question without such assumptions.

Let V and W be continuous random variables. The variable V is *stochastically larger* than W if,

$$\mathbb{P}(V \geq x) \geq \mathbb{P}(W \geq x)$$

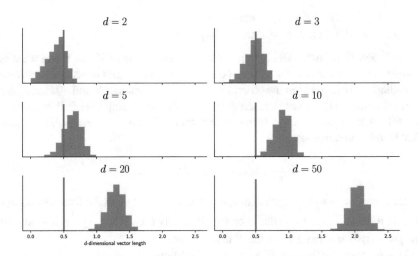

Fig. 3.39 Each panel shows the histogram of lengths of uniformly distributed d dimensional random vectors. The population to the left of the dark vertical line are those that are contained in the inscribed hypersphere. This shows that fewer points are contained in the hypersphere with increasing dimension

Fig. 3.40 The black line density function is stochastically larger than the gray one

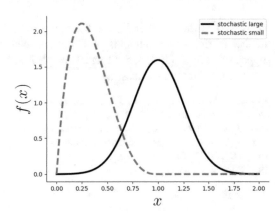

for all $x \in \mathbb{R}$ with strict inequality for at least one x. This can be expressed using CDFs as the following:

$$F_V(x) \leq F_W(x)$$

The term *stochastically smaller* means the obverse of this. For example, the black line density function shown in Fig. 3.40 is stochastically larger than the gray one.

The Mann-Whitney-Wilcoxon Test The Mann-Whitney-Wilcoxon test approaches the following alternative hypotheses:

- $H_0 : F(x) = G(x)$ for all x versus
- $H_a : F(x) < G(x)$, F stochastically larger than G.

where F and G are two CDFs. Suppose we have two datasets X and Y and we want to know if they are drawn from the same underlying probability distribution or if one is stochastically larger than the other. There are n_x elements in X and n_y elements in Y. If we combine these two datasets and rank them, then, under the null hypothesis, any data element should be as likely as any other to be assigned any particular rank. That is, the combined set Z,

$$Z = \{X_1, \ldots, X_{n_x}, Y_1, \ldots, Y_{n_y}\}$$

contains $n = n_x + n_y$ elements. Thus, any assignment of n_y ranks from the integers $\{1, \ldots, n\}$ to $\{Y_1, \ldots, Y_{n_y}\}$ should be equally likely (i.e., $\mathbb{P} = \binom{n}{n_y}^{-1}$). Importantly, this property is independent of the F distribution.

That is, we can define the U statistic as the following:

$$U_X = \sum_{i=1}^{n_x} \sum_{j=1}^{n_y} \mathbb{I}(X_i \geq Y_j)$$

where $\mathbb{I}(\cdot)$ is the usual indicator function. For an interpretation, this counts the number of times that elements of Y outrank elements of X. For example, let us suppose that $X = \{1, 3, 4, 5, 6\}$ and $Y = \{2, 7, 8, 10, 11\}$. We can get a this in one move using Numpy broadcasting:

```
>>> x = np.array([ 1,3,4,5,6 ])
>>> y = np.array([2,7,8,10,11])
>>> U_X = (y <= x[:,None]).sum()
>>> U_Y = (x <= y[:,None]).sum()
>>> (U_X, U_Y)
(4, 21)
```

Note that

$$U_X + U_Y = \sum_{i=1}^{n_x} \sum_{j=1}^{n_y} \mathbb{I}(Y_i \geq X_j) + \mathbb{I}(X_i \geq Y_j) = n_x n_y$$

because $\mathbb{I}(Y_i \geq X_j) + \mathbb{I}(X_i \geq Y_j) = 1$. We can verify this in Python,

```
>>> (U_X+U_Y) == len(x)*len(y)
True
```

Now that we can compute the U_X statistic, we have to characterize it. Let us consider U_X. If H_0 is true, then X and Y are identically distributed random variables. Thus all $\binom{n_x+n_y}{n_x}$ allocations of the X-variables in the ordered combined sample are equally likely. Among these, there are $\binom{n_x+n_y-1}{n_x}$ allocations that have a Y variable as the largest observation in the combined sample. For these, omitting this largest observation does not affect U_X because it would not have been counted anyway. The other $\binom{n_x+n_y-1}{n_x-1}$ allocations have an element of X as the largest observation. Omitting this observation reduces U_X by n_y.

With all that, suppose $N_{n_x,n_y}(u)$ be the number of allocations of X and Y elements that result in $U_X = u$. Under H_0 situation of equally likely outcomes, we have

$$p_{n_x,n_y}(u) = \mathbb{P}(U_X = u) = \frac{N_{n_x,n_y}(u)}{\binom{n_x+n_y}{n_x}}$$

From our previous discussion, we have the recursive relationship:

$$N_{n_x,n_y}(u) = N_{n_x,n_y-1}(u) + N_{n_x-1,n_y}(u - n_y)$$

After dividing all of this by $\binom{n_x+n_y}{n_x}$ and using the $p_{n_x,n_y}(u)$ notation above, we obtain the following:

$$p_{n_x,n_y}(u) = \frac{n_y}{n_x+n_y}p_{n_x,n_y-1}(u) + \frac{n_x}{n_x+n_y}p_{n_x-1,n_y}(u - n_y)$$

where $0 \le u \le n_x n_y$. To start this recursion, we need the following initial conditions:

$$p_{0,n_y}(u_x = 0) = 1$$

$$p_{0,n_y}(u_x > 0) = 0$$

$$p_{n_x,0}(u_x = 0) = 1$$

$$p_{n_x,0}(u_x > 0) = 0$$

To see how this works in Python,

```python
>>> def prob(n,m,u):
...         if u<0: return 0
...         if n==0 or m==0:
...             return int(u==0)
...         else:
...             f = m/float(m+n)
...             return (f*prob(n,m-1,u) +
...                     (1-f)*prob(n-1,m,u-m))
...
```

These are shown in Fig. 3.41 and approach a normal distribution for large n_x, n_y, with the following mean and variance:

$$\mathbb{E}(U) = \frac{n_x n_y}{2} \tag{3.16}$$

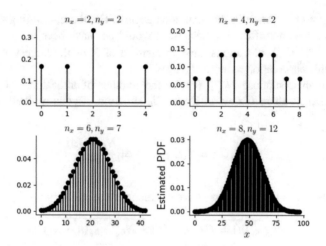

Fig. 3.41 The normal approximation to the distribution improves with increasing n_x, n_y

$$\mathbb{V}(U) = \frac{n_x n_y (n_x + n_y + 1)}{12} \tag{3.17}$$

The variance becomes more complicated when there are ties.

Example We are trying to determine whether or not one network configuration is faster than another. We obtain the following round-trip times for each of the networks:

```
>>> X=np.array([50.6,31.9,40.5,38.1,39.4,
...             35.1,33.1,36.5,38.7,42.3])
>>> Y=np.array([28.8,30.1,18.2,38.5,44.2,
...             28.2,32.9,48.8,39.5,30.7])
```

Because there are too few elements to use the `scipy.stats.mannwhitneyu` function (which internally uses the normal approximation to the U-statistic), we can use our custom function above, but first we need to compute the U_X statistic using Numpy:

```
>>> U_X = (Y <= X[:,None]).sum()
```

For the p-value, we want to compute the probability that the observed U_X statistic at least as great as what was observed:

```
>>> print(sum(prob(10,10,i) for i in range(U_X,101)))
0.08274697438784127
```

This is close to the usual five percent p-value threshold, so it is possible at a slightly higher threshold to conclude that the two sets of samples do *not* originate from the same underlying distribution. Keep in mind that the usual five percent threshold is just a guideline. Ultimately, it is up to the analyst to make the call.

Proving Mean and Variance for U-Statistic To prove Eq. 3.13.7, we assume there are no ties. One way to get at the result $\mathbb{E}(U) = n_x n_y/2$,

$$\mathbb{E}(U_Y) = \sum_j \sum_i \mathbb{P}(X_i \leq Y_j)$$

because $\mathbb{E}(\mathbb{I}(X_i \leq Y_j)) = \mathbb{P}(X_i \leq Y_j)$. Further, because all the subscripted X and Y variables are drawn independently from the same distribution, we have

$$\mathbb{E}(U_Y) = n_x n_y \mathbb{P}(X \leq Y)$$

and also

$$\mathbb{P}(X \leq Y) + \mathbb{P}(X \geq Y) = 1$$

because those are the two mutually exclusive conditions. Because the X variables and Y variables are drawn from the same distribution, we have $\mathbb{P}(X \leq Y) = \mathbb{P}(X \geq Y)$, which means $\mathbb{P}(X \leq Y) = 1/2$ and therefore $\mathbb{E}(U_Y) = n_x n_y/2$. Another way to get the same result is to note that, as we showed earlier, $U_X + U_Y = n_x n_y$. Then, taking the expectation of both sides noting that $\mathbb{E}(U_X) = \mathbb{E}(U_Y) = \mathbb{E}(U)$ gives

$$2\mathbb{E}(U) = n_x n_y$$

which gives $\mathbb{E}(U) = n_x n_y/2$.

Getting the variance is trickier. To start, we compute the following:

$$\mathbb{E}(U_X U_Y) = \sum_i \sum_j \sum_k \sum_l \mathbb{P}(X_i \geq Y_j \wedge X_k \leq Y_l)$$

Of these terms, we have $\mathbb{P}(Y_j \leq X_i \leq Y_j) = 0$ because these are continuous random variables. Let us consider the terms of the following type, $\mathbb{P}(Y_i \leq X_k \leq Y_l)$. To reduce the notational noise, let us rewrite this as $\mathbb{P}(Z \leq X \leq Y)$. Writing this out gives

$$\mathbb{P}(Z \leq X \leq Y) = \int_{\mathbb{R}} \int_Z^{\infty} (F(Y) - F(Z)) f(y) f(z) dy dz$$

where F is the cumulative density function and f is the probability density function $(dF(x)/dx = f(x))$. Let us break this up term by term. Using some calculus for the term:

$$\int_Z^{\infty} F(Y) f(y) dy = \int_{F(Z)}^1 F dF = \frac{1}{2}(1 - F(Z))$$

Then, integrating out the Z variable from this result, we obtain the following:

$$\int_{\mathbb{R}} \frac{1}{2} \left(1 - \frac{F(Z)^2}{2} \right) f(z) dz = \frac{1}{3}$$

Next, we compute

$$\int_{\mathbb{R}} F(Z) \int_{Z}^{\infty} f(y) dy f(z) dz = \int_{\mathbb{R}} (1 - F(Z)) F(Z) f(z) dz$$

$$= \int_{\mathbb{R}} (1 - F) F \, dF = \frac{1}{6}$$

Finally, assembling the result, we have

$$\mathbb{P}(Z \leq X \leq Y) = \frac{1}{3} - \frac{1}{6} = \frac{1}{6}$$

Also, terms like $\mathbb{P}(X_k \geq Y_i \wedge X_m \leq Y_i) = \mathbb{P}(X_m \leq Y_i \leq X_k) = 1/6$ by the same reasoning. That leaves the terms like $\mathbb{P}(X_k \geq Y_i \wedge X_m \leq Y_l) = 1/4$ because of mutual independence and $\mathbb{P}(X_k \geq Y_i) = 1/2$. Now that we have all the terms, we have to assemble the combinatorics to get the final answer.

There are $n_y(n_y - 1)n_x + n_x(n_x - 1)n_y$ terms of type $\mathbb{P}(Y_i \leq X_k \leq Y_l)$. There are $n_y(n_y-1)n_x(n_x-1)$ terms like $\mathbb{P}(X_k \geq Y_i \wedge X_m \leq Y_l)$. Putting this all together, this means that

$$\mathbb{E}(U_X U_Y) = \frac{n_x n_y (n_x + n_y - 2)}{6} + \frac{n_x n_y (n_x - 1)(n_y - 1)}{4}$$

To assemble the $\mathbb{E}(U^2)$ result, we need to appeal to our earlier result:

$$U_X + U_Y = n_x n_y$$

Squaring both sides of this and taking the expectation gives

$$\mathbb{E}(U_X^2) + 2\mathbb{E}(U_X U_Y) + \mathbb{E}(U_Y^2) = n_x^2 n_y^2$$

Because $\mathbb{E}(U_X^2) = \mathbb{E}(U_Y^2) = \mathbb{E}(U)$, we can simplify this as the following:

$$\mathbb{E}(U^2) = \frac{n_x^2 n_y^2 - 2\mathbb{E}(U_X U_Y)}{2}$$

$$\mathbb{E}(U^2) = \frac{n_x n_y (1 + n_x + n_y + 3n_x n_y)}{12}$$

Then, since $\mathbb{V}(U) = \mathbb{E}(U^2) - \mathbb{E}(U)^2$, we finally have

$$\mathbb{V}(U) = \frac{n_x n_y (1 + n_x + n_y)}{12}$$

Kolmogorov-Smirnov Test When you suspect that data come from a specific distribution, whose CDF you can calculate, the Kolmogorov-Smirnov test can help test whether or not this is the case. The Kolmogorov-Smirnov test calculates the maximum absolute difference between the empirical cumulative distribution function (CDF) and the suspected distribution.

$$D_n = \sup_x |F_n(x) - F(x)|$$

where F_n is the empirical CDF for n-data points and purported $F(x)$. The empirical CDF is defined as the following:

$$F_n(x) = \frac{1}{n} \sum_{i=1}^{n} \mathbb{I}(x_i \leq x)$$

where $\mathbb{I}$ is the indicator function. Amazingly, $\sqrt{n} D_n$ converges to the following Kolmogorov distribution for large n, for *any* CDF:

$$f(x) = -8 \sum_{k=1}^{\infty} (-1)^k k^2 x \exp\left\{-2k^2 x^2\right\} \mathbb{I}_{(0,\infty)}(x)$$

with corresponding cumulative density function

$$F(x) = \left[1 + 2\sum_{k=1}^{\infty} (-1)^k \exp\left\{-2k^2 x^2\right\}\right] \mathbb{I}_{(0,\infty)}(x)$$

where $\mathbb{I}_{(0,\infty)}(x)$ is the indicator function over the $(0, \infty)$ interval to enforce that $x > 0$. The Kolmogorov-Smirnov test is implemented in `scipy.stats.kstest`. It turns to that accurate calculation of the Kolmogorov distribution for all values is numerically difficult, so `scipy.stats.kstest` uses different approximations for input different values.

```
>>> import numpy as np
>>> @np.vectorize
... def fK(x):
...     'Kolmogorov PDF distribution'
...     k=np.arange(1,100)
...     return -8*np.sum((-1)**k*k**2 *x*np.exp(-2*k**2*x**2))
...
>>> @np.vectorize
```

```
...  def FK(x):
...      'Kolmogorov CDF distribution'
...      k = np.linspace(1,100)
...      return np.sqrt(2*np.pi)/x*np.sum(np.exp(-(2*k-1)**2
...                                *np.pi**2/(8*x**2)))
...
>>> def estcdf(a):
...      'empirical CDF'
...      x = np.sort(a)
...      y = np.arange(len(x))/float(len(x))
...      return x,y
...
```

Consider the following example of $n = 100$ samples from a χ_3^2 distribution. We use the estcdf function to create an estimated CDF for comparison with the analytical CDF from the same distribution. Figure 3.42 shows the two curves and the maximum absolute difference between them, which is distributed proportion to the Kolmogorov distribution.

```
>>> from scipy import stats
>>> n=100
>>> chirv = stats.chi2(3)
>>> a = chirv.rvs(n)
>>> xe,ye = estcdf(a)
>>> tdist = stats.chi2(3)
>>> Dn=np.abs(tdist.cdf(xe)-ye)
>>> print(Dn.max()) # KS statistic below
0.1394954506688874
>>> stats.kstest(a,tdist.cdf) # two-sided KS test in
          scipy.stats
KstestResult(statistic=0.14949545066888742, pvalue=
0.0204598580916997)
```

We can write a short simulation to repeat this exercise and compare the histograms from the simulation with the analytical Kolmogorov distribution, as shown in Fig. 3.43.

```
>>> # simulation
>>> def run_simulation(n=100,ntrials=1000):
...      o = []
...      for i in range(ntrials):
...          a = chirv.rvs(n)
...          xe,ye = estcdf(a)
...          o.append(np.max(np.abs(tdist.cdf(xe)-ye)))
...      Dns = np.array(o)*np.sqrt(n)
...      return Dns
```

Fig. 3.42 Empirical CDF vs. analytical χ_3^2 distribution and maximum absolute difference between them

Fig. 3.43 Comparing simulated histograms with Kolmogorov distribution

3.14 Survival Analysis

3.14.1 Survival Curves

The problem is to estimate the length of time units (e.g., subjects, individuals, components) that exist in a cohort over time. For example, consider the following data. The rows are the days in a 7-day period, and the columns are the individual units. For example, these could be five people on a life raft. Those who survive for a given day are indicated by 1 and those that do not by 0.

```
>>> import pandas as pd
>>> d = pd.DataFrame(index=range(1,8),
...                  columns=['A','B','C','D','E'],
...                  data=1)
>>> d.loc[3:,'A']=0
>>> d.loc[6:,'B']=0
>>> d.loc[5:,'C']=0
>>> d.loc[4:,'D']=0
```

Fig. 3.44 The red squares indicate a dead subject, and the blue, a living subject

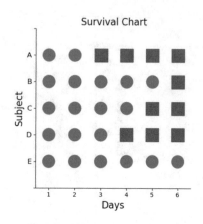

Survival Chart

```
>>> d.index.name='day'
>>> d.T
day   1   2   3   4   5   6   7
A     1   1   0   0   0   0   0
B     1   1   1   1   1   0   0
C     1   1   1   1   0   0   0
D     1   1   1   0   0   0   0
E     1   1   1   1   1   1   1
```

Importantly, survival is a one-way street—once a subject is *dead*, then that subject cannot return to the experiment. This is important because survival analysis is also applied to component failure or other topics where this fact is not so obvious. The following chart shows the survival status of each of the subjects for all 7 days. The blue circles indicate that the subject is alive, and the red squares indicate death of the subject (see Fig. 3.44). The survival probability is plotted in Fig. 3.45.

There is another important recursive perspective on this calculation. Imagine there is a life raft containing $[A, B, C, D, E]$. Everyone survives until day two when A dies. This leaves four in the life raft $[B, C, D, E]$. Thus, from the perspective of day one, the survival probability is the probability of surviving just up until day two and then surviving day two, $\mathbb{P}_S(t \geq 2) = \mathbb{P}(t \notin [0, 2)|t < 2)\mathbb{P}_S(t = 2) = (1)(4/5) = 4/5$. In words, this means that surviving past the second day is the product of surviving the second day itself and not having a death up to that point (i.e., surviving up to that point). Using this recursive approach, the survival probability for the third day is $\mathbb{P}_S(t \geq 3) = \mathbb{P}_S(t > 3)\mathbb{P}_S(t = 3) = (4/5)(3/4) = 3/5$. Recall that just before the third day, the life raft contains $[B, C, D, E]$ and on the third day we have $[B, C, E]$. Thus, from the perspective of just before the third day, there are four survivors in the raft, and on the third day, there are three $3/4$. Using this recursive argument generates the same plot and comes in handy with censoring.

Fig. 3.45 The survival
probability decreases by day

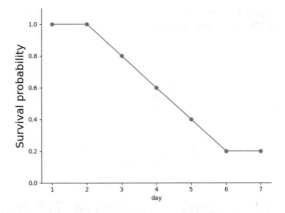

3.14.2 Censoring and Truncation

Censoring occurs when a subject leaves (right censoring) or enters (left censoring)
the study. There are two general types of right censoring. The so-called Type I right
censoring is when a subject randomly drops out of the study. This random dropout is
another statistical effect that has to be accounted for in estimating survival. Type II
right censoring occurs when the study is terminated when enough specific random
events occur.

Likewise, left censoring occurs when a subject enters the study prior to a certain
date, but exactly when this happened is unknown. This happens in study designs
involving two separate studies stages. For example, a subject might enroll in the first
selection process but be ineligible for the second process. Specifically, suppose a
study concerns drug use and certain subjects have used the drug before the study but
are unable to report exactly when. These subjects are left censored. Left truncation
(a.k.a. staggered entry, delayed entry) is similar except the date of entry is known.
For example, a subject that starts taking a drug after being initially left out of the
study.

Right censoring is the most common, so let us consider an example. Let us
estimate the survival function given the following survival times in days:

$$\{1, 2, 3^+, 4, 5, 6^+, 7, 8\}$$

where the censored survival times are indicated by the plus symbol. As before, the
survival time at the 0th day is $8/8 = 1$; the first day is $7/8$; the second day $=$
$(7/8)(6/7)$. Now, we come to the first right censored entry. The survival time for the
third day is $(7/8)(6/7)(5/5) = (7/8)(6/7)$. Thus, the subject who dropped out is not
considered *dead* and cannot be counted as such but is considered just *absent* as far
as the functional estimation of the probabilities goes. Continuing for the fourth day,
we have $(7/8)(6/7)(5/5)(4/5)$, the fifth day, $(7/8)(6/7)(5/5)(4/5)(3/4)$, the sixth
(right censored) day $(7/8)(6/7)(5/5)(4/5)(3/4)(3/3)$, and so on (see Table 3.3).

Table 3.3 Survival
probabilities with censoring

Day	$\mathbb{P}_S$
0	1
1	7/8
2	3/4
3+	3/4
4	3/5
5	9/20
6+	9/20
7	9/40
8	0

3.14.3 Hazard Functions and Their Properties

Generally, the *survival function* is a continuous function of time $S(t) = \mathbb{P}(T > t)$ where T is the event time (e.g., time of death). Note that the cumulative density function, $F(t) = \mathbb{P}(T \leq t) = 1 - S(t)$ and $f(t) = \frac{dF(t)}{dt}$, is the usual probability density function. The so-called *hazard function* is the instantaneous rate of failure at time t:

$$h(t) = \frac{f(t)}{S(t)} = \lim_{\Delta t \to 0} \frac{\mathbb{P}(T \in (t, t + \Delta t] | T \geq t)}{\Delta t}$$

Note that is a continuous-limit version of the calculation we performed above. In words, it says given the event time $T \geq t$ (subject has survived up to t), what is the probability of the event occurring in the differential interval Δt for a vanishingly small Δt. Note that this is not the usual derivative-slope from calculus because there is no difference term in the numerator. The hazard function is also called the *force of mortality*, *intensity rate*, or the *instantaneous risk*. Informally, you can think of the hazard function as encapsulating the two issues we are most concerned about: deaths and the population at risk for those deaths. Loosely speaking, the probability density function in the numerator represents the probability of a death occurring in a small differential interval. However, we are not particularly interested in unqualified deaths, but only deaths that can happen to a specific at-risk population. Returning to our lifeboat analogy, suppose there are 1000 people in the lifeboat and the probability of anybody falling off the lifeboat is 1/1000. Two things are happening here: (1) the probability of the bad event is small, and (2) there are a lot of subjects over which to spread the probability of that bad event. This means that the hazard rate for any particular individual is small. On the other hand, if there are only two subjects in the life raft and the probability of falling off is 3/4, then the hazard rate is high because not only is the unfortunate event probable, the risk of that unfortunate event is shared by only two subjects.

It is a mathematical fact that

$$h(t) = \frac{-d \log S(t)}{dt}$$

This leads to the following interpretation:

$$S(t) = \exp\left(-\int_0^t h(u)du\right) := \exp(-H(t))$$

where $H(t)$ is the *cumulative hazard function*. Note that $H(t) = -\log S(t)$. Consider a subject whose survival time is 5 years. For this subject to have died at the fifth year, it had to be alive during the fourth year. Thus, the *hazard* at 5 years is the failure rate per-year, conditioned on the fact that the subject survived until the fourth year. Note that this is *not* the same as the unconditional failure rate per year at the fifth year, because the unconditional rate applies to all units at time zero and does not use information about survival up to that point gleaned from the other units. Thus, the *hazard function* can be thought of as the point-wise unconditional probability of experiencing the event, scaled by the fraction of survivors up to that point.

Example To get a sense of this, let us consider the example where the probability density function is exponential with parameter λ, $f(t) = \lambda \exp(-t\lambda)$, $\forall t > 0$. This makes $S(t) = 1 - F(t) = \exp(-t\lambda)$, and then the hazard function becomes $h(t) = \lambda$, namely, a constant. To see this, recall that the exponential distribution is the only continuous distribution that has no memory:

$$\mathbb{P}(X \le u + t | X > u) = 1 - \exp(-\lambda t) = \mathbb{P}(X \le t)$$

This means no matter how long we have been waiting for a death to occur, the probability of a death from that point onward is the same—thus the hazard function is a constant.

3.14.4 Expectations

Given all these definitions, it is an exercise in integration by parts to show that the expected life remaining is the following:

$$\mathbb{E}(T) = \int_0^\infty S(u)du$$

This is equivalent to the following:

$$\mathbb{E}(T | t = 0) = \int_0^\infty S(u)du$$

and we can likewise express the expected remaining life at t as the following:

$$\mathbb{E}(T \,|\, T \geq t) = \frac{\int_t^\infty S(u)du}{S(t)}$$

3.14.5   Parametric Regression Models

Because we are interested in how study parameters affect survival, we need a model that can accommodate regression in exogenous (independent) variables ($\mathbf{x}$).

$$h(t|\mathbf{x}) = h_o(t) \exp(\mathbf{x}^T \boldsymbol{\beta})$$

where $\boldsymbol{\beta}$ are the regression coefficients and $h_o(t)$ is the baseline instantaneous hazard function. Because the hazard function is always nonnegative, the effects of the covariates enter through the exponential function. These kinds of models are called *proportional hazard rate models*. If the baseline function is a constant (λ), then this reduces to the *exponential regression model* given by the following:

$$h(t|\mathbf{x}) = \lambda \exp(\mathbf{x}^T \boldsymbol{\beta})$$

3.14.6   Cox Proportional Hazards Model

The tricky part about the above proportional hazard rate model is the specification of the baseline instantaneous hazard function. In many cases, we are not so interested in the absolute hazard function (or its correctness), but rather a comparison of such hazard functions between two study populations. The Cox model emphasizes this comparison by using a maximum likelihood algorithm for a partial likelihood function. There is a lot to keep track of in this model, so let us try the mechanics first to get a feel for what is going on.

Let j denote the jth failure time, assuming that failure times are sorted in increasing order. The hazard function for subject i at failure time j is $h_i(t_j)$. Using the general proportional hazards model, we have

$$h_i(t_j) = h_0(t_j) \exp(z_i \beta) := h_0(t_j)\psi_i$$

To keep it simple, we have $z_i \in \{0, 1\}$ that indicates membership in the experimental group ($z_i = 1$) or the control group ($z_i = 0$). Consider the first failure time, t_1, for subject i failing is the hazard function $h_i(t_1) = h_0(t_1)\psi_i$. From the definitions, the probability that subject i is the one who fails is the following:

$$p_1 = \frac{h_i(t_1)}{\sum h_k(t_1)} = \frac{h_0(t_1)\psi_i}{\sum h_0(t_1)\psi_k}$$

where the summation is over all surviving units up to that point. Note that the baseline hazard cancels out and gives the following:

$$p_1 = \frac{\psi_i}{\sum_k \psi_k}$$

We can keep computing this for the other failure times to obtain $\{p_1, p_2, \ldots p_D\}$. The product of all of these is the *partial likelihood*, $L(\psi) = p_1 \cdot p_2 \cdots p_D$. The next step is to maximize this partial likelihood (usually logarithm of the partial likelihood) over β. There are a lot of numerical issues to keep track of here. Fortunately, the Python `lifelines` module can keep this all straight for us.

Let us see how this works using the `Rossi` dataset that is available in lifelines.

```
>>> from lifelines.datasets import load_rossi
>>> from lifelines import CoxPHFitter
>>> rossi_dataset = load_rossi()
```

The Rossi dataset concerns prison recidivism. The `fin` variable indicates whether or not the subjects received financial assistance upon discharge from prison.

- week: week of first arrest after release, or censoring time.
- arrest: the event indicator, equal to 1 for those arrested during the period of the study and 0 for those who were not arrested.
- fin: a factor, with levels yes if the individual received financial aid after release from prison, and no if he did not; financial aid was a randomly assigned factor manipulated by the researchers.
- age: in years at the time of release.
- race: a factor with levels black and other.
- wexp: a factor with levels yes if the individual had full-time work experience prior to incarceration and no if he did not.
- mar: a factor with levels married if the individual was married at the time of release and not married if he was not.
- paro: a factor coded yes if the individual was released on parole and no if he was not.
- prio: number of prior convictions.
- educ: education, a categorical variable coded numerically, with codes 2 (grade 6 or less), 3 (grades 6 through 9), 4 (grades 10 and 11), 5 (grade 12), or 6 (some post-secondary).
- emp1—emp52: factors coded yes if the individual was employed in the corresponding week of the study and no otherwise.

```
>>> rossi_dataset.head()
   week  arrest  fin  age  race  wexp  mar  paro  prio
0    20       1    0   27     1     0    0     1     3
```

1	17	1	0	18	1	0	0	1	8
2	25	1	0	19	0	1	0	1	13
3	52	0	1	23	1	1	1	1	1
4	52	0	0	19	0	1	0	1	3

Now, we just have to set up the calculation in lifelines, using the scikit-learn style. The lifelines module handles the censoring issues.

```
>>> cph = CoxPHFitter()
>>> cph.fit(rossi_dataset,
...              duration_col='week',
...              event_col='arrest')
<lifelines.CoxPHFitter: fitted with 432 total observations,
318 right-censored observations>
>>> cph.print_summary() # access the results using cph.summary
<lifelines.CoxPHFitter: fitted with 432 total observations,
318 right-censored observations>
             duration col = 'week'
                event col = 'arrest'
     baseline estimation = breslow
   number of observations = 432
number of events observed = 114
  partial log-likelihood = -658.75
         time fit was run = 2022-02-08 21:37:54 UTC

---
```

	coef	exp(coef)	se(coef)	coef lower 95%
covariate				
fin	-0.38	0.68	0.19	-0.75
age	-0.06	0.94	0.02	-0.10
race	0.31	1.37	0.31	-0.29
wexp	-0.15	0.86	0.21	-0.57
mar	-0.43	0.65	0.38	-1.18
paro	-0.08	0.92	0.20	-0.47
prio	0.09	1.10	0.03	0.04

	coef upper 95%	exp(coef) lower 95%	exp(coef) upper 95%
covariate			
fin	-0.00	0.47	1.00
age	-0.01	0.90	0.99
race	0.92	0.75	2.50
wexp	0.27	0.57	1.30
mar	0.31	0.31	1.37
paro	0.30	0.63	1.35
prio	0.15	1.04	1.16

	z	p	-log2(p)
covariate			
fin	-1.98	0.05	4.40
age	-2.61	0.01	6.79
race	1.02	0.31	1.70

Fig. 3.46 This shows the
fitted coefficients from the
summary table for each
covariate

```
wexp          -0.71    0.48            1.06
mar           -1.14    0.26            1.97
paro          -0.43    0.66            0.59
prio           3.19  <0.005           9.48
---
Concordance = 0.64
Partial AIC = 1331.50
log-likelihood ratio test = 33.27 on 7 df
-log2(p) of ll-ratio test = 15.37
```

The values in the summary are plotted in Fig. 3.46.

The Cox proportional hazards model object from `lifelines` allows us to
predict the survival function for an individual with given covariates, assuming that
the individual just entered the study. For example, for the first individual (i.e., row)
in the `rossi_dataset`, we can use the model to predict the survival function for
that individual.

```
>>> cph.predict_survival_function(rossi_dataset.iloc[0,:]).head()
           0
1.0    0.998
2.0    0.995
3.0    0.993
4.0    0.990
5.0    0.988
```

This result is plotted in Fig. 3.47.

3.15 Expectation Maximization

Expectation maximization (EM) is a powerful algorithm for handling cases where
the variables of interest are hidden or *latent* in the data available. Let us denote
the visible variable as v and the hidden variable as h. For example, consider a

Fig. 3.47 The Cox proportional hazards model can predict the survival probability for an individual based on their covariates

model $p_\theta(v, h)$ where we only have data for v. For maximum likelihood we want to maximize the log-likelihood given the visible data to estimate the parameter θ. Let us consider h as a discrete random variable (common for mixture distributions). One idea is to maximize the log-likelihood of v and integrate out h as in the following:

$$\sum_v \log p_\theta(v) = \sum_v \log \sum_h p_\theta(v_i, h)$$

Sometimes this is called the incomplete log-likelihood. Now that the missing h issue has been settled, we can proceed with our usual maximum likelihood approach by maximizing this over θ, which can become very complicated as the dimension of the problem increases. Expectation maximization uses an alternative approach. We can rewrite the summand above as the following:

$$\log p_\theta(v) = \log \sum_h p_\theta(v, h)$$

so that for any probability distribution $r(h|v)$, we have

$$\log p_\theta(v) = \log \sum_h r(h|v) \left(\frac{p_\theta(v, h)}{r(h|v)} \right)$$

We can rewrite this as

$$\log p_\theta(v) = \log \mathbb{E}_{r(h|v)} \left(\frac{p_\theta(v, h)}{r(h|v)} \right)$$

So that with concave log, we can use Jensen's inequality 2.11 in the following:

$$\log p_\theta(v) \geq \mathbb{E}_{r(h|v)}(\log\left(\frac{p_\theta(v, h)}{r(h|v)}\right))$$

and then expand this as the following:

$$\log p_\theta(v) \geq \sum_h r(h|v) \log\left(\frac{p_\theta(v, h)}{r(h|v)}\right)$$

and then summing this up is the log-likelihood:

$$\ell(\theta) = \sum_v \log p_\theta(v_i) \geq \sum_v \sum_h r(h|v) \log\left(\frac{p_\theta(v, h)}{r(h|v)}\right)$$

We define the lower bound on the right as

$$q = \sum_v \sum_h r(h|v) \log\left(\frac{p_\theta(v, h)}{r(h|v)}\right)$$

Our usual strategy is to maximize the likelihood $\ell(\theta)$, but now we want to maximize the lower bound q. Because we arbitrarily picked $r(h|v)$, the next obvious step is to pick another r distribution that increases the lower bound. Specifically, we want to maximize the following objective function:

$$\sum_h r(h|v) \log\left(\frac{p_\theta(v, h)}{r(h|v)}\right) = \sum_h r(h|v) \log p_\theta(v, h) - \sum_h r(h|v) \log r(h|v)$$

subject to the constraints $\sum_h r(h|v) = 1$ and $r(h|v) \geq 0$. We can do this by expanding the logarithm and composing the Lagrangian:

$$\mathcal{L} = \sum_h r(h|v) \log p_\theta(v, h) - \sum_h r(h|v) \log r(h|v) - \sum_h \lambda_z r(h|v) - v(1 - \sum_h r(h|v))$$

Then, using unconstrained maximization,

$$\frac{d\mathcal{L}}{dr} = \log p_\theta(v, h) - \log r(h|v) - 1 - v = 0$$

and solving for the optimimum $r^*(h|v)$ and normalizing $\sum_h r^*(h|v) = 1$ gives the final result:

$$r^*(h|v) = p_\theta(h|v)$$

This maximization process is called the *Expectation* step in the expectation maximization algorithm. The *Maximization* step works on the θ parameters,

$$\arg\max_{\theta} \sum_v \sum_h r^*(h|v) \log p_\theta(v, h)$$

Example To see how this all works together, let us consider the following problem from [4]. The hidden variable is $h \in \{1, 2\}$, and the visible variable is $v \in \mathbb{R}$ with parameter $\theta \in \mathbb{R}$

$$p_\theta(v|h) = \frac{1}{\sqrt{\pi}} e^{-(v-\theta h)^2}$$

The prior probability on h is $p(h = 1) = p(h = 2) = \frac{1}{2}$. We only observe values of v and not h, and we want to estimate θ. From our result above for the expectation step, we know that to optimize the lower bound, we need the following density for some initial θ_0:

$$p_\theta(h|v) = \frac{p_\theta(v|h)p(h)}{p(v)}$$

Note that because h takes only two values and that $\sum_h p_\theta(h|v) = 1$, we only need to keep track of one of the h values. We can get $p(v)$ by summing over $p(h)$ as shown in Fig. 3.48.

$$p(v) = p_\theta(v|h = 1)\frac{1}{2} + p_\theta(v|h = 2)\frac{1}{2} = \frac{e^{-(v-\theta)^2} + e^{-(v-2\theta)^2}}{2\sqrt{\pi}}$$

Then, to do the expectation step, we compute

$$p_{\theta_0}(h = 1|v) = \frac{p_{\theta_0}(v|h = 1)p(h = 1)}{p(v)} = \frac{e^{-(v-\theta_0)^2}}{e^{-(v-\theta_0)^2} + e^{-(v-2\theta_0)^2}}$$

Fig. 3.48 There are two lumps here because of the ambiguity in the hidden variable

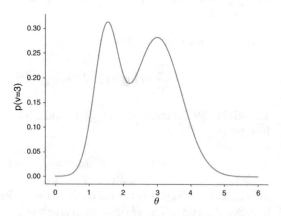

Note that θ_0 is from the prior iteration. The maximization step is then

$$\arg\max_{\theta} \sum_{v} \sum_{h} p_{\theta_0}(h|v) \log p_{\theta}(v, h)$$

For the maximization over θ, simplifying out the irrelevant terms,

$$\arg\max_{\theta} \sum_{v} \sum_{h} p_{\theta_0}(h|v) \log p_{\theta}(v|h)$$

This is the usual maximum likelihood problem weighted by the result of the expectation step using the previous value of θ_0. We can reduce the notation and emphasize the relevant terms by writing the inner summation as the following:

$$\arg\max_{\theta} \sum_{v} \mathbb{E}_{p_{\theta_0}(h|v)}(-(v - h\theta)^2)$$

Maximizing this using calculus gives the following for a single v:

$$\theta^* = v\mathbb{E}_{p_{\theta_0}(h|v)}(h) / \mathbb{E}_{p_{\theta_0}(h|v)}(h^2)$$

which gives the iteration

$$\theta_{k+1} = v \frac{e^{3\theta_k^2} + 2e^{2\theta_k v}}{e^{3\theta_k^2} + 4e^{2\theta_k v}} = v\left(1 - \frac{2}{e^{(3\theta_k - 2v)\theta_k} + 4}\right)$$

The following Python code implements the EM algorithm:

```python
>>> import numpy as np
>>> theta=2 # starting point
>>> v=3 # visible data
>>> for i in range(5):
...     theta = v*(1-2/(np.exp((3*theta-2*v)*theta)+4))
...     print(f'theta={theta}')
...
theta=1.7999999999999998
theta=1.6173826651486614
theta=1.5563791505956408
theta=1.5458135899717353
theta=1.5442871279738428
```

The following is the lower bound for the algorithm.

$$\mathbb{E}_{p_{\theta_0}(h|v)}(-(v - h\theta)^2) = -\frac{e^{\theta_o^2 - 4v\theta_o + v^2}\left((v - 2\theta)^2 e^{2v\theta_o} + e^{3\theta_o^2}(v - \theta)^2\right)}{e^{(v - 2\theta_o)^2} + e^{(v - \theta_o)^2}}$$

At convergence, when $\theta = \theta_o$, this simplifies to the following:

Fig. 3.49 The lower bound
of for the EM algorithm

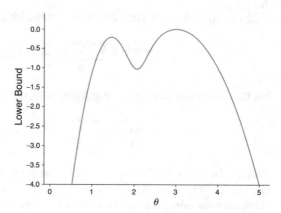

$$\mathbb{E}_{p_\theta(h|v)}(-(v - h\theta)^2) = -\frac{e^{\theta^2+v^2-4\theta v}\left(e^{3\theta^2}(v - \theta)^2 + e^{2\theta v}(v - 2\theta)^2\right)}{\left(e^{(v-2\theta)^2} + e^{(v-\theta)^2}\right)}$$

which is shown in Fig. 3.49. Notice that the maximum is achieved when $\theta = 3$, but
that is *not* found by the EM algorithm when its starting value is closer to the local
maximum at $\theta = 1.5$. Because the EM algorithm is greedy, once it locks into the
local minimum, it can never escape and find the global maximum.

Example: Censoring Sometimes collected data are thresholded due to limitations
in the data collection apparatus, for example. For example, if y_i are the samples then
some $z_i = \min(y_i, a)$. One approach is to drop the thresholded values and then just
use the other y_i for estimation. However, the thresholded values z_i can still help with
the estimation. Consider $Y \sim \mathcal{N}(\mu, 1)$. The complete log-likelihood for n samples
is the following:

$$\ell(\mu) = -\frac{1}{2}\sum_{i=1}^{m}(y_i - \mu)^2 - \frac{1}{2}\sum_{i=m+1}^{n}(z_i - \mu)^2$$

We need the following definitions for the standard normal distribution with the
following corresponding probability density function, $\phi(x)$:

$$\phi(x) = \frac{e^{-\frac{x^2}{2}}}{\sqrt{2\pi}}$$

with corresponding cumulative density function, $\Phi(x)$ The expectation step uses the
truncated normal distribution. To simplify this, we can take the derivative first and
then do the expectation step. This will make it easier to do the maximization step
later. This is legitimate due to a technicality.

$$\frac{d}{d\mu}\ell(\mu) = \sum_{i=1}^{m}(y_i - \mu) + \sum_{i=m+1}^{n}\mathbb{E}_{\mu_o}(z_i - \mu) = 0$$

This simplifies as

$$J = m\bar{y} - m\mu + (n-m)\mathbb{E}_{\mu_o}[(z - \mu)]$$

where

$$\mathbb{E}_{\mu_o}[(z - \mu)] = \int_{a}^{\infty}(z - \mu)\phi(z - \mu_o)dz$$

To proceed, we need the following fact:

$$\int_{a}^{\infty}(x - \mu)\phi(x - \mu)dx = \frac{e^{-\frac{(a-\mu)^2}{2}}}{\sqrt{2\pi}} = \phi(a - \mu)$$

Then,

$$\mathbb{E}_{\mu_o}[(z - \mu)] = \int_{a}^{\infty}(z - \mu_o + \mu_o - \mu)\phi(z - \mu_o)dz$$

$$= \int_{a}^{\infty}(z - \mu_o)\phi(z - \mu_o)dz + \int_{a}^{\infty}(\mu_o - \mu)\phi(z - \mu_o)dz$$

$$= \phi(a - \mu_o) + (\mu_o - \mu)(1 - \Phi(a - \mu_o))$$

Now, we can rewrite J above as the following,

$$J = m\bar{y} - m\mu + (n-m)(\phi(a - \mu_o) + (\mu_o - \mu)(1 - \Phi(a - \mu_o))) = 0$$

Then, solving for μ gives the update step.

$$\mu_{k+1} = \frac{m\bar{y} + (n-m)\phi(a - \mu_k) + (n-m)\mu_k(1 - \Phi(a - \mu_k))}{m + (n-m)(1 - \Phi(a - \mu_k))}$$

Let us try evaluating this using Python. First, let us assemble the modules:

```
>>> import numpy as np
>>> from scipy import stats
```

Now, we construct the terms we need for the update equation:

```
>>> phi = stats.norm().pdf
>>> Phi = stats.norm().cdf
>>> a = 4.5 # cutoff threshold
```

Let us create some samples. Remember that the parameter of this generative distribution is what we want to estimate.

```
>>> nrm = stats.norm(4) # mean is 4
>>> samples = nrm.rvs(25)
```

Next, we assemble the terms we need for the update equation:

```
>>> y = samples[samples<a]
>>> z = samples[samples>=a] # clipped
>>> m = len(y)
>>> n = len(y)+len(z)
>>> ym = np.mean(y)

>>> mu = 3 # initialize
>>> for i in range(5):
...     print(f'mu={mu}')
...     mu = ((m * ym) + (n - m)*phi(a-mu)
...           + (n-m)*mu*(1-Phi(a-mu)) )/(m+(n-m)*(1-Phi(a-mu)))
...
mu=3
mu=3.609396048783591
mu=3.7007889552321247
mu=3.72018353521635
mu=3.72448619620872
```

Let us compare this result to just estimating using only the y_i samples (i.e., the unclipped ones) and using all of the samples.

```
>>> print(f'mu={mu:3.2f},ym={ym:3.2f},all={np.mean(samples):3.2f}')
mu=3.73,ym=3.56,all=3.99
```

Note that the μ is greater than $\bar{y}$ so that just using the y_i data and excluding the clipped data results in an underestimation of the parameter. This means that the algorithm has derived important value from the clipped data in terms of estimating the underlying parameter. A good exercise would be to try more samples and change the cutoff threshold to see how this changes.

Example: Telephone Plans Suppose there are n regions that are surveyed, but for the first m regions, respondents are asked to select between four plans, and in the remaining $n - m$ regions, the respondents chose among an additional fifth plan.[8] The response for the ith participant is modeled as a multinomial distribution, $\mathcal{M}(1, (p_1, p_2, p_3, p_4, p_5))$. This means that the sum of all respondents in the ith region is $T_i \sim \mathcal{M}(n_i, (p_1, p_2, p_3, p_4, p_5))$, but this is only for those $n - m$ regions for which the respondents had five options. The distribution for the other m regions that had only four option is modeled by the following multinomial distribution for the i_{th} region:

---

[8] This example is inspired from p.179 in [38] and is presented here in excruciating detail with Python code.

$$\binom{n_i + x_i}{T_{i,1}, T_{i,2}, T_{i,3}, T_{i,4}, x} p_1^{T_{i,1}} p_2^{T_{i,2}} p_3^{T_{i,3}} p_3^{T_{i,4}} p_4^{T_{i,4}} p_5^{x_i}$$

The notation above means the following:

$$\binom{n}{k_1, k_2, \cdots, k_m} = \frac{n!}{k_1! k_2! \cdots k_m!}$$

Conceptually, there are x_i more hypothetical participants, beyond the n_i participants who answered only among four options. The hidden variable is x_i, which is the number of participants who *would* have filled out the fifth option. The process to apply the EM algorithm is the same. The complete likelihood is the following:

$$L(\mathbf{x}, \mathbf{T}) = \prod_{i=1}^{m} \binom{n_i + x_i}{T_{i,1}, T_{i,2}, T_{i,3}, T_{i,4}, x_i} p_1^{T_{i,1}} p_2^{T_{i,2}} p_3^{T_{i,3}} p_3^{T_{i,4}} p_4^{T_{i,4}} p_5^{x_i} \times$$

$$\prod_{i=m+1}^{n} \binom{n_i}{T_{i,1}, \cdots, T_{i,5}} p_1^{T_{i,1}} p_2^{T_{i,2}} p_3^{T_{i,3}} p_3^{T_{i,4}} p_4^{T_{i,4}} p_5^{T_{i,5}}$$

We must marginalize x_i out to obtain the incomplete likelihood. The key step is to recall the negative binomial distribution sums to one as in the following:

$$\sum_{k=0}^{\infty} \binom{k+r-1}{k} (1-p)^r p^k = 1$$

which gives the useful equation

$$\sum_{k=0}^{\infty} \binom{k+r-1}{k} p^k = \frac{1}{(1-p)^r}$$

Focusing on the key part of the summation to marginalize and using the above equation, we obtain the following:

$$\sum_{x_i=0}^{\infty} \binom{n_i + x_i}{x_i} p_5^{x_i} = \frac{1}{(1 - p_5)^{n_i + 1}}$$

This now gives the incomplete likelihood:

$$L(\mathbf{T}) = \prod_{i=1}^{m} \frac{n_i!}{T_{i1}! T_{i2}! T_{i3}! T_{i4}!} \frac{p_1^{T_{i,1}} p_2^{T_{i,2}} p_3^{T_{i,3}} p_3^{T_{i,4}} p_4^{T_{i,4}}}{(1 - p_5)^{n_i + 1}} \times$$

$$\prod_{i=m+1}^{n} \binom{n_i}{T_{i,1}, T_{i,2}, T_{i,3}, T_{i,4}, T_{i,5}} p_1^{T_{i,1}} p_2^{T_{i,2}} p_3^{T_{i,3}} p_3^{T_{i,4}} p_4^{T_{i,4}} p_5^{T_{i,5}}$$

To set up the expectation step, we have to find the distribution of the hidden variable given the visible variable by dividing $L(\mathbf{x}, \mathbf{T})$ by $L(\mathbf{T})$ to obtain the following:

$$\mathbb{P}(\mathbf{x}|\mathbf{T}) = \prod_{i=1}^{m} \binom{n_i + x_i}{n_i} \pi_5^{x_i} (1 - \pi_5)^{n_i+1}$$

which is a product of m negative binomial distributions and π_5 is the current value for p_5. With all that established, we are ready to compute the lower bound for the expectation step:

$$\mathbb{E}_{x_i \sim \mathbb{P}(\mathbf{x_i}|\mathbf{T})} (\log L(\mathbf{x}, \mathbf{T}))$$

The logarithm of the complete likelihood is the following:

$$\log L(\mathbf{x}, \mathbf{T}) = \sum_{i=1}^{m} \log \left\{ \binom{n_i + x_i}{T_{i,1}, T_{i,2}, T_{i,3}, T_{i,4}, x_i} p_1^{T_{i,1}} p_2^{T_{i,2}} p_3^{T_{i,3}} p_3^{T_{i,4}} p_4^{T_{i,4}} p_5^{x_i} \right\} +$$

$$\sum_{i=m+1}^{n} \log \left\{ \binom{n_i}{T_{i,1}, T_{i,2}, T_{i,3}, T_{i,4}, T_{i,5}} p_1^{T_{i,1}} p_2^{T_{i,2}} p_3^{T_{i,3}} p_3^{T_{i,4}} p_4^{T_{i,4}} p_5^{T_{i,5}} \right\} =: A + B$$

$$(3.18)$$

Separating out the terms with and without x_i, we have

$$A = \sum_{i=1}^{m} \log \left\{ \frac{(n_i + x_i)!}{T_{i,1}! T_{i,2}! T_{i,3}! T_{i,4}! x_i!} p_1^{T_{i,1}} p_2^{T_{i,2}} p_3^{T_{i,3}} p_3^{T_{i,4}} p_4^{T_{i,4}} p_5^{x_i} \right\}$$

$$= \sum_{i=1}^{m} \log \left\{ \frac{(n_i + x_i)!}{T_{i,1}! T_{i,2}! T_{i,3}! T_{i,4}! x_i!} \right\} + \log \left\{ p_1^{T_{i,1}} p_2^{T_{i,2}} p_3^{T_{i,3}} p_3^{T_{i,4}} p_4^{T_{i,4}} p_5^{x_i} \right\}$$

$$= \sum_{i=1}^{m} \log \left\{ (n_i + x_i)!/x_i! \right\} - \log \left\{ T_{i,1}! T_{i,2}! T_{i,3}! T_{i,4}! \right\} + x_i \log(p_5)$$

$$+ \log \left\{ p_1^{T_{i,1}} p_2^{T_{i,2}} p_3^{T_{i,3}} p_3^{T_{i,4}} p_4^{T_{i,4}} \right\}$$

Breaking out terms with x_i, we have the following expectation:

$$\mathbb{E}_{x_i \sim \mathbb{P}(\mathbf{x_i}|\mathbf{T})} \{ x_i \log(p_5) \} = \frac{\pi_5 (n_i + 1) \log(p_5)}{1 - \pi_5}$$

Furthermore,

$$\mathbb{E}_{x_i \sim \mathbb{P}(\mathbf{x_i}|\mathbf{T})}\left\{\log\left(\frac{(n_i + x_i)!}{x_i!}\right)\right\} = \mathbb{E}_{x_i \sim \mathbb{P}(\mathbf{x_i}|\mathbf{T})}\left\{\log\left((n_i + x_i) \times \cdots (x_i + 1)\right)\right\}$$

$$= \mathbb{E}_{x_i \sim \mathbb{P}(\mathbf{x_i}|\mathbf{T})}\left(\sum_{k=0}^{n_i-1}(n_i + x_i - k)\right)$$

$$= \frac{(\pi_5 + 1)\, n_i\, (n_i + 1)}{2\,(1 - \pi_5)}$$

Assembling this again gives

$$\mathbb{E}_{x_i \sim \mathbb{P}(\mathbf{x_i}|\mathbf{T})}(A) = \sum_{i=1}^{m} \frac{(\pi_5 + 1)\, n_i\, (n_i + 1)}{2\,(1 - \pi_5)} - \log\left\{T_{i,1}! T_{i,2}! T_{i,3}! T_{i,4}!\right\} + \frac{\pi_5\,(n_i + 1)\log(p_5)}{1 - \pi_5}$$

$$+ \log\left\{p_1^{T_{i,1}} p_2^{T_{i,2}} p_3^{T_{i,3}} p_3^{T_{i,4}} p_4^{T_{i,4}}\right\}$$

Thus, putting everything together for the expected log-likelihood gives the following objective that we have to maximize over p_i:

$$\mathbb{E}_{x_i \sim \mathbb{P}(\mathbf{x_i}|\mathbf{T})}(A + B) = \sum_{i=1}^{m} \frac{(\pi_5 + 1)\, n_i\, (n_i + 1)}{2\,(1 - \pi_5)} - \log\left\{T_{i,1}! T_{i,2}! T_{i,3}! T_{i,4}!\right\}\frac{\pi_5\,(n_i + 1)\log(p_5)}{1 - \pi_5}$$

$$+ \log\left\{p_1^{T_{i,1}} p_2^{T_{i,2}} p_3^{T_{i,3}} p_3^{T_{i,4}} p_4^{T_{i,4}}\right\}$$

$$+ \sum_{i=m+1}^{n} \log\left\{\binom{n_i}{T_{i,1}, T_{i,2}, T_{i,3}, T_{i,4}, T_{i,5}} p_1^{T_{i,1}} p_2^{T_{i,2}} p_3^{T_{i,3}} p_3^{T_{i,4}} p_4^{T_{i,4}} p_5^{T_{i,5}}\right\} =: J(\mathbf{p})$$

$$(3.19)$$

Also, we have to enforce the constraint that $\sum_{i=1}^{5} p_i = 1$ in the optimization. Using calculus to take the partial derivatives of the objective. Here is the partial derivative for p_1.

$$\partial_{p_1} J = \frac{-1}{(1 - p_1 - p_2 - p_3 - p_4)} \frac{\pi_5}{1 - \pi_5} \sum_{i=1}^{m}(n_i + 1)$$

$$+ \frac{1}{p_1}\sum_{i=1}^{m} T_{i,1} + \frac{1}{p_1}\sum_{i=m+1}^{n} T_{i,1}$$

$$- \frac{1}{(1 - p_1 - p_2 - p_3 - p_4)}\sum_{i=m+1}^{n} T_{i,5} = 0 \qquad (3.20)$$

With the following definitions:

$$n_m = \sum_{i=1}^{m} n_i$$

$$c_j = \frac{\sum_{i=1}^{n} T_{i,j}}{\frac{\pi_5 (n_m + m)}{1 - \pi_5} + \sum_{i=m+1}^{n} T_{i,5}}$$

In general, solving this for p_i gives the following iteration:

$$p_1^{(k+1)} = \frac{c_1^{(k)} (1 - p_2^{(k)} - p_3^{(k)} - p_4^{(k)})}{c_1^{(k)} + 1}$$

$$p_2^{(k+1)} = \frac{c_2^{(k)} (1 - p_1^{(k)} - p_3^{(k)} - p_4^{(k)})}{c_2^{(k)} + 1}$$

$$p_3^{(k+1)} = \frac{c_3^{(k)} (1 - p_1^{(k)} - p_2^{(k)} - p_4^{(k)})}{c_3^{(k)} + 1}$$

$$p_4^{(k+1)} = \frac{c_4^{(k)} (1 - p_1^{(k)} - p_2^{(k)} - p_3^{(k)})}{c_4^{(k)} + 1}$$

$$p_5^{(k+1)} = (1 - p_1^{(k+1)} - p_2^{(k+1)} - p_3^{(k+1)} - p_4^{(k+1)})$$

The iteration starts with the initial condition π_5 and then updates the c_j terms and then the p_5 term.

```
>>> df  # NaN means no data available for Plan 5
Plan      1  2  3  4     5
Region
0         2  4  2  2    NaN
1         4  4  1  1    NaN
2         2  1  5  2    NaN
3         3  4  0  3    NaN
4         1  2  6  1    NaN
5         3  2  1  4    NaN
6         3  1  2  4    NaN
7         0  3  4  3    NaN
8         3  3  2  2    NaN
9         2  3  3  2    NaN
10        2  3  3  2    NaN
11        2  4  2  2    NaN
12        2  2  3  3    NaN
13        1  2  3  4    NaN
14        5  1  2  2    NaN
15        4  0  4  0   12.0
16        5  0  1  2   12.0
17        3  2  5  5    5.0
```

```
18      4  5  3  3    5.0
19      0  2  3  3   12.0
```

Note that only in the last few rows are there values for the fifth plan.

```
>>> m, n, p5 = 15, 20, 0.5
>>> p1 = p2 = p3 = p4 = (1-p5)/4
>>> W = df.sum(axis=0)
>>> nm = df.sum(axis=1)[:m].sum()
>>> c = W/(p5*(nm+m)/(1-p5)+df.loc[m:,5].sum())
```

The next block shows the iteration:

```
>>> from collections import defaultdict
>>> out = defaultdict(list)
>>> for i in range(1,12):
...         p1=c[1]/(1+c[1])*(1-p2-p3-p4)
...         p2=c[2]/(1+c[2])*(1-p1-p3-p4)
...         p3=c[3]/(1+c[3])*(1-p1-p2-p4)
...         p4=c[4]/(1+c[4])*(1-p1-p2-p3)
...         p5=1-p1-p2-p3-p4
...         c=W/(p5*(nm+m)/(1-p5)+df.loc[(m):,5].sum())
...         print('p1=%0.3f, p2=%0.3f, p3=%0.3f, p4=%0.3f,
...         p5=%0.3f'%(p1,p2,p3,p4,p5))
...         out[i].extend([p1,p2,p3,p4,p5])
...
p1=0.122, p2=0.116, p3=0.132, p4=0.121, p5=0.509
p1=0.120, p2=0.114, p3=0.130, p4=0.119, p5=0.517
p1=0.119, p2=0.112, p3=0.129, p4=0.117, p5=0.523
p1=0.118, p2=0.111, p3=0.128, p4=0.116, p5=0.527
p1=0.117, p2=0.110, p3=0.127, p4=0.115, p5=0.531
p1=0.116, p2=0.110, p3=0.126, p4=0.115, p5=0.533
p1=0.116, p2=0.109, p3=0.125, p4=0.114, p5=0.535
p1=0.116, p2=0.109, p3=0.125, p4=0.114, p5=0.537
p1=0.115, p2=0.109, p3=0.125, p4=0.113, p5=0.538
p1=0.115, p2=0.109, p3=0.124, p4=0.113, p5=0.538
p1=0.115, p2=0.108, p3=0.124, p4=0.113, p5=0.539
```

This is the estimate for p_5 just using the last rows for which data are available.

```
>>> (df.loc[m:,5]/df.loc[m:,:].sum(axis=1)).mean()
0.4599999999999996
```

Compare this to the estimate from EM for p_5 below (Fig. 3.50).

```
>>> print('p1=%0.3f, p2=%0.3f, p3=%0.3f, p4=%0.3f,
...       p5=%0.3f'%(p1,p2,p3,p4,p5))
p1=0.115, p2=0.108, p3=0.124, p4=0.113, p5=0.539
```

If we perform the exact same analysis but starting with p5=0.9, we get the result in Fig. 3.51 with the final estimate below.

```
>>> print('p1=%0.3f, p2=%0.3f, p3=%0.3f, p4=%0.3f,
...       p5=%0.3f'%(p1,p2,p3,p4,p5))
p1=0.091, p2=0.085, p3=0.097, p4=0.087, p5=0.640
```

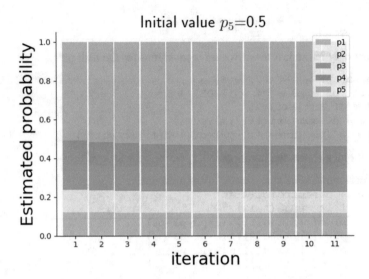

Fig. 3.50 EM iterations for multinomial probabilities starting with $p_5 = 0.5$

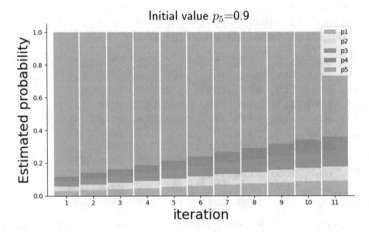

Fig. 3.51 EM iterations for multinomial probabilities starting with $p_5 = 0.9$. Compare to Fig. 3.50

These are different because EM is sensitive to initial values.

3.16   Survey Sampling

The purpose of survey sampling is to estimate statistical quantities based on a n-sized subsample of a larger N population. For example, we want to estimate the number of cigarette smokers in a given city. We could telephone everybody in the city and ask if they smoke, but that would be expensive and laborious. Instead, suppose we go to a busy downtown street corner and ask people if they smoke. Then, based on the responses, we could easily compute the proportion of smokers to non-smokers, but then how does this result reflect on the entire population? Furthermore, what if the streetcorner we chose just happens to be in front of a health-food store? How does the health food store effect the number of potential smokers who answered your survey? That is, more generally, how can other ambient information influence the final estimate of the survey across the entire city? These are the questions survey sampling is designed to address.

The population under survey consists of individual *units*. Each unit u_i has an attribute (or set of attributes) denoted by Y_i. The goal is to statistically characterize Y_i over the entire finite population N using a subsample of size $n < N$, denoted $\mathcal{S}_n$. The *sampling design* is the probability of any particular $\mathcal{S}_n$ and this probability ultimately determines the statistical properties that characterize our conclusions about $\mathcal{S}_n$ in terms of the larger N population. The *inclusion probability* under the sampling design is the probability that any particular unit appears in $\mathcal{S}_n$, $\mathbb{P}\left(u_i = U_j\right)$. There are two primary categories of sampling: *replacement* and *nonreplacement*. Under replacement, a particular unit can appear multiple times in $\mathcal{S}_n$ which cannot happen with nonreplacement.

The inclusion probabilities are key to analysis. For example, we have

$$\bar{y}_s = \frac{1}{n} \sum_{i=1}^{n} y_i$$

where $u_i \in \mathcal{S}_n$. Because of linearity of the expectation and sampling with replacement means $\mathbb{P}(y_i = Y_j) = \frac{1}{N}$, we also have $\mathbb{E}(y_i) = \overline{Y}$ and therefore

$$\mathbb{E}(\bar{y}_s) = \overline{Y}$$

To explore this, let us consider a small simulation of heights where $N = 100$ and $n = 10$.

```
>>> import numpy as np
>>> from scipy import stats
>>> heights = stats.norm(5,1).rvs(100)
```

Now, let us select a random sample *with* replacement:

```
>>> sample = np.random.choice(heights,10)
>>> sample.mean() # ybar above
4.795029626941257
```

If we keep randomly sampling like this, we obtain the mean over all sampled sets:

```
>>> sample = np.random.choice(heights,(10,100))
>>> sample.mean(axis=0).mean()
5.0721303754908185
```

The `axis=0` means we first took the mean over the `10` samples for each of the `100` sample selections. Compare this to the result using all N samples:

```
>>> heights.mean()
5.095579243655205
```

There is naturally some difference here. To quantify this, we compute the variance of $\bar{y}_s$:

$$\mathbb{V}(\bar{y}_s) = \frac{1}{n^2} \sum_{i=1}^{n} \mathbb{V}(y_i) = \frac{\sigma_Y^2}{n}$$

To see this, note that

$$\sigma_Y^2 := \frac{1}{N} \sum_{i=1}^{N} (Y_i - \overline{Y})^2$$

$$= \frac{1}{N} \sum_{i=1}^{N} \left(Y_i^2 - 2Y_i\overline{Y} + \overline{Y}^2 \right)$$

$$= \overline{Y}^2 + \frac{1}{N} \sum_{i=1}^{N} \left(Y_i^2 - 2Y_i\overline{Y} \right)$$

$$= \overline{Y}^2 + \frac{1}{N} \sum_{i=1}^{N} Y_i^2 - 2\frac{1}{N}\overline{Y} \sum_{i=1}^{N} Y_i$$

$$= \overline{Y}^2 + \frac{1}{N} \sum_{i=1}^{N} Y_i^2 - 2\overline{Y}^2$$

$$\sigma_Y^2 = -\overline{Y}^2 + \frac{1}{N} \sum_{i=1}^{N} Y_i^2$$

From which we can identify the following term:

$$\mathbb{E}(y_s^2) = \sum_{i=1}^{N} \frac{1}{N} Y_i^2 = \sigma_Y^2 + \overline{Y}^2$$

Then, we need $\mathbb{E}\left(\bar{y}_s^2\right)$,

$$\mathbb{E}\left(\bar{y}_s^2\right) = \mathbb{E}\left(\frac{1}{n^2} \left(\sum_{i=1}^{n} y_i \right)^2 \right) = \frac{1}{n^2} \left(\sum_{i=1}^{n} \mathbb{E}\left(y_i^2\right) + \sum_{i=1}^{n} \sum_{j\neq i}^{n} \mathbb{E}\left(y_i\right) \mathbb{E}\left(y_j\right) \right)$$

$$= \frac{1}{n^2} \left(\sum_{i=1}^{n} \mathbb{E}\left(y_i^2\right) + \sum_{i=1}^{n} \sum_{j \neq i}^{n} \mathbb{E}\left(y_i\right) \mathbb{E}\left(y_j\right) \right)$$

$$= \frac{1}{n} \left(\sigma_Y^2 + \overline{Y}^2 \right) + \frac{n^2 - n}{n^2} \overline{Y}^2$$

$$= \frac{1}{n} \left(\sigma_Y^2 + \overline{Y}^2 \right) + \overline{Y}^2 - \frac{1}{n} \overline{Y}^2$$

$$= \frac{1}{n} \sigma_Y^2 + \overline{Y}^2$$

Finally we have our result

$$\mathbb{V}(\bar{y}_s) = \mathbb{E}(\bar{y}_s^2) - \overline{Y}^2 = \frac{\sigma_Y^2}{n} \tag{3.21}$$

which means we can reduce the variance by taking larger n. Of course, we don't know σ_Y^2 ahead of time, so we have to estimate it with the sample variance over $\mathcal{S}_n$:

$$s_y^2 = \frac{1}{n-1} \sum_{i=1}^{n} (y_i - \bar{y}_s)^2$$

To prove this is an unbiased estimator, we can use the conditional variance formula,

$$\mathbb{V}(A) = \mathbb{V}_B(\mathbb{E}(A|B)) + \mathbb{E}_B(\mathbb{V}(A|B))$$

where $A = Y$ and $B = \mathcal{S}_n$. Then using our previous result,

$$\mathbb{V}_{\mathcal{S}_n}(\mathbb{E}(Y|\mathcal{S}_n)) = \mathbb{V}_{\mathcal{S}_n}(\bar{y}_s) = \frac{\sigma_Y^2}{n}$$

and

$$\mathbb{E}_{\mathcal{S}_n}(\mathbb{V}(y|\mathcal{S}_n)) = \mathbb{E}_{\mathcal{S}_n} \left(\frac{1}{n} \left(\sum_{i=1}^{n} (y_i - \bar{y}_s)^2 \right) \right) = \frac{n-1}{n} \mathbb{E}(s_y^2)$$

Putting it together in the conditional variance formula gives the following:

$$\sigma_Y^2 = \frac{\sigma_Y^2}{n} + \frac{n-1}{n} \mathbb{E}(s_y^2)$$

which gives,

$$\mathbb{E}(s_y^2) = \sigma_Y^2$$

which says that the expectation of the adjusted population variance over the sample S_n is the variance over the entire population. Because we don't know the σ_Y^2 ahead of time, we have to estimate it from the sample

$$\widehat{\mathbb{V}(\bar{y}_s)} = \frac{s_y^2}{n}$$

which is justified by the prior result.

Sampling Without Replacement Everything we have derived so far as for the case of sampling *with* replacement. For sampling *without* replacement, all $\binom{N}{n}$ outcomes are equally probable so we have

$$\mathbb{P}(y_i \in S_n) = \frac{\binom{N-1}{n-1}}{\binom{N}{n}} = \frac{n}{N}$$

So we have

$$\mathbb{E}(y_i) = \sum_{j=1}^{N} \frac{1}{N} Y_j = \overline{Y}$$

We also have

$$\mathbb{E}(y_i^2) = \frac{1}{N} \sum_{j=1}^{N} Y_j^2$$

The nonreplacement makes the covariance trickier than for replacement case. For simplicity and without loss of generality, we can assume $\overline{Y} = 0$. We have the following probability for both Y_k and Y_l to be in the n-sample:

$$\mathbb{P}(y_j \in S_n \bigwedge y_i \in S_n) = \frac{\binom{N-2}{n-2}}{\binom{N}{n}} = \frac{n(n-1)}{N(N-1)}$$

Then, given particular y_i and y_j, we have the following probability:

$$\mathbb{P}(y_j = Y_k \bigwedge y_i = Y_l) = \frac{1}{N(N-1)}$$

So, for the expectation, we have the following:

$$\mathbb{E}(y_i y_j) = \sum_{l=1}^{N} \sum_{k \neq l}^{N} Y_k Y_l \mathbb{P}(y_j = Y_k \bigwedge y_i = Y_l)$$

By substitution for the probability of the pair, we obtain

$$
\mathbb{E}(y_i\, y_j) = \frac{1}{N(N-1)} \sum_{l=1}^{N} \sum_{k\neq l}^{N} Y_k Y_l = \frac{1}{N(N-1)} \sum_{l=1}^{N} Y_l \sum_{k\neq l}^{N} Y_k
$$

$$
= \frac{1}{N(N-1)} \sum_{l=1}^{N} Y_l \left(N\overline{Y} - Y_l \right)
$$

$$
= \frac{1}{N(N-1)} \sum_{l=1}^{N} Y_l \left(0 - Y_l \right)
$$

$$
= \frac{-1}{N(N-1)} \sum_{l=1}^{N} Y_l^2 = \frac{-\sigma_Y^2}{N-1}
$$

where we used the $\overline{Y} = 0$ assumption on the penultimate line above.

With the individual y_i terms settled, we can get to the main event of computing the mean of $\overline{y}_s$. The mean follows directly from linearity and the above result for $\mathbb{E}(y_i)$:

$$
\mathbb{E}\left(\overline{y}_s\right) = \mathbb{E}\left(\frac{1}{n}\sum_{i=1}^{n} y_i\right) = \frac{1}{n}\sum_{i=1}^{n} \mathbb{E}(y_i) = \overline{Y}
$$

For the variance, using the $\overline{Y} = 0$ assumption, we have

$$
\mathbb{V}\left(\overline{y}_s\right) = \mathbb{E}\left(\overline{y}_s^2\right)
$$

Now we can write out the $\mathbb{E}\left(\overline{y}_s^2\right)$ term as

$$
\mathbb{E}\left[\left(\frac{1}{n}\sum_{i=1}^{n} y_i\right)^2\right] = \frac{1}{n^2}\sum_{i=1}^{n} \mathbb{E}\left(y_i^2\right) + \frac{1}{n^2}\sum_{i=1}^{n}\sum_{j\neq i}^{n} \mathbb{E}\left(y_i\, y_j\right)
$$

$$
= \frac{\sigma_Y^2}{n} + \frac{\sigma_Y^2}{n^2}\sum_{i=1}^{n}\sum_{j\neq i}^{n} \frac{-1}{N-1}
$$

$$
= \frac{\sigma_Y^2}{n} - \frac{\sigma_Y^2}{n}\frac{(n-1)}{N-1}
$$

$$
\mathbb{V}\left(\overline{y}_s\right) = \frac{\sigma_Y^2}{n}\frac{N-n}{N-1}
$$

Recall that the adjusted population variance is

$$S_Y^2 = \frac{1}{N-1} \sum_{i=1}^{N} (Y_i - \bar{Y})^2 = \sigma_Y^2 \frac{N}{N-1}$$

So, by substituting this into the prior equation, we obtain

$$\mathbb{V}(\bar{y}_s) = \left(\frac{1}{n} - \frac{1}{N}\right) S_Y^2$$

We can once again use the conditional variance formula:

$$\sigma_Y^2 = \left(1 - \frac{1}{N}\right) S_Y^2 = \mathbb{V}_{\mathcal{S}_n} \left(\mathbb{E}(y \mid S_n)\right) + \mathbb{E}_{\mathcal{S}_n} \left(\mathbb{V}(y \mid S_n)\right)$$

$$\left(1 - \frac{1}{N}\right) S_Y^2 = \mathbb{V}_{\mathcal{S}_n} \left(\bar{y}_{\mathcal{S}_n}\right) + \mathbb{E}_{\mathcal{S}_n} \left(\frac{1}{n} \sum_{i=1}^{n} (y_i - \bar{y}_s)^2\right)$$

$$\left(1 - \frac{1}{N}\right) S_Y^2 = \left(\frac{1}{n} - \frac{1}{N}\right) S_Y^2 + \mathbb{E}_{\mathcal{S}_n} \left(\frac{1}{n} \sum_{i=1}^{n} (y_i - \bar{y}_s)^2\right)$$

$$\left(1 - \frac{1}{N}\right) S_Y^{2'} = \left(\frac{1}{n} - \frac{1}{N}\right) S_Y^2 + \frac{n-1}{n} \mathbb{E}_{\mathcal{S}_n} \left(s_y^2\right)$$

By clearing the terms above, we can identify

$$\mathbb{E}_{\mathcal{S}_n} (s_y^2) = S_Y^2$$

The punchline here is that the expectation of the adjusted population variance of the n-sample is the adjusted population variance for the full population. As with the sampling with replacement case, we still need an estimator for the variance of $\mathbb{V}(\hat{y}_s)$

$$\widehat{\mathbb{V}(\bar{y}_s)} = \frac{s_y^2}{n} \left(1 - \frac{n}{N}\right)$$

which is justified by the prior result. The factor $\left(1 - \frac{n}{N}\right)$ is usually called the *finite population correction*.

Example: Estimating Variance Using Finite Population Correction To make the above calculations concrete, consider the following data:

```
>>> data = np.array([-1]*100 + [0]*100 +[1]*100)
>>> N, n = len(data), 20
```

Here is a random sample without replacement:

```
>>> sample = np.random.choice(data,n,replace=False)
>>> ys = np.mean(sample)
>>> ys
0.0
```

Let us estimate the $\mathbb{V}(\bar{y}_s)$ by drawing multiple samples:

```
>>> x = [np.mean(np.random.choice(data,n,replace=False)) for
         i in range(1000)]
>>> np.var(x), (1-n/N)*np.var(sample)/n
(0.033226977500000004, 0.023333333333333334)
```

The last line shows the estimated variance by using the simulation compared to using the finite population correction factor on a *single* sample. These two numbers are pretty close, so we can estimate the variance of the sample mean $\bar{y}_s$ using single sample which, practically speaking, is all we have.

Example: Sorties Suppose we have a 100-mile-wide region over which we fly sorties exactly north-south. For each flight, we record the number of deer observed alongside the mile-wide view of the aircraft. We have a sample (without replacement) of 20 flights. We want to estimate the number of deer in this region. For this, the *unit* is the individual flight. The *feature y* is the number of deer observed, and $N = 100$ is the total number of possible units available (number of 1-mile wide flights). Given the following data:

```
>>> y = [9,12,1,1,10,9,9,12,13,12,6,4,12,11,1,5,2,12,12,4]
```

According to our results, we have the following sample mean and variance, y_s and s_y^2:

```
>>> ys = np.mean(y)
>>> sy2 = np.var(y,ddof=1)
>>> ys, sy2
(7.85, 19.18684210526316)
```

To estimate the variance of the sample mean, we compute

```
>>> n, N = 20, 100
>>> var_ys = (1/n - 1/N)*sy2
>>> var_ys
0.7674736842105264
```

To estimate the total deer population using our survey, we can compute the following:

```
>>> total_pop_ = ys * N
>>> total_pop_
785.0
```

With corresponding estimated variance, we can write out the complete result:

```
>>> var_total_pop_  =  N**2 * var_ys
>>> print(f'total = {total_pop_} with estimated standard
          error = {np.sqrt(var_total_pop_):3.2f}')
total = 785.0 with estimated standard error = 87.61
```

Note the subtle choice of what constitutes a unit for this example and how N was therefore known a priori.

3.16.1 Unequal Sampling with Replacement for Weighted Totals

So far, we have only considered the *unweighted* case of estimating

$$\overline{Y} = \frac{1}{N} \sum_{i=1}^{N} Y_i$$

but what about the more general *weighted* case

$$\overline{Y} = \sum_{i=1}^{N} p_i Y_i$$

with $\sum_{i=1}^{n} p_i = 1$ using a $\mathcal{S}_n$ sample *with* replacement? Sampling with unequal probability p_i facilitates this more general weighted sum. To see this, consider the following:

$$\mathbb{E}(y_i) = \sum_{j=1}^{N} \mathbb{P}(y_i = Y_j) Y_j$$

If we take the inclusion probability as the following:

$$p_j := \mathbb{P}(y_i = Y_j)$$

then, the expectation over the unweighted sample $\mathcal{S}_n$ becomes the following:

$$\mathbb{E}(\overline{y}_s) = \frac{1}{n} \sum_{i=1}^{n} \mathbb{E}(y_i) = \frac{1}{n} \sum_{i=1}^{n} \sum_{j=1}^{N} p_j Y_j = \overline{Y}$$

Because all of the corresponding derivations for the with replacement case follow through, we have the following:

$$\widehat{\mathbb{V}(\bar{y}_s)} = \frac{s_y^2}{n}$$

Thus, by using unequal sampling, we can estimated weighted averages as $\overline{Y}$. Conversely, this means that unequal sampling leads to estimating $\overline{Y}$ as a weighted sum.

3.16.2 Unequal Sampling for Unweighted Totals

We want unweighted total but are forced to using unequal sampling with replacement for practical reasons. In other words, we want to estimate $Y_T = \sum_{i=1}^N Y_i$ using an unequal probability sampling with the following probability:

$$p_j = \mathbb{P}\left(y_i = Y_j\right)$$

This is reducible to the previous case by substituting $Z_i = Y_i/p_i$. Thus, for the unequal probability sample, S_n, we have the usual estimator:

$$\bar{z}_s = \frac{1}{n}\sum_{i=1}^n z_i = \frac{1}{n}\sum_{i=1}^n \frac{y_i}{p_i}$$

The corresponding estimator for the variance is the same as before

$$\widehat{\mathbb{V}\left(\bar{z}_s\right)} = \frac{s_z^2}{n}$$

where

$$s_z^2 = \frac{1}{n-1}\sum_{i=1}^n \left(\frac{y_i}{p_i} - \bar{z}_s\right)^2$$

which is known as the Hansen-Hurwitz estimator.

Example: Unequal Sampling for Unweighted Total Let us generate some simulated data:

```
>>> N, n = 100, 50
>>> Y = np.arange(1,N+1)
>>> pY = np.array([1,]*50 + [2,]*50) # setup weights
>>> pY = pY/pY.sum() # normalize
```

We want to estimate the sum of the elements in Y:

```
>>> Y.sum() # want to estimate this with sample mean
5050
```

By re-weighting and defining the Z_i terms, we can use unequal probability sampling:

```
>>> Z = Y/pY # re-weight Y-values
>>> sample = np.random.choice(Z,size=n,p=pY)
>>> sample.mean(), Y.sum() # pretty close
(5206.5, 5050)
```

And the for the variance,

```
>>> sample.var()/n # Hansen-Hurwitz estimate
49387.905
```

Because we are using simulated data, we can estimate the variance using simulation

```
>>> nsamples = 5_000
>>> np.random.choice(Z, size=(n,nsamples), p=pY).mean(axis=0).var()
63367.63790796
```

which is pretty close to the Hansen-Hurwitz result.

3.16.3   Unequal Sampling Without Replacement

As in the previous section, we want to estimate the total Y_T, but here we sample without replacement. Consider the following Horvitz-Thompson estimator:

$$Y_{HT} = \sum_{i=1}^{n} \frac{y_i}{\pi_i}$$

where $\pi_i = \mathbb{P}(Y_i \in \mathcal{S}_n)$. If we define $a_j = 1$ if $Y_j \in \mathcal{S}_n$ with probability π_j and zero otherwise, then we can write this over all N of the elements of Y_j as follows:

$$Y_{HT} = \sum_{j=1}^{N} \frac{Y_j}{\pi_j} a_j$$

Note that

$$\mathbb{E}(a_j) = \pi_j$$

with corresponding variance,

$$\mathbb{V}(a_j) = \pi_j(1 - \pi_j)$$

With these definition, we can compute the following:

$$\mathbb{E}\left(Y_{HT}\right) = \sum_{j=1}^{N} \mathbb{E}\left(a_j\right) \frac{Y_j}{\pi_j} = \sum_{j=1}^{N} Y_j = Y_T$$

which means that Y_{HT} is an unbiased estimator of Y_T. Let us define $\pi_{ij} = \mathbb{P}\left(Y_i \in \mathcal{S}_n \wedge Y_j \in \mathcal{S}_n\right)$. We want to compute the variance using

$$\mathbb{V}\left(Y_{HT}\right) = \mathbb{E}\left(Y_{HT}^2\right) - \mathbb{E}\left(Y_{HT}\right)^2$$

To get the second term, we can write the mean as the following:

$$\mathbb{E}\left(Y_{HT}\right) = Y_T = \sum_{i=1}^{N} \pi_i \left(\frac{Y_i}{\pi_i}\right)$$

and then square it as shown:

$$\mathbb{E}\left(Y_{HT}\right)^2 = \left(\sum_{i=1}^{N} \pi_i \left(\frac{Y_i}{\pi_i}\right)\right)^2 = \sum_{i=1}^{N} \pi_i^2 \left(\frac{Y_i}{\pi_i}\right)^2 + \sum_{i=1}^{N}\sum_{j\neq i}^{N} \pi_i \pi_j \left(\frac{Y_i}{\pi_i}\right)\left(\frac{Y_j}{\pi_j}\right)$$

Writing out the first term,

$$\mathbb{E}\left(Y_{HT}^2\right) = \mathbb{E}\left[\left(\sum_{i=1}^{N} a_i \frac{Y_i}{\pi_i}\right)^2\right] = \sum_{i=1}^{N} \pi_i \left(\frac{Y_i}{\pi_i}\right)^2 + \sum_{i=1}^{N}\sum_{j\neq i}^{N} \pi_{ij} \left(\frac{Y_i}{\pi_i}\right)\left(\frac{Y_j}{\pi_j}\right)$$

Combining the two terms gives the variance:

$$\mathbb{V}\left(Y_{HT}\right) = \sum_{i=1}^{N}\left[\pi_i \left(\frac{Y_i}{\pi_i}\right)^2 - \pi_i^2 \left(\frac{Y_i}{\pi_i}\right)^2\right] + \sum_{i=1}^{N}\sum_{j\neq i}^{N}\left[\pi_{ij}\left(\frac{Y_iY_j}{\pi_i\pi_j}\right) - \pi_i\pi_j\left(\frac{Y_iY_j}{\pi_i\pi_j}\right)\right]$$

$$= \sum_{i=1}^{N}\left(\pi_i - \pi_i^2\right)\left(\frac{Y_i}{\pi_i}\right)^2 + \sum_{i=1}^{N}\sum_{j\neq i}^{N}\left(\pi_{ij} - \pi_i\pi_j\right)\left(\frac{Y_iY_j}{\pi_i\pi_j}\right)$$

To combine the result, we add and then subtract the diagonal terms to get rid of the $j \neq i$ in the sum:

$$\mathbb{V}\left(Y_{HT}\right) = \sum_{i=1}^{N}\left(\pi_i - \pi_i^2\right)\left(\frac{Y_i}{\pi_i}\right)^2 + \sum_{i=1}^{N}\left(\sum_{j=1}^{N}\left(\pi_{ij} - \pi_i\pi_j\right)\left(\frac{Y_iY_j}{\pi_i\pi_j}\right)\right) -$$

$$\sum_{i=1}^{N} \left(\pi_i - \pi_i^2\right) \left(\frac{Y_i}{\pi_i}\right)^2$$

$$= \sum_{i=1}^{N} \sum_{j=1}^{N} \left(\pi_{ij} - \pi_i \pi_j\right) \left(\frac{Y_i Y_j}{\pi_i \pi_j}\right)$$

To estimate this variance using $\mathcal{S}_n$, we can use the following:

$$\widehat{\mathbb{V}(Y_{HT})} = \sum_{i=1}^{n} \sum_{j=1}^{n} \left(1 - \frac{\pi_i \pi_j}{\pi_{ij}}\right) \frac{y_i y_j}{\pi_i \pi_j}$$

which is attributed to Horvitz and Thompson. Note that there is an alternative *Yates-Grundy* form of this estimate. The variance estimator is unbiased as long as $\pi_{ij} > 0$ but is unstable (especially for small samples) and can become negative in certain cases. Practically speaking, it can be difficult to derive the second-order probabilities π_{ij} which may involve calculating $\binom{n}{2}$ possible pairs in the sample.

3.16.4 *Probability Proportional to Size (PPS) Cluster Sampling*

For practical reasons, it is often more convenient to sample using clusters as opposed to directly sampling individual units. Consider N clusters $C_1, C_2, \ldots, C_N$ of cardinality $X_1, X_2, \ldots, X_N$. Suppose n clusters are selected with replacement with probability proportional to cardinality of each cluster

$$p_j = \frac{X_j}{\sum_{i=1}^{N} X_j} = \frac{X_j}{X_T}$$

where X_T is the total over all clusters and Y_j is the total of y in the jth cluster, C_j, and we want to estimate Y_T. Then, the Hansen-Hurwitz unbiased population estimator is the following:

$$\widehat{Y}_T = \frac{1}{n} \sum_{i=1}^{n} \frac{y_i}{p_i} = \frac{X_T}{n} \sum_{i=1}^{n} \frac{y_i}{x_i}$$

Note that $\frac{y_i}{x_i}$ is the mean of the values of $y_{(k)}$ in the C_i cluster. Recall that the unbiased estimator of the population mean, $\overline{Y}$ is the following:

$$\overline{Y} = \sum_{i=1}^{N} p_j \overline{Y}_j$$

From our previous result, we have

$$\mathbb{V}\left(\hat{Y}_T\right) = \frac{X_T^2}{n} \sum_{j=1}^{N} p_j \left(\overline{Y}_j - \overline{Y}\right)^2$$

with corresponding estimator for the variance

$$\widehat{\mathbb{V}\left(\hat{Y}_T\right)} = \frac{X_T^2}{n} \frac{1}{n-1} \sum_{i=1}^{n} \left(\overline{y}_i - \overline{y}_s\right)^2$$

where $\overline{y}_s$ is the sample mean of the cluster means.

Example: Probability Proportional to Size (PPS) Cluster Sampling Suppose we have four hotels (i.e., clusters) and we want to estimate the total number of pet-friendly hotel rooms in town. The following table shows the number of rooms and number of pet-friendly rooms in each hotel. The Ym_i in the table shows the mean of pet-friendly rooms per hotel, and the p_i is the probability of selecting each hotel which is the proportion of rooms available in that hotel over the total number of available hotel rooms. Thus, each hotel is randomly selected based upon the total number of available rooms there.

```
>>> rd
   num_rooms   num_pets    p_i    Ym_i
0         57         18  0.291   0.316
1         71         31  0.362   0.437
2         13          5  0.066   0.385
3         55         26  0.281   0.473
```

Our survey sample consists of two hotels selected with replacement according to p_i,

```
>>> sample = np.random.choice(rd.Ym_i, 2, p=rd['p_i'])
>>> sample
array([0.31578947, 0.43661972])
>>> rd.num_rooms.sum()*sample.mean(), rd.num_pets.sum()
(73.73610081541882, 80)
```

The last line shows the Hansen-Hurwitz estimator for the number of total pet-friendly rooms compared to the actual number of pet-friendly rooms available. The estimated variance of the total number of pet-friendly rooms available is given below:

```
>>> ((sample - sample.mean())**2).sum()*rd.num_rooms.sum()**2/2
140.2179007484884
```

Because there are few available hotels overall and sampling with replacement means that the same sample can appear more than once, it is possible for the estimated variance to equal zero.

3.16.5　Stratified Random Sampling

The practical motivation behind stratified random sampling is to ensure that no one group dominates the overall estimate without accounting for potentially smaller groups' contribution. For example, if a region is dominated by a few large hospitals, and we want to estimate the number of available beds using PPS methods, then we may have repeated samples of the large hospitals, just because these have the most beds overall. By using stratified random sampling, we can have *large*, *medium*, and *small* groups and then sample each group independently, which reduces the dominating effect of the large hospitals in this estimation. The sampling population is split into separate groups, and sampling is performed separately within each group (i.e., stratum).

Convenient and practically workable strata may already be defined by geographic region, administrative criteria (e.g., voting districts), or some other factor like age or gender. The main disadvantage of stratified sampling is that it requires every enumeration unit to be identified by stratum before sampling.

Instead of using pre-existing and convenient strata, you could create strata by splitting the population with respect to another feature that is thought to be associated with the feature of interest. In our hospital example, we could stratify hospitals with respect to floor space, data which may be easily available from tax records and which should be highly associated with number of beds because beds consume floor space. On the other hand, if we want to estimate the average daily cost of a hospital bed per patient, then the geographic region of the hospital may be more important, supposing that wealthy areas can charge patients more than poor areas.

Let us consider a numerical example using some penguin data from seaborn.

```
>>> import seaborn as sns
>>> df =
        sns.load_dataset('penguins')[['island','body_mass_g',
        'species']]
>>> df.head()
     island  body_mass_g species
0  Torgersen       3750.0  Adelie
1  Torgersen       3800.0  Adelie
2  Torgersen       3250.0  Adelie
3  Torgersen         NaN  Adelie
4  Torgersen       3450.0  Adelie
```

We want to use stratified sampling to estimate the body mass of the penguins by stratifying on the islands they live on. As shown, some islands are more highly populated with penguins than others, and some islands have heavier penguins than others (see Fig. 3.52):

```
>>> df.groupby('island').count()['species']
island
Biscoe    168
Dream     124
```

Fig. 3.52 The heavier
penguins live on Biscoe
island

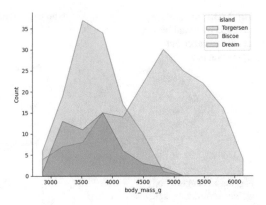

```
Torgersen      52
Name: species, dtype: int64
```

For this example, we want to sample 30 penguins total by stratifying across the islands. In the following, the groupby partitions the dataframe by island, and then the apply function uses the sample method to collect ten samples of body_mass_g in each partition with replacement (replace=True). The next groupby partitions the result by level=0, which corresponds to the island name, and the remainder computes the average weight of the penguins. Because this is a simulation, we can do this 1000 times and collect the result in the stratified variable.

```
>>> # generate samples by stratifying on island
>>> nsamples = 30
>>> stratified =[df.groupby('island')
...                 .apply(lambda i:i.sample(nsamples//3,
...                                    replace=True))['body_mass_g']
...                 .groupby(level=0).sum().sum()/nsamples
...               for i in range(1000)]
```

As a comparison, we generate 1000 random samples that ignore the islands and collect the result in the unstratified variable:

```
>>> # generate unstratified samples ignoring islands
>>> unstratified =[df['body_mass_g'].sample(nsamples,
...                                    replace =True).mean()
...               for i in range(1000)]
```

We next assemble the two sets of samples into a single dataframe, sampled_data,

```
>>> sampled_data=pd.DataFrame(dict(stratified=stratified,
...                                unstratified=unstratified))
```

that we can compare using Seaborn histplot

```
>>> ax = sns.histplot(data=sampled_data,
...                 element='poly',
...                 kde=True)
```

Fig. 3.53 The samples stratified by island have a smaller variance but significant bias

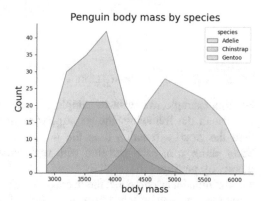

Fig. 3.54 Penguin body weight stratified by species

which shows that the samples stratified by island create estimates of smaller variance but with significant bias. The variance (i.e., spread) of the `unstratified` samples is greater but with less bias (see Fig. 3.53).

To continue with this example, let us redo this exercise but now stratifying on species instead of island. Figure 3.54 shows that the Gentoo penguins are heavier. Note that this `species` stratification variable is closely related to body weight.

Now, we repeat the process as before and generate `1000` simulated estimates and Fig. 3.55. Note that the bias for the stratified estimates of body weight is significantly reduced because destratification variable (`species`) is strongly correlated with the study variable `body_mass_g`.

```
>>> # generate samples by stratifying on species
>>> stratified =[df.groupby("species")
...                 .apply(lambda i:i.sample(nsamples//3,
...                                         replace
        =True))['body_mass_g']
...                 .groupby(level=0).sum().sum()/nsamples
...                 for i in range(1000)]
>>> sampled_data = pd.DataFrame(dict(stratified=stratified,
...                                 unstratified=unstratified))
```

Discussion Survey sampling is a large and mature area of statistics, so here we have just tried to highlight the key issues. Note that we have only considered estimating

Fig. 3.55 The bias for the stratified estimates of body weight is reduced significantly when the stratification variable (species) is strongly correlated with the study variable (body_mass_g). Compare to Fig. 3.53

the mean and total using survey sampling but other statistics such as the median and ratio metrics are also possible, albeit with much more complicated calculations for variance. Recall that the sampling design is the $\mathbb{P}(S_n)$ and the inclusion probability is π_i; these are obviously related, but it turns out that Hanurav's algorithm (1966) established the correspondence between these two sampling considerations because he proved that any sampling design results in a π_i and vice versa. Thus, there is no meaningful differentiation between these two.

We did not discuss whether or not estimators are minimum variance unbiased estimators. In the most formal sense, specific estimators should be paired with specific sampling strategies to ensure minimum variance properties, but most survey sampling strategies are dominated by practical considerations. Thus, the choice of desired metric and the corresponding estimator for that metric are both design parameters for the analyst.

It is typical for sampling results to include weights which reflect the sampling design. For example, $w_i = 1/\pi_i$ for all survey units reported. This makes it straightforward to compute totals and averages by just using these unit weights. It is common to collect partial responses for all or certain elements of a survey. Missing responses require special handling and use methods related to missing data imputation (see Sect. 3.18).

3.17   Log-linear Models

Log-linear models are a generalization of generalized linear models (GLMs) and facilitate regression analyses of contingency tables with complex interactions between the variables while accounting for the structure of the contingency table. Specifically, GLMs require you to select input/output variables to perform the regression and then focus on interpreting the resulting regression coefficients. For example, logistic regression is a special case of log-linear models where a single binary value response variable is a function of the covariates, but log-linear models do not need to choose any one variable as the response to the others. This is important in situations where several factors simultaneously interact and the cause and effect relationship is unknown.

We have already covered the Fisher exact test (see Sect. 3.5.6) and the odds ratio for two-by-two contingency tables in Sect. 3.5.7. Our goal is to generalize this analysis to larger two- and three-dimensional contingency tables. For contingency table analysis, we are primarily concerned with multinomial and Poisson probability sampling models. Poisson models are easier to extend to multidimensional contingency tables and are fundamental to log-linear models.

Unfortunately, as of the time of this writing, there is no established Python module that implements log-linear models, although Statsmodels does consider the two-by-two contingency table. Consequently, we illustrate the main features of this theory using Xarray and Pandas methods. We build on our prior examples of contingency table analysis and show how log-linear models provide more flexibility and extensions for analysis. Log-linear models comprise a vast statistical theory, so our coverage here is by no means comprehensive, and we refer you to references [1, 2, 10, 39, 41, 43, 45], and [51] for deeper study. The main struggle with log-linear models is notational, with different approaches to the topic using differing notations. We have tried to pick a middle road for notation that aligns with the corresponding illustrative Python code. It is easy to get lost in the fog of mathematical symbols, so this section has many worked out numerical examples that you can experiment with to cement your conceptual understanding. We provide intuitive motivation and references for the key mathematical results that are too technical and distracting to be included here.

3.17.1 Poisson and Multinomial Models

Contingency tables are used to summarize categorical data and are naturally characterized by multinomial distributions. Recall that the form of the multinomial distribution is a product of its parameters. The great idea with log-linear models is to use logarithms to expand these distributions and thereby enable a wide range of categorical data models to use with these distributions. It is technically difficult to work with the multinomial distribution in high dimensions, but there is a maximum likelihood equivalence between the multinomial and the Poisson distribution that facilitates log-linear modeling. Suppose we start with the following Poisson log-linear model:

$$\log \mu = \Psi \beta$$

where Ψ is a binary matrix. The Poisson distribution parameter is μ, so this model says that the logarithm of the μ parameters is a subset of the β vector variables picked off by the Ψ matrix. Consider the jth row:

$$\log \mu_j = \boldsymbol{\psi}_j^T \boldsymbol{\beta} = \phi_j$$

This is the Poisson log-likelihood:

$$L = -\sum_{i=1}^{n} \log x_i! + \sum_{i=1}^{n} x_i \log \mu_i - \sum_{i=1}^{n} \mu_i$$

Substituting ϕ_i, we now have

$$L = -\sum_{i=1}^{n} \log x_i! + \sum_{i=1}^{n} x_i \phi_i - \tau$$

where $\tau = \sum_{j=1}^{n} \mu_j$. Note that

$$N \log \tau = N \log \left(\sum_{j=1}^{n} \mu_j \right)$$

with $N = \sum_{j=1}^{n} x_i$. We can add and subtract $N \log \tau$ into the log-likelihood to obtain

$$L = -\sum_{i=1}^{n} \log x_i! + \sum_{i=1}^{n} \left(x_i \phi_i - x_i \log \left(\sum_{j=1}^{n} \mu_j \right) \right) + N \log \tau - \tau$$

With the following:

$$p_i = \frac{\mu_i}{\sum_{j=1}^{n} \mu_j} = \frac{\exp(\phi_i)}{\sum_{j=1}^{n} \exp(\phi_j)}$$

we can substitute into the log-likelihood:

$$L = -\sum_{i=1}^{n} \log x_i! + \sum_{i=1}^{n} \left(x_i \phi_i - x_i \log \frac{\exp(\phi_i)}{p_i} \right) + N \log \tau - \tau$$

$$= -\sum_{i=1}^{n} \log x_i! + \sum_{i=1}^{n} (x_i \phi_i - x_i \phi_i + x_i \log p_i) + N \log \tau - \tau$$

$$= -\sum_{i=1}^{n} \log x_i! + \sum_{i=1}^{n} x_i \log p_i + N \log \tau - \tau$$

Compare this result to the log-likelihood for the multinomial distribution:

$$\log\left\{\frac{n!}{\prod_i x_i}\right\} + \sum_{i=1}^{n} x_i \log p_i$$

Keep in mind that for solving the log-likelihood to obtain the maximum likelihood estimator, we only care about terms that contain the parameter (p_i in this case). The τ terms at the end reflect the *unconditional* Poisson distribution where the total count is not constrained, so the distribution for the sum of $N = \sum_{i=1}^{n} x_i$ is also random with Poisson distribution with parameter $\tau = \sum_{j=1}^{n} \mu_j$ and thus the trailing terms are the log-likelihood for N. For computing the maximum likelihood, these trailing terms are ignored, and the solutions for this Poisson log-likelihood are identical to those for the multinomial log likelihood. This is important because it is much easier to reason about the Poisson model, especially for high-dimensional contingency tables.

3.17.2   Log-linear Models

Consider a Poisson log-linear model for counts in a two-dimensional contingency table with n_r rows and n_c columns. We have already seen multiple methods for assessing independence for two-by-two contingency tables but here want to examine the same issues using log-linear models. Suppose that the data in the table are independent counts according to a Poisson distribution with parameter $\mu_{i,j}$, then

$$\mu_{i,j} = \mu \phi_i \psi_j$$

where ϕ_i, ψ_j are positive constants with $\sum_i \phi_i = \sum_i \psi_i = 1$ and $\mu = \sum_{i,j} \mu_{i,j}$. The statement of independence here means that $\mu_{i,j}$ is a product of terms that depend only on i and j separately, not jointly. Keep in mind that there is no loss of generality by focusing on the $\mu_{i,j}$ terms instead of the individual cell probabilities $p_{i,j}$ because $\mu_{i,j} = N p_{i,j}$. The Poisson log-linear model is the following:

$$\log \mu_{i,j} = \beta_0 + \beta_i^A + \beta_j^B$$

where A and B denote the categorical variables in the two-dimensional table. The problem is that this model is seriously over-parameterized, especially for a 2×2 table. The binary matrix shown is only of rank three which means the model is not *identifiable* because more than one set of parameter values can produce the same model.

$$\begin{pmatrix} \log \mu_{1,1} \\ \log \mu_{1,2} \\ \log \mu_{2,1} \\ \log \mu_{2,2} \end{pmatrix} = \begin{pmatrix} 1\,1\,0\,1\,0 \\ 1\,1\,0\,0\,1 \\ 1\,0\,1\,1\,0 \\ 1\,0\,1\,0\,1 \end{pmatrix} \begin{pmatrix} \beta_0 \\ \beta_1^A \\ \beta_2^A \\ \beta_1^B \\ \beta_2^B \end{pmatrix}$$

By setting $\beta_1^A = \beta_1^B = 0$, we can reduce the matrix above to three columns.

$$\begin{pmatrix} \log \mu_{1,1} \\ \log \mu_{1,2} \\ \log \mu_{2,1} \\ \log \mu_{2,2} \end{pmatrix} = \begin{pmatrix} 1\,0\,0 \\ 1\,0\,1 \\ 1\,1\,0 \\ 1\,1\,1 \end{pmatrix} \begin{pmatrix} \beta_0 \\ \beta_2^A \\ \beta_2^B \end{pmatrix}$$

We can plug this model into the log-likelihood equations above to obtain the Following:

$$\frac{\partial L}{\partial \beta_i^A} = x_{i,\bullet} - \sum_{j=1}^{n_c} \exp\left(\beta_0 + \beta_i^A + \beta_j^B\right) = x_{i,\bullet} - \mu_{i,\bullet}$$

and also

$$\frac{\partial L}{\partial \beta_j^B} = x_{\bullet,j} - \sum_{i=1}^{n_r} \exp\left(\beta_0 + \beta_i^A + \beta_j^B\right) = x_{\bullet,j} - \mu_{\bullet,j}$$

Thus, the maximum likelihood solution is the following:

$$\hat{\mu}_{i,j} = x_{i,\bullet} x_{\bullet,j} / N$$

where $N = \sum_{i,j} x_{i,j} = \mu$. Given the matrix above, we can solve for the $\hat{\beta}$ terms using the $\hat{\mu}$ terms:

$$\hat{\beta}_0 = \log \hat{\mu}_{1,1}$$

$$\widehat{\beta_2^A} = \log \hat{\mu}_{2,2} - \log \hat{\mu}_{1,2}$$

$$\widehat{\beta_2^B} = \log \hat{\mu}_{2,2} - \log \hat{\mu}_{2,1}$$

As before, the quality of the model can be assessed with the Pearson statistic (see Sect. 3.5.7) for testing the goodness of fit

$$X^2 = \sum_{i=1}^{n_c} \sum_{j=1}^{n_r} \frac{(x_{i,j} - \hat{\mu}_{i,j})^2}{\hat{\mu}_{i,j}}$$

with χ^2 distribution, $(n_r - 1)(n_c - 1)$ degrees of freedom.

It may have occurred to you that we could have chosen another set of conditions to achieve identifiability and that these choices are motivated by the interpretability of the terms. For example, proceeding directly from the maximum likelihood solution,

$$\log \hat{\mu}_{i,j} = -\log n + \log x_{i,\bullet} + \log x_{\bullet,j}$$

This means that log-linear independence boils down to terms that depend on i and j separately, and in this case,

$$\hat{\beta}_0 = -\log N$$
$$\widehat{\beta_2^A} = \log x_{2,\bullet}$$
$$\widehat{\beta_2^B} = \log x_{\bullet,2}$$

Notice that the terms only depend on the marginals of the second row and column and the β_0 term is the log of the sum of all the elements in the contingency table. Because the terms depend only on the marginals, they cannot capture any between-cell interactions. Now, let us consider a more general model that accounts for possible interactions between the rows and columns.

$$\log \mu_{i,j} = \beta_0 + \beta_i^A + \beta_j^B + \gamma_{i,j}^{AB}$$

Once again, we have to supply additional constraints because there are too many free variables. Consider the equations for a 2×2 table,

$$\begin{pmatrix} \log \mu_{1,1} \\ \log \mu_{1,2} \\ \log \mu_{2,1} \\ \log \mu_{2,2} \end{pmatrix} = \begin{pmatrix} 1 1 0 1 0 1 0 0 0 \\ 1 1 0 0 1 0 1 0 0 \\ 1 0 1 1 0 0 0 1 0 \\ 1 0 1 0 1 0 0 0 1 \end{pmatrix} \begin{pmatrix} \beta_0 \\ \beta_1^A \\ \beta_2^A \\ \beta_1^B \\ \beta_2^B \\ \gamma_{1,1}^{AB} \\ \gamma_{1,2}^{AB} \\ \gamma_{2,1}^{AB} \\ \gamma_{2,2}^{AB} \end{pmatrix}$$

The matrix rank equals four so we can set $\beta_1^A = \beta_1^B = 0$ as before and also set $\gamma_{i,j}^{AB} = 0$ except for $\gamma_{2,2}^{AB}$, for all i, j.

$$\begin{pmatrix} \log \mu_{1,1} \\ \log \mu_{1,2} \\ \log \mu_{2,1} \\ \log \mu_{2,2} \end{pmatrix} = \begin{pmatrix} 1\,0\,0\,0 \\ 1\,0\,1\,0 \\ 1\,1\,0\,0 \\ 1\,1\,1\,1 \end{pmatrix} \begin{pmatrix} \beta_0 \\ \beta_2^A \\ \beta_2^B \\ \gamma_{2,2}^{AB} \end{pmatrix}$$

Plugging this model into the log-likelihood equations and solving for the model parameters gives the following:

$$\widehat{\gamma_{2,2}^{AB}} = \log \frac{x_{1,1} x_{2,2}}{x_{2,1} x_{1,2}}$$

The term on the right is the log odds ratio because $\hat{p}_{i,j} = x_{i,j}/N$. We can rewrite this as the following:

$$\exp \left(\widehat{\gamma_{2,2}^{AB}} \right) = \frac{\hat{p}_{1,1} \hat{p}_{2,2}}{\hat{p}_{2,1} \hat{p}_{1,2}}$$

This model for the two-by-two contingency table adds $(n_r - 1)(n_c - 1) = 1$ parameters, so it is the *saturated* (i.e., overfitted) model. Interestingly, as compared to the independent model, we estimated only one additional parameter $\gamma_{2,2}^{AB}$ to achieve saturation, and this turned out to be the odds ratio we have previously discussed. The following are the rest of the estimated terms for this model:

$$\hat{\beta}_0 = \log x_{1,1}$$

$$\widehat{\beta_2^A} = \log x_{2,1} - \log x_{1,1}$$

$$\widehat{\beta_2^B} = \log x_{1,2} - \log x_{1,1}$$

As opposed to the independence model, note that the terms do *not* depend on the marginals, but on the table cell values $(x_{i,j})$ individually.

Example: Tumors in Mice Consider the following data concerning tumor development in mice exposed to a fungicide [51].

```
>>> import numpy as np
>>> import xarray as xr
>>> data =np.array([[4,5],[12,74]])
>>> xa = xr.DataArray(data,coords={'i':['exposed','control'],
...                                'j':['tumors','no tumors']}
...                  )
>>> xa
<xarray.DataArray (i: 2, j: 2)>
array([[ 4,   5],
       [12, 74]])
Coordinates:
  * i           (i) <U7 'exposed' 'control'
  * j           (j) <U9 'tumors' 'no tumors'
```

First, we consider the expected table under the independence model:

```
>>> expected = xa.sum('j')*xa.sum('i')/xa.sum()
>>> expected
<xarray.DataArray (i: 2, j: 2)>
array([[ 1.51578947,  7.48421053],
       [14.48421053, 71.51578947]])
Coordinates:
  * i         (i) <U7 'exposed' 'control'
  * j         (j) <U9 'tumors' 'no tumors'
```

As the dimension of the contingency tables increases, it will be more difficult to attach a meaningful interpretation to each of the determined coefficients in the regression, but the included terms allow us to control the within-model interactions and then rate the fit of the overall model.

3.17.3 I × J × K Log-linear Models

As we begin our discussion of high-dimensional contingency tables, let us first convince ourselves that our prior analysis of three-dimensional contingency tables can be captured using log-linear models. Recall that we considered several different ways to formulate the notion of independence, either jointly or conditionally. Now, let us consider how these prior formulations can be represented as log-linear models.

Below is our data from the previous section on contingency tables

```
>>> import pandas as pd
>>> import numpy as np
>>> import xarray as xr

>>> data = np.array([[16,7,15,34,5,3],
...                   [1,1,3,8,1,3]]).reshape((2,3,2))
>>> xa = xr.DataArray(data, coords = {
...                     'behavior(i)':['deviant','non-deviant'],
...                     'adversity(j)':['low','medium','high'],
...                     'risk(k)': ['N','R']
...                     }
...                  ).astype(np.int64)
>>> xa
<xarray.DataArray (behavior(i):2, adversity(j):3, risk(k):2)>
array([[[16,  7],
        [15, 34],
        [ 5,  3]],

       [[ 1,  1],
        [ 3,  8],
        [ 1,  3]]])
Coordinates:
  * behavior(i)     (behavior(i)) <U11 'deviant' 'non-deviant'
  * adversity(j)    (adversity(j)) <U6 'low' 'medium' 'high'
  * risk(k)         (risk(k)) <U1 'N' 'R'
```

along with the following hypothesis:

$$H_1 : p_{i,j,k} = p_{i,\bullet,\bullet} p_{\bullet,j,k}$$

As shown previously, we can compute the expected table as the following:

```
>>> expected =
         xa.sum(['adversity(j)','risk(k)'])*xa.sum('behavior(i)')/xa.sum()
>>> expected
<xarray.DataArray (behavior(i):2, adversity(j):3, risk(k):2)>
array([[[14.02061856,  6.59793814],
        [14.84536082, 34.63917526],
        [ 4.94845361,  4.94845361]],

       [[ 2.97938144,  1.40206186],
        [ 3.15463918,  7.36082474],
        [ 1.05154639,  1.05154639]]])
Coordinates:
  * behavior(i)    (behavior(i)) <U11 'deviant' 'non-deviant'
  * adversity(j)   (adversity(j)) <U6 'low' 'medium' 'high'
  * risk(k)        (risk(k)) <U1 'N' 'R'
```

Note that in our discussion of this data, we did not discuss maximum likelihood estimation for this expected table, but we used the maximum likelihood equations to derive the coefficients in the log-linear model. Keeping with this approach, let us configure the maximum likelihood equations for this particular problem and solve for the maximum likelihood estimator of this expected table.

Using Sympy, we define the appropriate variables:

```
>>> import sympy as S
>>> muv = S.symbols('mu1:3(1:4)(1:3)', positive=True)
>>> xv = S.symbols('x1:3(1:4)(1:3)', positive=True)
>>> bAv = S.symbols('bA1:3', real=True)
>>> bBCv = S.symbols('bBC1:4(1:3)', real=True)
```

For this next step, we want to pack the Sympy variables into Numpy arrays that are easier to manipulate:

```
>>> mu = np.array(muv).reshape(2,3,2)
>>> xd = np.array(xv).reshape(2,3,2)
>>> bBC = np.array(bBCv).reshape(3,2)
>>> subsdict = {bAv[1]:0}
>>> values = xa.values.flatten().tolist()
```

Now, we construct the log-likelihood function as in the following, using nested loops to make sure and assign the correct values from the Xarray xa into the equations:

```
>>> L = 0
>>> for i in range(2):
...     for j in range(3):
...         for k in range(2):
...             L += (bAv[i]+bBC[j,k])*xd[i,j,k]-S.exp(bAv[i]+bBC[j,k])
...             subsdict[xd[i,j,k]] = values.pop(0)
...
```

The next step is the maximize the log-likelihood using calculus as in the following:

```
>>> t = L.subs(subsdict)
>>> sols, = S.solve([t.diff(i) for i in t.free_symbols])
>>> sols
{bA1: log(80/17), bBC11: log(289/97), bBC12: log(136/97),
bBC21: log(306/97), bBC22: log(714/97), bBC31: log(102/97),
bBC32: log(102/97)}
```

Because we want to compare this to our existing `expected` table, we can extract the terms as follows:

```
>>> mudict = {}
>>> for i in range(2):
...        for j in range(3):
...            for k in range(2):
...                mudict[mu[i,j,k]]=(S.exp((bAv[i]+bBC[j,k])
...                                    .subs(sols)
...                                    .subs(subsdict))
...                                    .evalf())
...
>>> mudict
{mu111: 14.0206185567010, mu112: 6.59793814432990,
mu121: 14.8453608247423, mu122: 34.6391752577320,
mu131: 4.94845360824742, mu132: 4.94845360824742,
mu211: 2.97938144329897, mu212: 1.40206185567010,
mu221: 3.15463917525773,
mu222: 7.36082474226804, mu231: 1.05154639175258,
mu232: 1.05154639175258}
```

These terms match one-for-one with the values shown in the Xarray `expected` above. We glossed over certain technicalities above. For example, we did not check the second derivative when we maximized the log-likelihood equations. It turns out that the Poisson distribution we are using for this model, and, in fact, the exponential family that contains the Poisson distribution, does not require it. Nonetheless, this is a very tedious procedure that can only get worse as the number of dimensions increases.

What we want is a way to compute the maximum likelihood estimator for the expected table without all this work. Fortunately, Birch criteria provide exactly this in two parts. First, for all nonzero entries of each marginal subtable that are nonzero, these marginal subtables are sufficient statistics for $\mu_{i,j,k}$. Second, all sufficient marginal subtables must be equal to the corresponding marginal subtables of their estimated means for the estimated means to be the maximum likelihood estimators. This is a consequence of the fact that sufficient statistics for this exponential family *are* the maximum likelihood estimators. Note that Birch criteria provide a way to test whether or not the estimated means in the table are the maximum likelihood estimators, but *not* how to construct these estimators. Here is an example of a marginal subtable that matches the maximum likelihood estimator:

$$\hat{\mu}_{i,\bullet} = x_{i,\bullet}$$

This criterion is the basic result that powers the iterative proportional fitting algorithm.

3.17.4   *Iterative Proportional Fitting*

The Birch criteria give us a way of testing whether or not the estimator is maximum likelihood estimates, but it does not tell us how to construct such an estimate. Some maximum likelihood estimates may *not* have a closed form and may require iterated solutions, which is what the *iterative proportional fitting algorithm* (IPF) provides. The algorithm is based upon the fact that

$$1 = \frac{n_{i,j,\bullet}}{\hat{\mu}_{i,j,\bullet}} = \frac{n_{i,\bullet,j}}{\hat{\mu}_{i,\bullet,j}} = \frac{n_{\bullet,i,j}}{\hat{\mu}_{\bullet,i,j}}$$

which leads to these identities:

$$\hat{\mu}_{i,j,k} = \frac{n_{i,j,\bullet}}{\hat{\mu}_{i,j,\bullet}} \hat{\mu}_{i,j,k}$$

$$\hat{\mu}_{i,j,k} = \frac{n_{i,\bullet,j}}{\hat{\mu}_{i,\bullet,j}} \hat{\mu}_{i,j,k}$$

$$\hat{\mu}_{i,j,k} = \frac{n_{\bullet,i,j}}{\hat{\mu}_{\bullet,i,j}} \hat{\mu}_{i,j,k}$$

Thus the iteration becomes

$$\hat{\mu}_{i,j,k}^{[3t+1]} = \frac{n_{i,j,\bullet}}{\hat{\mu}_{i,j,\bullet}} \hat{\mu}_{i,j,k}^{[3t]}$$

$$\hat{\mu}_{i,j,k}^{[3t+2]} = \frac{n_{i,\bullet,j}}{\hat{\mu}_{i,\bullet,j}} \hat{\mu}_{i,j,k}^{[3t+1]}$$

$$\hat{\mu}_{i,j,k}^{[3t+3]} = \frac{n_{\bullet,i,j}}{\hat{\mu}_{\bullet,i,j}} \hat{\mu}_{i,j,k}^{[3t+2]}$$

The initial cases are usually taken to be

$$\hat{\mu}_{i,j,k}^{[0]} = 1 \ \forall i, j, k$$

Birch's results enable us to derive estimates of the expected cell counts for all of the elementary cells from the configurations of sums, without having to go through the intermediate step of estimating the individual regression terms. The code below makes this algorithm concrete, but before we can use it, we have to identify the log-linear models that it facilitates.

3.17.5   Hierarchical Models

We need some definitions to proceed. Non-comprehensive models are those that do not have at least one parameter for every dimension in the table. For example, for a three-dimensional model

$$\log \mu_{i,j,k} = \beta_0 + \beta_i^A + \beta_j^B + \beta_k^C$$

setting any of the parameters to zero results in a non-comprehensive model. Conversely, a comprehensive model has at least the main effect (i.e., any of the terms above except for β_0) in every dimension.

Hierarchical log-linear models are the easiest to fit and explain and require that high-order interactions always accompany lower-order interactions. For example, the following is *not* a hierarchical log-linear model:

$$\log \mu_{i,j,k,l} = \beta_{i,j,k}^{ABC}$$

However, by including all subsets of interaction terms, we obtain a hierarchical model:

$$\log \mu_{i,j,k,l} = \beta_0 + \beta_i^A + \beta_j^B + \beta_k^C + \beta_l^D + \beta_{j,k}^{BC} + \beta_{i,j}^{AB} + \beta_{j,l}^{BD} + \beta_{i,k}^{AC} + \beta_{i,j,k}^{ABC}$$

The notation for keeping track of all of these terms is tricky, so this model can be specified by its *generating class* which is a shorthand for specifying the maximum interactions as brackets in a set. For the model above, the generating class would be {[ABC], [BD]}. The [ABC] element of the generating class requires that [AB], [BC], and [AC] be included in the hierarchical model. This model is comprehensive because it includes the [A], [B], [C], and [D] terms.

Because of the Poisson/multinomial distribution, maximum likelihood estimates can be derived from the sufficient statistics in the model. We can use the Birch's results with the generating class notation to identify the maximum likelihood estimates using the marginal subtables. For example, given the generating class {[ABC], [BD]}, the marginal subtable $n_{i,j,k,\bullet}$ is needed to identify the maximum likelihood estimate for [ABC]. Likewise, the generating class [BD] marginal subtable, $n_{\bullet,j,\bullet,l}$. Some examples can make this notation easier to understand.

Consider the two-dimensional log-linear model with the generating class {[A], [B]}

$$\log \mu_{i,j} = \beta_0 + \beta_i^A + \beta_j^B$$

This means that $\log \mu_{i,j}$ can be written as a function of i and j, separately. The marginal subtables are $n_{\bullet,j}$ and $n_{i,\bullet}$. Using the Birch criteria, we have

$$\hat{\mu}_{\bullet,j} = n_{\bullet,j}$$

and

$$\hat{\mu}_{i,\bullet} = n_{i,\bullet}$$

and thus our familiar result

$$\hat{\mu}_{i,j} = n_{i,\bullet} n_{\bullet,j} / n_{\bullet,\bullet}$$

Keep in mind that the marginal totals have to be greater than zero, according to Birch. Next, consider the comprehensive three-dimensional log-linear model with generating class $\{[AB], [C]\}$:

$$\log \mu_{i,j,k} = \beta_0 + \beta_i^A + \beta_j^B + \beta_k^C + \beta_{i,j}^{AB}$$

That is, the log-linear model is a function of (i, j) and k, so using Birch's results again, we have

$$\hat{\mu}_{i,j,k} = n_{i,j,\bullet} n_{\bullet,\bullet,k} / n_{\bullet,\bullet,\bullet}$$

That satisfies the conditions for the model. An alternative, three-dimensional model for the generating class $\{[AB], [BC]\}$ is the following:

$$\log \mu_{i,j,k} = \beta_0 + \beta_i^A + \beta_j^B + \beta_k^C + \beta_{i,j}^{AB} + \beta_{j,k}^{BC}$$

for which we have via Birch's results

$$\hat{\mu}_{i,j,k} = n_{i,j,\bullet} n_{\bullet,j,k} / n_{\bullet,\bullet,\bullet}$$

For the following model with the generating class $\{[AB], [AC], [BC]\}$, we have

$$\log \mu_{i,j,k} = \beta_0 + \beta_i^A + \beta_j^B + \beta_k^C + \beta_{i,j}^{AB} + \beta_{j,k}^{BC} + \beta_{i,k}^{AC}$$

with sufficient marginal subtables $n_{i,j,\bullet}$, $n_{i,\bullet,k}$, and $n_{\bullet,j,k}$, but in this case we don't have a closed form solution for $\hat{\mu}_{i,j,k}$ and must rely on an iterative solution that satisfies the following equations:

$$\hat{\mu}_{i,j,\bullet} = n_{i,j,\bullet}$$
$$\hat{\mu}_{i,\bullet,k} = n_{i,\bullet,k}$$
$$\hat{\mu}_{\bullet,j,k} = n_{\bullet,j,k}$$

Example: Conditional Independence IPF To see this in action using iterative proportional fitting, let us return to the personality-cholesterol data from before

```
>>> data = np.array([[716,79],
...                  [207,25],
...                  [819,67],
...                  [186,22]]).reshape((2,2,2))
>>> xa = xr.DataArray(data,
...             coords = {'personality(i)': ['A','B'],
...                       'cholesterol(j)': ['normal','high'],
...                       'blood_pressure(k)': ['normal','high'],
...                      }
...             ).astype(np.int64)
>>> xa
<xarray.DataArray (personality(i): 2, cholesterol(j): 2,
blood_pressure(k): 2)>
array([[[716,   79],
        [207,   25]],

       [[819,   67],
        [186,   22]]])
Coordinates:
  * personality(i)      (personality(i)) <U1 'A' 'B'
  * cholesterol(j)      (cholesterol(j)) <U6 'normal' 'high'
  * blood_pressure(k)   (blood_pressure(k)) <U6 'normal' 'high'
```

with the same hypothesis that cholesterol and blood pressure are independent given personality type:

$$p_{i,j,k} = p_{i,j,\bullet} p_{i,\bullet,k} / p_{i,\bullet,\bullet}$$

Using our generating class notation, this is written $\{[AB][AC]\}$. Now, we set up the initial conditions The, iterative proportional fitting algorithm:

```
>>> mu = xa*0+1 # initialize
```

Then, we can do the iteration:

```
>>> for i in range(5):
...     mu = (xa.sum('blood_pressure(k)')/
...           mu.sum('blood_pressure(k)'))*mu # [AB]
...     mu = (xa.sum('cholesterol(j)')/
...           mu.sum('cholesterol(j)'))*mu    # [AC]
...
>>> mu.transpose('personality(i)',
...              'cholesterol(j)',
...              'blood_pressure(k)') # realign for comparison
<xarray.DataArray (personality(i): 2, cholesterol(j): 2, blood
 _pressure(k): 2)>
array([[[714.49367089,   80.50632911],
        [208.50632911,   23.49367089]],

       [[813.9213894 ,   72.0786106 ],
        [191.0786106 ,   16.9213894 ]]])
Coordinates:
  * personality(i)      (personality(i)) <U1 'A' 'B'
```

```
*  blood_pressure(k)    (blood_pressure(k))  <U6 'normal' 'high'
*  cholesterol(j)       (cholesterol(j))     <U6 'normal' 'high'
```

Note that the iteration stops when the values for mu stop changing significantly, but I left that logic out of the loop to reduce clutter. It turns out the IPF converges very quickly and is guaranteed to converge. The result for mu here is exactly the same as we got for using the formula in the contingency table approach, which is encouraging, given that we made no claims to maximum likelihood properties of that formula.

Example: Complete Independence IPF While we have the personality-cholesterol data at hand, let us re-visit the complete independence model, $\{[A][B][C]\}$:

```
>>> mu0 = xa*0+1 # initialize
>>> for i in range(5):
...     mu0 = (xa.sum(['cholesterol(j)','blood_pressure(k)'])/
...            mu0.sum(['cholesterol(j)','blood_pressure(k)']))*mu0 # [A]
...     mu0 = (xa.sum(['personality(i)','blood_pressure(k)'])/
...            mu0.sum(['personality(i)','blood_pressure(k)']))*mu0 # [B]
...     mu0 = (xa.sum(['personality(i)','cholesterol(j)'])/
...            mu0.sum(['personality(i)','cholesterol(j)']))*mu0     # [C]
...
>>> mu0.transpose('personality(i)','cholesterol(j)', 'blood_pressure(k)')
<xarray.DataArray (personality(i): 2, cholesterol(j): 2, blood_pressure(k): 2)>
array([[[739.88436419,   74.06518791],
        [193.66396207,   19.38648583]],

       [[788.15335387,   78.89709403],
        [206.29831987,   20.65123223]]])
Coordinates:
  * blood_pressure(k)    (blood_pressure(k))  <U6 'normal' 'high'
  * cholesterol(j)       (cholesterol(j))     <U6 'normal' 'high'
  * personality(i)       (personality(i))     <U1 'A' 'B'
```

Again, this result exactly matches our previous contingency table analysis (see Sect. 3.5.7).

3.17.6 Deviance

The ability to compute the maximum likelihood expected contingency table using IPF is great, but now that we can consider many different kinds of models, we need to compare them against each other systematically, and for that, we return to the *deviance* (also known as KL-divergence in machine learning) that we used for generalized linear models, which is commonly denoted as G^2 for log-linear models. For example, given the two models for the personality-cholesterol data above, we can compute the deviance between the expected table for conditional independence and the original data as follows:

```
>>> float(2*(xa*np.log(xa/mu)).sum())  # conditional independence
2.062585010431811
```

We can likewise compute the deviance between the expected table for the complete independence model and the original data as follows:

```
>>> float(2*(xa*np.log(xa/mu0)).sum())  # complete independence
8.722939543258185
```

Because the conditional independence has a smaller deviance from the actual data than the complete independence model, we can conclude that it fits the data better, but we must also consider the degrees of freedom in each case to really judge between then. Typically, we want a small deviance with many degrees of freedom. For the conditional independence model, we have two degrees of freedom, and for the complete independence model, we have four. Degrees of freedom can be tricky to calculate, so the next section goes into the details. Due to properties of the deviance we can directly compute

```
>>> float(2*(mu0*np.log(mu0/mu)).sum())
6.66702094574363
```

which is the difference of the two deviance values above. The χ_2^2 threshold for 95% significance for two degrees of freedom is the following:

```
>>> from scipy import stats
>>> stats.chi2(2).isf(.05)
5.991464547107983
```

Thus, we can conclude that the difference between the conditional independence and complete independence model is statistically significant because it exceeds this threshold.

3.17.7 Degrees of Freedom

The degrees of freedom is the difference between the number of parameters that *could* be estimated minus the parameters that the model requires. Consider the generating class $\{[A][B][C]\}$ which corresponds to the following complete independence model for a three-dimensional contingency table:

$$\beta_0 + \beta_i^A + \beta_j^B + \beta_k^C$$

The β_0 is one parameter to estimate, and the other terms give $n_r - 1$, $n_c - 1$, and $n_l - 1$ parameters, from left to right, where n_l is the size of the third dimension (sometimes called *layers*). Remember that each of the marginals have to sum to one, so each of the dimensions has one parameter that is completely determined by the rest. The total degrees of freedom in this case is

$$n_r n_c n_l - (1 + (n_r - 1) + (n_c - 1) + (n_l + 1))$$

For example, for $n_c = n_r = n_l = 2$, we have four degrees of freedom. Consider the $\{[AB][C]\}$ model:

$$\beta_0 + \beta_i^A + \beta_j^B + \beta_k^C + \gamma_{i,j}^{AB}$$

In this case, we gain another $(n_r - 1)(n_c - 1)$ parameters to estimate because the new term represents a valid two-dimensional probability distribution that has to total to one as do its marginals. The total degrees of freedom is therefore

$$n_r n_c n_l - (1 + (n_r - 1) + (n_c - 1) + (n_l - 1) + (n_r - 1)(n_c - 1))$$

For the $n_c = n_r = n_l = 2$ case, this computes out to be three. Now, consider the $\{[AB][BC]\}$ model of conditional independence:

$$\beta_0 + \beta_i^A + \beta_j^B + \beta_k^C + \gamma_{i,j}^{AB} + \gamma_{j,k}^{BC}$$

This is the same number of parameters of the previous case except for the additional $(n_c - 1)(n_l - 1)$ terms. Thus, the total degrees of freedom is therefore

$$n_r n_c n_l - (1 + (n_r - 1) + (n_c - 1) + (n_l - 1) + (n_r - 1)(n_c - 1)) + (n_l - 1)(n_c - 1))$$

and in the $n_c = n_r = n_l = 2$ case, we have two degrees of freedom. For the $\{[AB][BC][AC]\}$ case, we have $(n_r - 1)(n_l - 1)$ additional parameters compared to the previous case which gives one degree of freedom when $n_c = n_r = n_l = 2$. The final fully saturated case is $\{[ABC]\}$ which adds the $(n_r - 1)(n_c - 1)(n_l - 1)$ terms which makes for zero total degree of freedom. Hopefully, you can see the general pattern here which applies to larger than three-dimensional contingency tables.

Example: Four-Dimensional Mouse Tension Data Consider the following four dimensional contingency table for a study to determine the relationship between two different drugs and muscle tension in mice. For each mouse, a muscle was identified, and its tension was measured. A randomly chosen drug was given to the mouse, and the muscle tension was measured again. The muscle was then tested to identify which type of muscle it was. The weight of the muscle was also measured.

```
>>> data = np.array([[3,21],[23,11],
...                  [22,32],[4,12],
...                  [3,10],[41,21],
...                  [45,23],[6,22]])
>>> x = xr.DataArray(data.reshape((2,2,2,2)),
...                  coords = {'T': ['hi','lo'],
...                            'W': ['hi','lo'],
...                            'M': ['1','2'],
...                            'D': ['1','2']
...                            },
...                  attrs = {'description':'muscle tension data',
```

```
...                              'T': 'muscle tension',
...                              'W': 'muscle weight',
...                              'M': 'muscle type',
...                              'D': 'drug type'
...                            }
...                  )
>>> x
<xarray.DataArray (T: 2, W: 2, M: 2, D: 2)>
array([[[[ 3, 21],
         [23, 11]],

        [[22, 32],
         [ 4, 12]]],


       [[[ 3, 10],
         [41, 21]],

        [[45, 23],
         [ 6, 22]]]]])
Coordinates:
  * T          (T) <U2 'hi' 'lo'
  * W          (W) <U2 'hi' 'lo'
  * M          (M) <U1 '1' '2'
  * D          (D) <U1 '1' '2'
Attributes:
    description:  muscle tension data
    T:            muscle tension
    W:            muscle weight
    M:            muscle type
    D:            drug type
```

Let us consider the complete independence model $\{[T][W][M][D]\}$ where the letters correspond to each of the variables as described in the `attrs` attribute above. We can compute the expected contingency table using the formula

$$\hat{\mu}_{i,j,k,l} = x_{i,\bullet,\bullet,\bullet}x_{\bullet,j,\bullet,\bullet}x_{\bullet,\bullet,k,\bullet}x_{\bullet,\bullet,\bullet,l}/N^3$$

This is implemented in the code below using `itertools.combinations` and `functools.reduce`:

```
>>> from functools import reduce
>>> import itertools as it
>>> mu = reduce(lambda i,j:i*j, [x.sum(i)
...                              for i in it.combinations(x.coords,3)
...                              ])/x.sum()**3
>>> float(2*(x*np.log(x/mu)).sum())    # dof = 11
127.3506927454614
```

The number of degrees of freedom for this model is $16 - (1+1+1+1+1) = 11$. Next, let us consider the model of all two-way interactions

$$\{[TW][TM][WM][TD][WD][MD]\}$$

We need the IPF algorithm to compute this because there is no closed form estimator:

```
>>> mui = x*0+1 # initialize
>>> for i in range(6):
...     mui = x.sum(['T','W'])/mui.sum(['T','W']) * mui
...     mui = x.sum(['T','M'])/mui.sum(['T','M']) * mui
...     mui = x.sum(['W','M'])/mui.sum(['W','M']) * mui
...     mui = x.sum(['T','D'])/mui.sum(['T','D']) * mui
...     mui = x.sum(['W','D'])/mui.sum(['W','D']) * mui
...     mui = x.sum(['M','D'])/mui.sum(['M','D']) * mui
...
>>> float(2*(x*np.log(x/mui)).sum())  # dof = 5
47.66855634814131
```

The number of degrees of freedom for this model is six fewer in the previous complete independence model $11-6=5$. Next, let us consider the model of all three-way interactions

$$\{[TWM][TWD][TMD][WMD]\}$$

with one degree of freedom.

```
>>> mui = x*0+1
>>> for i in range(6):
...     mui = x.sum('T')/mui.sum('T') * mui
...     mui = x.sum('W')/mui.sum('W') * mui
...     mui = x.sum('M')/mui.sum('M') * mui
...     mui = x.sum('D')/mui.sum('D') * mui
...
>>> float(2*(x*np.log(x/mui)).sum()) # dof = 1
0.1112228895307559
```

This shows that we can get a smaller (i.e., better) deviance if we use more degrees of freedom (i.e., overfit the data).

3.17.8 Graphical Models

Using generating classes makes it easy to specify a wide variety of potential models, and using the deviance, we can compare them pair-wise, but now we can have *too* many models to choose from, and so our new problem is model interpretability. Graphical models add certain addition restrictions to our models, but the benefit is that we can draw the models as a network graph of nodes and edges and then interpret them graphically. A model is *graphical* if all two-factor terms generated by a higher-order interaction *includes* the higher-order interaction term. For example, the graphical model $\{[ABCD][CDE]\}$ is determined by the four-factor term $\{[ABCD]\}$ and the three-factor term $\{[CDE]\}$.

Fig. 3.56 Graphical model
for log-linear model
{[ABD], [BCD]}

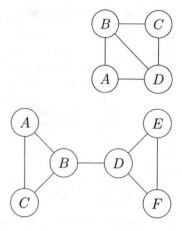

Fig. 3.57 Graphical model
for log-linear model
{[ABC], [BD], [DEF]}

All two-factor terms generated by {[ABCD]} are included in the model, namely,
[AB], [AC], [AD], [BC], [BD], [CD]. Likewise, {[CDE]} generates the following two-factor terms that are also included in the model, [CD], [CE], [DE].
Conversely, if a model includes two-factor terms like {[AB], [BC], [AC]}, then it
must include the [ABC] higher-order term in order to be a *graphical* model.

Because the saturated model is the graphical model containing all possible two-
factor terms, any log-linear model can be embedded in a graphical model. Thus, to
interpret a log-linear model, we want the smallest graphical model that contains that
log-linear model. For example, the non-graphical model {[AB][AC][AD][BC]} is
embedded in the graphical model {[ABC][AD]}, which has the conditional independence interpretation that BC is independent of D, given A. This interpretation
applies to the embedded non-graphical model, although the non-graphical model
involves additional constraints. Consider the log-linear model {[ABD], [BCD]}
shown in Fig. 3.56.

Recall that a *clique* is a graph with all nodes mutually connected. Another way
to look at log-linear models as graphical models is to consider the items of the
generating class a maximal clique (i.e., not contained in a bigger clique). With
this interpretation, it is easy to read off log-linear models and their independence
relations from network graphs. A *chain* is a sequence of edges that lead from one
node to another node. The main theorem for using chains in graphical models says
that if node sets $\mathcal{A}$, $\mathcal{B}$, and $\mathcal{C}$ are disjoint node subsets and (if and only if) every
chain between a node in $\mathcal{A}$ and a node in $\mathcal{B}$ contains at least one node in $\mathcal{C}$, then the
nodes in $\mathcal{A}$ and the nodes in $\mathcal{B}$ are independent given $\mathcal{C}$.

For example, for Fig. 3.57, we can read off the independence relationships
that A and C are independent of D, E, F given B, which can be symbolized as
$AC \perp DEF|B$. Also, $EF \perp ABC|D$. Marginal relationships (i.e., those that do
not involve all terms) can also be read off the graph. For example, $AC \perp DE|B$.

3.17.9   Model Selection

Choosing between so many possible log-linear models is tricky. There are two primary methods: backward selection and forward selection. With backward selection, you start with a saturated model and then incrementally delete terms with the least significant χ^2 statistic until all remaining terms have the prespecified minimum level of significance. The forward selection model is basically the same but in the other direction—start with the independence model, and keep adding terms until no term achieves a predetermined minimum level of significance. The advantage of either of these methods is that they are quick to implement, but that is primarily because they consider a relatively small subset of all possible models available. They are also both sensitive to the choice of the initial model and furthermore can miss intermediate good models entirely and sometimes require complicated reparameterization along the way.

In 1976, Brown proposed looking at *marginal* association and *partial* association measures as a way of navigating between multiple proposed models. Marginal association is the deviance obtained when all other terms are collapsed and the resulting marginalized table is measured against the independence model for that table. For example, the test of marginal association for $[ABC]$ is the differential deviance between $[AB][BC][AC]$ and $[ABC]$ and that for $[ABCD]$ is the differential deviance between $[ABC][ABD][ACD][BCD]$ and $[ABCD]$.

Partial association involves computing the differential deviance between the model with and without the term of interest (see example below). The idea is that when the partial association and the marginal association are significantly different for a given term, then this can indicate the presence of Simpson's paradox. Looking at these two measures of association also makes it obvious which terms are important in the final resulting model.

It is best to examine Brown's ideas by example. Consider the following data regarding perching behavior of lizards, organized and coded into the following categories: height above ground under/over 5 ft ($h \in \{0, 1\}$), diameter of branch under/over 2 inches ($d \in \{0, 1\}$), insolation sun/shade ($i \in \{0, 1\}$), time of day morning/midday/afternoon ($t \in \{0, 1, 2\}$), species anolis grahami/opalinus ($s \in \{0, 1\}$).

```
>>> X
<xarray.DataArray (h: 2, d: 2, i: 2, t: 3, s: 2)>
array([[[[[20,    2],
          [ 8,    1],
          [ 4,    4]],

         [[34,   11],
          [69,   20],
          [18,   10]]],


        [[[ 8,    3],
```

```
                [ 4,    1],
                [ 5,    3]],


            [[17,  15],
             [60,  32],
             [ 8,   8]]]],



          [[[[13,   0],
             [ 8,   0],
             [12,   0]],

            [[31,   5],
             [55,   4],
             [13,   3]]],


           [[[ 6,   0],
             [ 0,   0],
             [ 1,   1]],

            [[12,   1],
             [21,   5],
             [ 4,   4]]]]]])
Coordinates:
  * h        (h)  int64  0 1
  * d        (d)  int64  0 1
  * i        (i)  int64  0 1
  * t        (t)  int64  0 1 2
  * s        (s)  int64  0 1
```

To compute the marginal association for the $[dh]$ term, we sum over all the other terms to obtain the reduced table:

```
>>> m = x.sum(('i','t','s'))
>>> m
<xarray.DataArray (h: 2, d: 2)>
array([[201, 164],
       [144,  55]])
Coordinates:
  * h        (h)  int64  0 1
  * d        (d)  int64  0 1
```

Next, we use this table to compute the independence model (which we have done previously) to obtain

```
>>> indep = reduce(lambda i,j: i*m.sum(j) ,['h','d'],1)/m.sum()
>>> indep
<xarray.DataArray (d: 2, h: 2)>
array([[223.2712766, 121.7287234],
       [141.7287234,  77.2712766]])
```

```
Coordinates:
  * d          (d) int64 0 1
  * h          (h) int64 0 1
```

Finally, we compute the deviance between this independence model and the original table:

```
>>> float(2*(m*np.log(m/indep)).sum())
16.619383942523577
```

We can repeat this process for all the two and three factor terms to obtain the following summary including degrees of freedom and *p*-value (mpval):

```
>>> df
```

item	marginal_assoc	dof	mpval
dh	16.619	1	0.000
hs	26.856	1	0.000
hi	0.945	1	0.331
ht	2.317	2	0.314
ds	18.524	1	0.000
di	3.574	1	0.059
dt	3.735	2	0.155
is	6.472	1	0.011
st	6.368	2	0.041
it	47.969	2	0.000
dhs	0.474	1	0.491
dhi	0.181	1	0.670
dht	1.088	2	0.580
his	2.181	1	0.140
hst	0.401	2	0.818
hit	1.070	2	0.586
dis	0.305	1	0.581
dst	0.007	2	0.996
dit	2.373	2	0.305
ist	1.099	2	0.577

Note that the items with the *t* variable have an additional degree of freedom because there are three levels in the *t* category.

Partial association computes the differential deviance with and without a given factor. For example, to compute the partial association for the [*hd*] term, we have to first compute the model with *all* two factor terms included, namely, [*ht*][*dh*][*hi*][*hs*][*dt*][*it*][*st*][*di*][*ds*][*is*]. Then compute the same model *without* the [*hd*] factor and compute the deviance between these two models. The result is summarized next in the extended dataframe:

```
>>> df
```

item	marginal_assoc	dof	mpval	partial_assoc	ppval
dh	16.619	1	0.000	9.821	0.002
hs	26.856	1	0.000	22.010	0.000

hi	0.945	1	0.331	0.034	0.853
ht	2.317	2	0.314	2.532	0.282
ds	18.524	1	0.000	12.703	0.000
di	3.574	1	0.059	0.806	0.369
dt	3.735	2	0.155	2.682	0.262
is	6.472	1	0.011	7.562	0.006
st	6.368	2	0.041	11.453	0.003
it	47.969	2	0.000	48.896	0.000
dhs	0.474	1	0.491	0.318	0.573
dhi	0.181	1	0.670	0.393	0.531
dht	1.088	2	0.580	1.481	0.477
his	2.181	1	0.140	3.714	0.054
hst	0.401	2	0.818	1.648	0.439
hit	1.070	2	0.586	2.976	0.226
dis	0.305	1	0.581	0.040	0.841
dst	0.007	2	0.996	0.011	0.994
dit	2.373	2	0.305	2.497	0.287
ist	1.099	2	0.577	2.654	0.265

Examining the above dataframe shows that the two-factor terms $[it]$ and $[hs]$ have the highest marginal associations which indicates that the independence model is failing spectacularly for these factors and these should therefore be included in the final model. We can start with $\{[it][hs][ds]\}$ that has 37 degrees of freedom and gives a deviance of 59.26. This corresponds to the following graphical model in Fig. 3.58:

We can easily interpret that (i, t) and (h, s, d) are mutually independent and that $h \perp d|s$ so that given species, height, and branch diameter are independent. Proceeding in order by partial association, we can add the $[ts]$ factor for 35 degrees of freedom and deviance of 52.9. This gives the updated graphical model in Fig. 3.59.

The addition of the extra edge enables us to conclude that given t (time of day), i (insolation) is independent of s (species). Additionally, (it) is independent of h given s, also. The χ^2 p-value for this model is 0.045, which is on the edge of rejection using the standard 0.05 level. By adding the $[hd]$ term, we could get the corresponding p-value to 0.27 for 34 degrees of freedom. Looking at the values

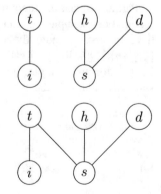

Fig. 3.58 Graphical model for log-linear model $\{[it][hs][ds]\}$

Fig. 3.59 Graphical model for log-linear model $\{[it][hs][ds][ts]\}$

for partial and marginal association, it does not look like adding three factor terms is worth it in terms of degrees of freedom or interpretability. Remember we seek the smallest sensible model that fits the data. The [hd] term is likely there because tree branches get thinner with height. The [it] term suggests that insolation and time-of-day interact, probably due to the diurnal cycle. The model further suggests that the two species have different preferences regarding branch heights, diameters, and time-of-day. These are all insights (among others) that would be hard to hypothesize by just looking at the raw five-dimensional table!

3.17.10   Table Raking

The IPF algorithm can help interpret contingency tables by forcing the marginals to add up to the same number (typically 100), which makes it easier to interpret the individual table cells. This is called *table raking*. The primary assumption is that any biases are limited to misrepresentation of the frequencies of the categories, not to the joint relationships between them. This means that if you somehow generated fake samples to fill in the marginal sums, the joint relationships between the variables would stay the same for the log-linear model, which is equivalent to stating that the main effect terms are incorrect, but all the interactions are correct.

Let us consider the following dataframe regarding political party classifications and the simple independence model:

```
>>> data = [[306,279,116],
...         [185,312,194],
...         [26,134,338]]
>>> x = xr.DataArray(data,
...       coords = {'i':['democrat','independent','republican'],
...                 'j':['liberal','moderate', 'conservative'], }
...       )
```

Now, we set up the usual IPF iteration:

```
>>> mui =x.copy() # initialize

>>> for i in range(6):
...       mui = 100/mui.sum('i') * mui
...       mui = 100/mui.sum('j') * mui # marginals forced to 100
...
>>> mui
<xarray.DataArray (i: 3, j: 3)>
array([[55.01212733, 32.45331276, 12.53455991],
       [36.74463004, 40.09540122, 23.15996874],
       [ 8.23156706, 27.44933262, 64.31910032]])
Coordinates:
  * i          (i) <U11 'democrat' 'independent' 'republican'
  * j          (j) <U12 'liberal' 'moderate' 'conservative'
```

Note that the marginals are now all approximately equal to 100.

```
>>> mui.sum('i')
<xarray.DataArray (j: 3)>
array([ 99.98832443,  99.9980466 , 100.01362897])
Coordinates:
  * j            (j) <U12 'liberal' 'moderate' 'conservative'
>>> mui.sum('j')
<xarray.DataArray (i: 3)>
array([100., 100., 100.])
Coordinates:
  * i            (i) <U11 'democrat' 'independent' 'republican'
```

The advantage is the new raked table is easier to interpret than the original data, and we can see the political independents are most strongly moderate compared to democrat and republican groups.

```
>>> mui.sel({'i': 'independent'})
<xarray.DataArray (j: 3)>
array([36.74463004, 40.09540122, 23.15996874])
Coordinates:
    i            <U11 'independent'
  * j            (j) <U12 'liberal' 'moderate' 'conservative'
```

3.18 Missing Data

Missing data is common and usually handled by simply omitting any incomplete cases and proceeding as usual with the remainder of the data. The problem with this approach is that it can lead to serious bias in the estimated parameters and cause pernicious downstream consequences. For example, for medical applications, a dense and complete set of features for a patient is expensive to collect, but ignoring missing data and keeping only those complete records means that the model is biased to those who could afford to pay for expensive diagnostics. This may be acceptable if the model is intended for use among affluent patients, with ready access to high-quality medical diagnostics, but what about everybody else? The ultimate result of such a model is poor performance among those only partially represented because of missing data.

Another common strategy to missing data is to fill in the missing values with a fixed number like zero. This again is another form of bias because it draws the model's parameters toward optimizing loss functions near the origin, which is especially bad if zero is not even in the domain of the data. Still another strategy is to fill the missing row elements in a given column with the mean or median of the non-missing values of that column. This again is another form of bias because it essentially asserts that the missing data were in fact derivable from the observed data, which begs the question of trying to collect so much data if a section of it is fully derivable from a subset. That is, the *potential* for new information *distinct* from the observed data is now lost in the missing elements.

Missing data has many subtle issues, so we must start by classifying these scenarios before we can figure out how to deal with them.

Categories of Missingness Suppose R is the binary matrix that stores the locations of the missing Y where 1 means the data is present and 0 absent; and ψ are parameters of the missing data model. The *missingness* model is the probabilistic mechanism that generates the R matrix.

Missing completely at random (MCAR) means that the probability of being missing depends only on ψ and not on the Y variables. This means that the probability of missing is the same for all elements. That is, the cause of the missing data is unrelated to the data itself. Mathematically, this means that the probability that data is missing is solely a function of the parameters of the missingness model (ψ), not any of the other variables, Y_{obs},

$$\mathbb{P}(R = 0 | Y_{obs}, Y_{mis}, \psi) = \mathbb{P}(R = 0 | \psi)$$

MAR (missing at random) means that the missingness probability may depend on observed information, Y_{obs}. MAR represents the probability of being missing equally only within strata defined by the observed data. For example, suppose we are measuring body weight, and, due to a mechanical problem with the measuring scale, we get missing data with a certain probability when a scale is placed upon a soft surface as opposed to a hard surface. If we note whether the measurement was taken on a hard or soft surface, and within each group we have MCAR, then for the whole experiment, we have MAR. The key idea is that we are observing the variable that partitions into MCAR. Mathematically this means that the probability an element is missing is not just a function of the missingness model, but also of the observed Y values.

$$\mathbb{P}(R = 0 | Y_{obs}, Y_{mis}, \psi) = \mathbb{P}(R = 0 | Y_{obs}, \psi)$$

MNAR (missing not at random) does not yield any simplifications, meaning that the probability of missing data depends on observed and unobserved information. MNAR is other than MCAR or MAR and includes the idea that the probability of missing data depends on unknown variables. For the scale example, it could be that the scale falls out of calibration with use on either surface and affects the probability of missingness. The calibration problem may be unknown to us.

Using the ψ variable above leaves the impression that this is something we need to estimate from the data. We would really prefer to ignore ψ because this deals with the issue of missingness as opposed to the embedded parameters θ of the Y_i data, which is our main objective. This question was articulated by Little and Rubin in 2002 with the concept of *ignorable*. That is, the missing data mechanism is ignorable for likelihood inference if (1) the data are MAR and (2) the joint parameter space of (ψ, θ) is the product of the parameter space of ψ and θ, respectively. This is another way of saying that there is no functional relationship ψ and θ.

The concept of ignorability is important for imputation models. Under ignorability,

$$\mathbb{P}(Y_{mis}|Y_{obs}, R) = \mathbb{P}(Y_{mis}|Y_{obs})$$

which implies that

$$\mathbb{P}(Y|Y_{obs}, R = 1) = \mathbb{P}(Y|Y_{obs}, R = 0)$$

This means that we can model the posterior distribution from the observed data and use this model to impute missing data. That is, if we don't have ignorability, then we have to include the R into the posterior distribution. Practically speaking, this means we have to develop two models for the posterior distribution, one for $R = 1$ and the other for $R = 0$. The MAR requirement above is important. For example, if we measure body weight between genders and ignorability is assumed, then missing data in a particular gender can be imputed from the other data in the *same* gender, not across different genders. It may be that ignorability is not appropriate, as in the case where missing data in one of the genders is based upon the measurement itself. That is, there may be some factor that increases the absent data in one of the genders based upon the measurement itself. This scenario requires careful data collection design and more advanced modeling techniques. The goal of multiple imputation is to obtain estimates of the parameter of interest that is valid for the entire population as if it had been available from the start. Because missingness fundamentally represents a lack of information, one can only say that the estimated population parameter under imputation should *on average* be equal to the target population parameter.

Example Let us consider a simple missingness model that illustrates these categories. Consider the following pair of jointly distributed Gaussian random variables where $(Y_1, Y_2) \sim \mathcal{N}(0, \rho = 0.5)$. In this case Y_2 is missing $(R_2 = 0)$ according to the following missing data model:

$$\mathbb{P}_M(R_2 = 0) = \psi_0 + \frac{\exp Y_1}{1 + \exp Y_1}\psi_1 + \frac{\exp Y_2}{1 + \exp Y_2}\psi_2$$

Under MCAR (Missing Completely at Random), we have $\psi = (0.5, 0, 0)$ which means that whether or not Y_2 is missing does not depend on either of the Y_i variables. Under MAR (missing at random), we have $\psi = (0, 1, 0)$ which means that the probability of missing Y_2 depends upon the observed variable Y_1. Importantly, there is no ψ_0, meaning that there is no arbitrariness of the missing Y_2. Under MNAR (Missing Not at Random), we have $\psi = (0, 0, 1)$, which means that the probability of missing Y_2 is a function of Y_2. The following code illustrates this missingness model:

```
>>> from scipy.stats import multivariate_normal, bernoulli
>>> from scipy.special import expit
>>> import pandas as pd
>>> rho = 0.5 # correlation coefficient
>>> rv = multivariate_normal([0,0],[[1,rho],[rho,1]])
>>> Ns = 2000
>>> df = pd.DataFrame(data=rv.rvs(Ns),columns=['Y_1','Y_2'])
>>> df['MCAR'] = bernoulli(0.5).rvs(Ns)
>>> df['MAR'] = bernoulli(expit(df.Y_1)).rvs()
>>> df['MNAR'] = bernoulli(expit(df.Y_2)).rvs()
>>> df.head()
      Y_1     Y_2  MCAR  MAR  MNAR
0 -0.504   0.030     0    0     1
1  0.263   0.959     1    1     1
2  0.844   1.451     0    1     0
3 -0.390  -0.830     1    1     0
4  0.432  -0.115     1    0     0
```

Now, we can compute the correlation matrix when omitting the missing cases for each category:

```
>>> df.loc[df.MCAR==1 ,['Y_1','Y_2']].corr()
        Y_1    Y_2
Y_1 1.000  0.486
Y_2 0.486  1.000

>>> df.loc[df.MAR==1 ,['Y_1','Y_2']].corr()
        Y_1    Y_2
Y_1 1.000  0.506
Y_2 0.506  1.000

>>> df.loc[df.MNAR==1 ,['Y_1','Y_2']].corr()
        Y_1    Y_2
Y_1 1.000  0.501
Y_2 0.501  1.000
```

Although mechanically similar, imputation is not prediction. The difference is that prediction is an attempt to re-create the lost data and imputation is trying to obtain statistically valid inferences from incomplete data. For example, to develop a prediction model, we can start with a complete dataset and then remove certain elements to artificially create missingness and then replace those elements according to the prediction model. We naturally choose the prediction model that minimizes the difference (e.g., minimizes the mean squared error between the two datasets) between the original complete data and the data that has been filled in with the predictions. The problem with this procedure is that the prediction model is optimized to the most probable values that are present and diminishes the uncertainty in the actual incomplete data. For example, it could be that in the real incomplete data, the missing data represents the majority of difficult cases and that the prediction model only optimizes for observed values, thus short-changing the uncertainty in the missing data, which is unable to influence the solution.

Standard Approaches Let us quickly review some standard approaches to data imputation.

- Listwise deletion: Eliminate all cases with missing data. This causes bias and loses valuable data. This is also known as complete-case only.
- Mean imputation: Replace missing data rows with the column mean. It is fast and easy to compute but will underestimate the variance and distort the relationship between variables. It will bias any other estimate besides the mean when the data are not MCAR.
- Regression imputation: Build a model for the observed data, and the fill in the missing values with predictions from that model. This gives unbiased estimates of the mean under MCAR and unbiased estimates under MAR *if* the factors that influence missingness are part of the regression model. The variability of the imputed data is systematically underestimated. This method produces predictive values that are realistic if the existing model is *perfect*, which implies that there was not any missing information in the first place.
- Stochastic regression imputation: Similar to regression imputation but adds noise to the model. It preserves regression weights and correlation between variables. The main idea is to randomly draw from the distribution of the residuals.
- Indicator method: Add a zero for each missing value and extend the regression model by a response indicator for that variable. Produces biased regression estimates even under MCAR.

The best known general method is multiple imputation (MI) which is a stochastic regression method.

3.18.1 Multiple Imputation

Multiple imputation is very general. The main idea is to compute multiple complete datasets for a single incomplete dataset and then aggregate the desired statistic across these data.

1. Start with incomplete dataset and replace missing values by generated plausible data drawn from a distribution specifically modeled for the missing entry. Each generated dataset is identical with respect to the observed values but differs in the imputed values.
2. Estimate the parameters of interest from each newly completed dataset separately. These are different for each completed dataset.
3. Pool the derived parameters and create a single consolidated estimate with corresponding variance. This variance combines the within-imputation (per completed dataset) variance and the new variance caused by the missing data (between-imputation variance).

The big advantage here is that multiple imputation separates the solution of the missing data problem from the solution of the completed data problem (i.e., estimating the desired statistical quantities). This contributes insight. The disadvantage is that it can be computationally expensive to impute for large datasets with many variables (i.e., many columns of data).

Fraction of Missing Information Consider K imputed datasets, and let e_i be the estimates of the parameter θ with U_i the corresponding estimated sample variances of these estimates. The multiple imputation estimate is the following:

$$\bar{e}_{MI} = \frac{1}{K} \sum e_i \tag{3.22}$$

The following measures the variance across the K imputed entries:

$$B_M = \frac{1}{K-1} \sum_{i=1}^{K} (e_i - \bar{e}_{MI})^2$$

The multiple imputation sampling variance is defined as the following,

$$T_M = \bar{U}_M + \left(1 + \frac{1}{K}\right) B_M \tag{3.23}$$

where

$$\bar{U}_M = \frac{1}{K} \sum_{i=1}^{K} U_i$$

is the average of the completed-data sampling variances. With these definitions, the ratio

$$r_M = \left(1 + \frac{1}{K}\right) \frac{B_M}{T_M}$$

is the proportionate increase in the variance due to missing data. This is called the fraction of missing information (FMI). Also,

$$a_M = \frac{T_M}{\bar{U}_M}$$

is the efficiency of the multiple imputation estimate relative to the observed-data estimate. The quantities are used to determine the number of imputations required and the impact of the missing data on downstream inferences and to judge the recovery of information from subjects with partial information via imputation. For example, if FMI is smaller than the proportion of subjects that would have been discarded in the complete-case-only analysis, the information has been recovered by including the partially observed subjects. This recovery of information depends on the parameter θ being estimated and the information in the observed subjects.

The t-distributed confidence intervals for θ have the following degrees of freedom:

$$\nu_M = (K - 1)/r_M^2 \tag{3.24}$$

for the confidence interval for the θ

$$\hat{\theta}_{MI} = \bar{e}_{MI} \pm t_{\alpha/2, \nu_M} \sqrt{T_M}$$

Theoretical Bayesian Justification of Multiple Imputation Let β be the parameters we want to estimate with $\mathbf{Y}_M$ and $\mathbf{Y}_O$, the missing and observed data, respectively. What we really want is $f\left(\beta, \mathbf{Y}_M | \mathbf{Y}_O\right)$. From the Bayesian perspective, the $\mathbf{Y}_M$ is a so-called nuisance parameter that is in the way of our objective to estimate β. We can expand this out into the following:

$$f\left(\beta, \mathbf{Y}_M | \mathbf{Y}_O\right) = f\left(\mathbf{Y}_M | \mathbf{Y}_O\right) f\left(\beta | \mathbf{Y}_M, \mathbf{Y}_O\right)$$

and then marginalize this over the $\mathbf{Y}_M$ to obtain the following:

$$f\left(\beta | \mathbf{Y}_O\right) = \mathbb{E}_{\mathbf{Y}_M | \mathbf{Y}_O}\left\{f\left(\beta | \mathbf{Y}_M, \mathbf{Y}_O\right)\right\}$$

Under standard regularity conditions, the order of integration can be exchanged to obtain the posterior mean and variance for the parameters of interest β

$$\mathbb{E}\left(\beta | \mathbf{Y}_O\right) = \mathbb{E}_{\mathbf{Y}_M | \mathbf{Y}_O}\left\{\mathbb{E}_{\beta | \mathbf{Y}_M, \mathbf{Y}_O}(\beta)\right\}$$

$$\mathbb{V}\left(\beta | \mathbf{Y}_O\right) = \mathbb{E}_{\mathbf{Y}_M | \mathbf{Y}_O}\left\{\mathbb{V}_{\beta | \mathbf{Y}_M, \mathbf{Y}_O}(\beta)\right\} + \mathbb{V}_{\mathbf{Y}_M | \mathbf{Y}_O}\left\{\mathbb{E}_{\beta | \mathbf{Y}_M, \mathbf{Y}_O}(\beta)\right\}$$

where we used the law of total variance for the variance above. All of this work is designed to produce something we can approximate using K samples drawn from the $f_{\mathbf{Y}_M | \mathbf{Y}_O}$ density

$$\mathbb{E}\left(\beta | \mathbf{Y}_O\right) \simeq \frac{1}{K} \sum_{k=1}^{K}\left\{\mathbb{E}_{\beta | \tilde{\mathbf{Y}}_{M,k}, \mathbf{Y}_O}(\beta)\right\} = \hat{\beta}$$

with corresponding

$$\mathbb{V}(\boldsymbol{\beta}|\mathbf{Y}_O) \simeq \frac{1}{K} \sum_{k=1}^{K} \mathbb{V}_{\boldsymbol{\beta}|\tilde{\mathbf{Y}}_{M,k},\mathbf{Y}_O}(\boldsymbol{\beta})$$

$$+ \frac{1}{K-1} \sum_{i=1}^{K} \left\{ \mathrm{E}_{\boldsymbol{\beta}|\tilde{\mathbf{Y}}_{M,k},\mathbf{Y}_O}(\boldsymbol{\beta}) - \hat{\boldsymbol{\beta}} \right\} \left\{ \mathrm{E}_{\boldsymbol{\beta}|\tilde{\mathbf{Y}}_{M,k},\mathbf{Y}_O}(\boldsymbol{\beta}) - \hat{\boldsymbol{\beta}} \right\}^T$$

Keep in mind that in order to compute the summation, we have to characterize the $f_{\mathbf{Y}_M|\mathbf{Y}_O}$. However, the full posterior density $f(\boldsymbol{\beta}, \mathbf{Y}_M|\mathbf{Y}_O)$ cannot generally be characterized by just the first two moments. This is where the Bernstein von-Mises theorem comes in. The Bernstein von-Mises theorem applies when the full data estimating equations are likelihood score equations. Briefly, the Bernstein von-Mises theorem states that the joint posterior distribution tends to the multivariate normal distribution as the sample size increases. Furthermore, as the sample size increases, the likelihood dominates the prior distribution so the mode of the likelihood (i.e., maximum likelihood estimator) can be used to obtain the required moments above. Because of the prior marginalization of $\mathbf{Y}_M$, only the resulting part of the posterior is being approximated in this way, which improves the approximation. More details in [7].

3.18.2 Canonical Example for Multiple Imputation

Normal Distribution with Known Variance (σ^2) In this section we will investigate the case for which the estimates for the mean and variance for the MI estimator are exact. Consider n samples for $Y \sim \mathcal{N}(\mu, \sigma)$ where σ is known a priori with n_O the number of values observed and n_M, the number of missing items. To construct each imputed dataset, we have to draw from the posterior density $\mathbf{Y}_M|\mathbf{Y}_O$. Given the observed items, we compute the unbiased estimator of μ

$$\hat{\mu} = \frac{1}{n_O} \sum_{i \in \mathcal{O}} Y_i$$

with corresponding variance

$$\mathbb{V}(\hat{\mu}) = \frac{\sigma^2}{n_O}$$

From the Bernstein von-Mises theorem, we know what the posterior distribution for this case is $f(\mu|\mathbf{Y}_O) \sim \mathcal{N}(\mu, \sigma^2)$ which we approximate as $\mathcal{N}(\hat{\mu}, \sigma^2)$. Nonetheless, in order to complete the imputation process, we have to design a posterior distribution for $\mathbf{Y}_M|\mathbf{Y}_O$ which we define as the following:

$$\mathbf{Y}_M | \mathbf{Y}_O \sim \mathcal{N}\left(\hat{\mu} \mathbf{1}_{n_M}, \sigma^2 \left(\mathbf{I}_{n_M} + \frac{1}{n_O} \mathbf{1}_{n_M} \mathbf{1}_{n_M}^T \right) \right) \tag{3.25}$$

For the kth iteration in MI, we have to draw the missing values and put them into the estimator for the mean:

$$\bar{Y}_k = \frac{1}{n} \left(\sum_{i \in \mathcal{M}} Y_{m,i} + n_O \hat{\mu} \right)$$

It is straightforward to show that

$$\mathbb{E}_{\mathbf{Y}_M | \mathbf{Y}_O}(\bar{Y}_k) = \hat{\mu}$$

Keep in mind that for fixed $\mathbf{Y}_O$, $\hat{\mu}$ is also fixed because it is a function of the observed data. Thus, we have the following:

$$\bar{Y}_k \sim \mathcal{N}(\hat{\mu}, \sigma^2 \frac{n_M}{n^2} + \sigma^2 \frac{n_M^2}{n^2 n_O})$$

Note that the second term in the variance comes from the off diagonal terms in the above covariance matrix for $\mathbf{Y}_M | \mathbf{Y}_O$. The MI estimator for the mean is the following:

$$\hat{\mu}_{MI} = \frac{1}{K} \sum_{k=1}^{K} \bar{Y}_k$$

Importantly, this is an unbiased estimator for $\hat{\mu}$,

$$\mathbb{E}_{\mathbf{Y}_M | \mathbf{Y}_O}(\hat{\mu}_{MI}) = \hat{\mu}$$

and the $\bar{Y}_k$ are mutually independent. Thus, we have the following:

$$\mathbb{V}_{\mathbf{Y}_M | \mathbf{Y}_O}(\hat{\mu}_{MI}) = \frac{1}{K} \frac{n_M \sigma^2}{n^2} + \frac{1}{K} \frac{n_M^2 \sigma^2}{n^2 n_O}$$

To remove the conditioning on $\mathbf{Y}_O$ and obtain the unconditional variance, we use the law of total variance,

$$\mathbb{V}(\hat{\mu}_{MI}) = \mathbb{E}(\mathbb{V}(\hat{\mu}_{MI} | \mathbf{Y}_O)) + \mathbb{V}(\mathbb{E}(\hat{\mu}_{MI} | \mathbf{Y}_O))$$

Plugging in the prior terms and recognizing that we are taking the expectations over the $Y_O \sim \mathcal{N}(\mu, \sigma^2)$, we obtain

$$\mathbb{V}(\hat{\mu}_{MI}) = \left(\frac{\sigma^2}{n_O} + \frac{1}{K} \frac{n_M \sigma^2}{n^2} + \frac{1}{K} \frac{n_M^2 \sigma^2}{n^2 n_O} \right) = \frac{\sigma^2}{n_O} \left(1 + \frac{1}{K} \frac{n_M}{n} \right)$$

The MI procedure provides estimates for the mean and variance for this estimator, as we noted previously. Taking the

$$\mathbb{E}\left(\bar{U}_M\right) = \frac{\sigma^2}{n}$$

and likewise for

$$\mathbb{E}\left(B_M\right) = \frac{n_M \sigma^2}{n n_O}$$

we obtain the following:

$$\mathbb{E}\left(T_M\right) = \mathbb{E}\left(\bar{U}_M\right) + \left(1 + \frac{1}{K}\right) \mathbb{E}\left(B_M\right) = \frac{\sigma^2}{n_O} \left(1 + \frac{1}{K} \frac{n_M}{n} \right)$$

which is exactly what we obtained earlier without using the MI estimators. This illustrates that MI is not inventing data out of nowhere, as these relationships arise automatically for this canonical situation. The subtle part is the design of the posterior distribution $\mathbf{Y}_M \sim \mathbf{Y}_O$ for sampling the missing items.

The more general case for unknown σ^2 follows the same process, except now the σ^2 terms for the distribution of the imputed values has to be drawn from a new distribution. A good choice for this, based upon the Fisher Cochran theorem, is the following:

$$\sigma^2 | \mathbf{Y}_O \sim \frac{(n_O - 1) S_O}{X^2}$$

where $X^2 \sim \chi^2_{n_O - 1}$ and S_O is the usual sample variance over the observed data. First, sample $v \sim \chi^2_{n_O - 1}$ and then compute a value for the variance as $\sigma_s^2 = (n_O - 1) S_O / v$ and use this for the variance for the posterior distribution $\mathbf{Y}_M | \mathbf{Y}_O$ in Eq. 3.25. This means use σ_s^2 instead of σ^2 and proceed as before. The key detail is to ensure a new sample for each of the MI trials generated using Eq. 3.25. Specifically, the mean used for each MI trial is a sample of $\mathcal{N}(\hat{\mu}, \sigma_s^2)$ and σ_s^2 is used for the posterior distribution in Eq. 3.25. The samples from this posterior distribution are used to fill the missing rows for the multiple imputation.

3.18.3 Worked Example for Multiple Imputation

There are many moving parts with multiple imputation, so let us code up a specific example in detail to see how everything we have discussed in this section fits together. For this example, we won't use the aggregation formulas that combine the final estimate from the individual imputed datasets, but instead examine the histograms of the estimated parameters generated by the imputations. This provides a better feeling for how the imputation process generates the parameters that are aggregated for the final estimate.

Consider the following missingness model:

$$P_M = \text{Bernoulli}\left(p = \mathcal{S}\left(-1 - \frac{1}{2}\delta - \frac{1}{2}\xi + 3\xi\delta\right)\right)$$

where $\mathcal{S}(\cdot)$ is the sigmoid function inverse of the logit function.

$$\mathcal{S}(x) = \frac{1}{1 + e^{-x}}$$

Notice that P_M is a probabilistic function of ξ, x and δ, which we observe in the data. This means that the probability model for missing data is different for each row because each row may have different values for ξ and δ. Given $x \sim \mathcal{N}(0, 1)$, we define the following:

$$\xi \sim \text{Bernoulli}\left(p = \mathcal{S}\left(\frac{1}{4} + \frac{3}{4}x\right)\right)$$

$$\delta \sim \text{Bernoulli}\left(p = \mathcal{S}\left(-\frac{1}{2} + \frac{1}{2}\xi + \frac{1}{2}x\right)\right)$$

The following code marks rows as missing according to P_M.

```
>>> from scipy.stats import norm
>>> def gen_data(nsamples=1000):
...     x = norm(0,1).rvs(nsamples)
...     # independent exposure variable
...     xi = bernoulli(expit(0.25+0.75*x)).rvs()
...     # binary dependent variable
...     d = bernoulli(expit(-0.5+0.5*xi+0.5*x)).rvs()
...     df = pd.DataFrame(dict(x=x,xi=xi,d=d))
...     # missingness model
...     missing = bernoulli(expit(-1-0.5*d-0.5*xi+3*d*xi)).rvs()
...     df['missing']=missing.astype(bool)
...     return df
...
```

Note that `missing==True` means that the entire row is absent. The next block creates a sample for this data:

```
>>> df = gen_data()
>>> df.head()
       x   xi   d   missing
0  -0.997   1   0      True
1  -0.220   1   0     False
2  -0.005   1   1      True
3  -0.178   1   0     False
4  -0.130   0   0     False
```

The following function simulates the data. Our goal is to compute the logistic regression of δ as a function of x, ξ. We want to understand the effect of missingness on the so-derived regression coefficients (including the intercept term).

Programming Tip

The Statsmodels smf module allows specifying regression formulas in a more natural format, similar to the R-language. For example, d ~ x + y means that the variable d is a function of x and y where the intercept term is implied.

```
>>> import statsmodels.formula.api as smf
>>> def run_simulation(n=1000):
...       df = gen_data(n)
...       # the intercept term is implied
...       r = smf.logit('d ~ xi + x',df).fit(disp=False)
...       rm = smf.logit('d ~ xi + x',df[~df.missing]).fit(disp=False)
...       return pd.DataFrame(dict(complete=r.params,nonmissing=rm.params))
...
```

For each run of the simulation, we collect the so-derived logistic regression parameters. Figure 3.60 shows the histogram of the intercept value for the logistic regression for the simulated data. The red vertical line indicates the true value (i.e., $-1/2$). Note the significant bias in the estimate for the case of missing data. Importantly, because the missingness model depends on the values of the columns themselves, it is not simply reducing the statistical power (i.e., shrinking the variance of the so-computed logistic regression parameters); it is *causing* this bias. Figures 3.61 and 3.62 show the corresponding histograms for the other estimated coefficients for the ξ and the x terms in the logistic regression.

```
>>> res = pd.concat([run_simulation() for i in range(500)])
>>> res.head()
            complete   nonmissing
Intercept     -0.868       -0.757
xi             0.841       -0.555
x              0.395        0.404
Intercept     -0.393       -0.282
xi             0.315       -0.888
```

Let us investigate the missingness model P_M analytically, so we can understand how it drives the bias we have observed. The subtle part is how the random variables

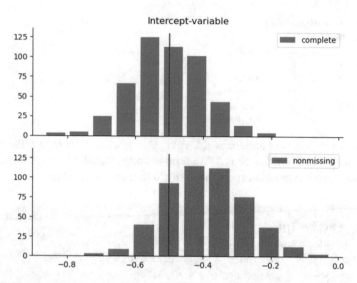

Fig. 3.60 The respective histograms show the derived intercept term for simulated data, without (top) and with missingness (bottom). Observe the bias in the data where the vertical line indicates the true value of the intercept

Fig. 3.61 The respective histograms show the derived coefficients for the ξ term for simulated data, without (top) and with missingness (bottom). Observe the bias in the data where the vertical line indicates the true value of the coefficient of ξ

interact. Keep in mind that each *row* of the data has its own distinct Bernoulli distribution based on the values along that row. First, let us consider the ξ variable:

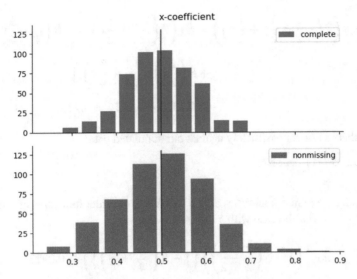

Fig. 3.62 The respective histograms show the derived coefficients for the x term for simulated data, without (top) and with missingness (bottom). Note the lack of previously observed bias in the data where the vertical line indicates the true value of the coefficient of x

$$\xi \sim \text{Bernoulli}\left(\mathcal{S}\left(\frac{1}{4} + \frac{3}{4}x\right)\right)$$

The probability that $\xi = 1$ is the following:

$$\mathbb{E}_X\left(\mathcal{S}\left(\frac{1}{4} + \frac{3}{4}x\right)\right) \approx 0.555$$

and check it against our simulated data.

```
>>> (df.xi==1).mean()
0.538
```

Given the following:

$$\delta \sim \text{Bernoulli}\left(\mathcal{S}\left(-\frac{1}{2} + \frac{1}{2}\xi + \frac{1}{2}x\right)\right)$$

the expected $\mathbb{P}(\delta = 1)$ is the following:

$$\mathbb{E}_{X,\xi}\left(S\left(-\frac{1}{2}+\frac{1}{2}\xi+\frac{1}{2}x\right)\right)=\mathbb{E}_X\left(S\left(-\frac{1}{2}+\frac{1}{2}x\right)(1-S\left(\frac{1}{4}+\frac{3}{4}x\right))\right.$$

$$\left.+S\left(\frac{1}{2}x\right)S\left(\frac{1}{4}+\frac{3}{4}x\right)\right)$$

$$\approx 0.1523 + 0.2972 = 0.4495$$

which checks out approximately against our simulated data:

```
>>> (df.d==1).mean()
0.416
```

To consider the joint probabilities of ξ, δ we have to consider four cases of $\xi \in \{0, 1\}$ and $\delta \in \{0, 1\}$. For the case with $\delta = 0 \wedge \xi = 0$,

$$\mathbb{E}_X\left(\left(1-S\left(\frac{x}{2}-\frac{1}{2}\right)\right)\left(1-S\left(\frac{3x}{4}+\frac{1}{4}\right)\right)\right) \approx 0.2923$$

For the case with $\delta = 0 \wedge \xi = 1$,

$$\mathbb{E}_X\left(S\left(\frac{3x}{4}+\frac{1}{4}\right)\left(1-S\left(\frac{x}{2}\right)\right)\right) \approx 0.2581$$

For the case with $\delta = 1 \wedge \xi = 0$,

$$\mathbb{E}_X\left(S\left(\frac{x}{2}-\frac{1}{2}\right)\left(1-S\left(\frac{3x}{4}+\frac{1}{4}\right)\right)\right) \approx 0.1523$$

For the case with $\delta = 1 \wedge \xi = 1$,

$$\mathbb{E}_X\left(S\left(\frac{3x}{4}+\frac{1}{4}\right)S\left(\frac{x}{2}\right)\right) \approx 0.2971$$

We can quickly check these values using our simulated data:

```
>>> df.groupby(['d','xi'])['x'].count().unstack()/len(df)
xi      0      1
d
0    0.316  0.268
1    0.146  0.270
```

To compute the probability of missing (P_M), we compute the logistic sigmoid for P_M each case of δ and ξ, arranged in Table 3.4 for clarity. Then, taking the $\mathbb{E}_X$ over these entries yields the following:

$$\begin{pmatrix} 0.0786242 & 0.047093 \\ 0.0277916 & 0.217241 \end{pmatrix}$$

Table 3.4 Terms to compute P_M

		δ	
		0	1
ξ	0	$\mathcal{S}(-1)\left(1-\mathcal{S}\left(\frac{x}{2}-\frac{1}{2}\right)\right)\left(1-\mathcal{S}\left(\frac{3x}{4}+\frac{1}{4}\right)\right)$	$\mathcal{S}\left(-\frac{3}{2}\right)\mathcal{S}\left(\frac{3x}{4}+\frac{1}{4}\right)\left(1-\mathcal{S}\left(\frac{x}{2}\right)\right)$
	1	$\mathcal{S}\left(-\frac{3}{2}\right)\mathcal{S}\left(\frac{x}{2}-\frac{1}{2}\right)\left(1-\mathcal{S}\left(\frac{3x}{4}+\frac{1}{4}\right)\right)$	$\mathcal{S}(1)\mathcal{S}\left(\frac{3x}{4}+\frac{1}{4}\right)\mathcal{S}\left(\frac{x}{2}\right)$

then, finally summing this up gives ≈ 0.3707. This checks out (approximately) against our simulated data:

```
>>> df.missing.mean()
0.347
```

Programming Tip
The Fortran library QUADPACK is available via `scipy.integrate` module and can compute the expectations above. For example, to compute $\mathbb{E}_X(\mathcal{S}(\frac{1}{4}+\frac{3}{4}x))$, we can do the following:

```
>>> from scipy.stats import norm
>>> from scipy.integrate import quad
>>> nrv = norm(0,1) # N(0,1) random variable
>>> f = lambda x: expit(1/4+3/4*x)*nrv.pdf(x) # integrand
>>> quad(f,-10,10) # no need to integrate all infinity
(0.5553083202603288, 4.393639370728342e-09)
```

The second term in the tuple above is the absolute error of the result. See the `scipy.integrate.quad` documentation for more details.

Figures 3.60, 3.61 and 3.62 show the bias of the logistic regression parameters created by missingness. The purpose of MI is to squeeze more out of the observed data by imputing data that reduces this bias. Our next step is to implement the analytical results in Sect. 3.18.2 by imputing the normally distributed variable with unknown variance:

```
>>> from scipy import stats
>>> from scipy.stats import multivariate_normal as mvn
>>> def imputeRows(Yo,m=100):
...     'Impute m-items with observed data, Yo.'
...     assert m>0
...     nO = len(Yo)
...     omu = Yo.mean() # observed mean
...     ovar = Yo.var() # observed variance
...     mns=[]
...     vars=[]
...     X2 = stats.chi2(nO-1) # chi2 distribution
...     v=(nO-1)*ovar/X2.rvs() # random variate
```

```
...      muhatk=stats.norm(omu,np.sqrt(v/nO)) # normal
         distribution
...      # Y_M|Y_O multivariate normal distribution
...      Ymi=mvn(np.ones(m)*muhatk.rvs(),
...                  v*np.ones((m,m))/nO+v*np.eye(m)).rvs()
...      # concatenate imputed data
...      Yc = np.r_[Ymi,Yo]
...      return Yc
...
```

Because δ and ξ are discrete, we partition the x-data into subsets based on these
pairs. For example,

```
>>> Xvals = df.query('d==0 and xi==0 and not missing')['x']
```

Then we impute x for the number of missing items in that subset:

```
>>> nO = Xvals.shape[0] # number observed in subset
>>> n = df.query('d==0 and xi==0').shape[0] # number total
        per subset
>>> Xc = imputeRows(Xvals,m=n-nO) # m is number imputed
>>> tf = pd.DataFrame(dict(x=Xc)).assign(xi=0,d=0) #
        construct completed data
>>> tf.head()
        x   xi   d
0  -2.133    0   0
1   0.138    0   0
2  -0.540    0   0
3  -0.363    0   0
4   0.672    0   0
```

This approach reduces the problem to that of Sect. 3.18.2 on a per-subset basis where
each subset is indexed by the discrete (ξ, δ) pair. Thus, the parameters required to
impute x are derived from the observed x data each subset. Then, the imputation
uses these parameters that fill in the missing data for each of the respective subsets.
The next function computes this for all of the subsets and then assembles the final
imputed data.

```
>>> def imputeDF(df):
...      def process_subset(df,d,xi):
...          'each (d,xi) pair is a subset'
...          Xvals = df.query(f'd=={d} and xi=={xi} and not
             missing')['x']
...          # number observed in subset
...          nO = Xvals.shape[0]
...          # number total per subset
...          n = df.query(f'd=={d} and xi=={xi}').shape[0]
...          Xc = imputeRows(Xvals,m=n-nO)
...          # pack into dataframe
...          return pd.DataFrame(dict(x=Xc)).assign(d=d,xi=xi)
...      tf =pd.concat([process_subset(df,i,j)
...                          for i in [0,1] for j in [0,1] ],
```

```
...                        ignore_index=True)
...          return tf
...
```

Now that we have a way of imputing the missing x items, we want to re-run our original simulation where we repeatedly performed the logistic regression and examined the population of so-created regression coefficients (see Figs. 3.60, 3.61 and 3.62). However, in this case, we impute the missing data for each run and then examine the population of logistic regression coefficients as shown in Figs. 3.63, 3.64 and 3.65 that show the corresponding histograms using the imputed data. Note how the bias in each case has been greatly reduced. This is the great value of imputing missing data.

```
>>> def run_MI_simulation(n=1000):
...          # simulate fresh data
...          df = gen_data(n)
...          # do full regression with complete data
...          r=smf.logit('d ~ xi + x',df).fit(disp=False)
```

Fig. 3.63 The respective histograms show the derived coefficients for the intercept term for simulated data, without (top), with missingness (middle), and with multiple imputation (bottom). The red vertical line is the true value. The top two plots are similar to Fig. 3.60 but for a different simulation run. Note that the bottom plot shows that missing imputation is able to correct the bias shown in the middle plot

Fig. 3.64 The respective histograms show the derived coefficients for the ξ term for simulated data, without (top), with missingness (middle), and with multiple imputation (bottom). The red vertical line is the true value. The top two plots are similar to Fig. 3.61, but for a different simulation run. Multiple imputation has reduced the bias observed in the top plot

```
...      # do logistic regression with observed data
...      rm=smf.logit('d ~ xi + x',df[~df.missing]).fit(disp=False)
...      # build imputed data
...      tf = imputeDF(df)
...      # do logistic regression on the imputed data
...      tm=smf.logit('d ~ xi + x',tf).fit(disp=False)
...      # assemble and return logistic regression coefficients
...      return pd.DataFrame(dict(complete=r.params,
...                               nonmissing=rm.params,
...                               mi=tm.params))
...
>>> # run simulation to gather logistic regression coefficients
>>> res=pd.concat([run_MI_simulation() for i in range(500)])
```

This example highlights the following central points:

- Multiple imputation's main advantage is that it does not require specification of the imputation model or special calculations.
- Multiple imputation yields good estimators for few imputations.

Fig. 3.65 The respective histograms show the derived coefficients for the x term for simulated data, without (top), with missingness (middle), and with missing imputation (bottom). The top two plots are similar to Fig. 3.62, but for a different simulation run. There is no bias in this case so multiple imputation is not helpful here

- Multiple imputation works best with estimators which are (asymptotically) normally distributed. For example, for logistic regression, the log-odds ratio scale is asymptotically normally distributed.

3.18.4 Multivariate Imputation by Chained Equations (MICE)

The prior example and the one before that used a known model to impute values for a univariate normally distributed random variable. For the general case, we can use multivariate imputation by chained equations (MICE). Recall when we discussed Gibbs sampling for Markov chain Monte Carlo, we marginalized a joint distribution and then sampled along each of the marginal distributions. This allowed us to sample from a complicated joint distribution when doing so directly was too difficult. Similarly, the main idea with MICE is to sample for missing variables conditional on the remaining covariates and use the samples to fill in for the missing values.

$$f(Y_{miss}|Y_{obs}, X_1, \ldots, X_n)$$

Because we don't know the joint distribution f, we cannot directly marginalize to obtain samples for Y_{miss}. Here we denote X_i as those covariates that are complete (i.e., have no missing values) and Y_i for other variables that are subject to missingness. We signify Y_{-j} as the set of Y_i variables *except* for j.

Let us assume the columns can be ordered by the number of missing values per column such that the first Y column has the least missing values. Concretely, the algorithm starts by identifying the Y_1 value with the fewest missing values and then regressing X_i onto Y_1. This generates regression parameters $\hat{\beta}_1$. The next step is to draw from $f_{\hat{\beta}_1}(Y_1|X_i)$ and use those values to fill in for the missing values in Y_1. Note that each filled-in value is *different* because it is a random draw from the regressed conditional distribution. By the end of this step, we have a freshly completed Y_1 column with no missing values. Next, we take the next Y_2 and regress again using the X_i variables and also the newly completed Y_1 column. Again, this generates regression parameters $\hat{\beta}_2$, and again those are used to sample $f_{\hat{\beta}_2}(Y_2|Y_1, X_i)$ and fill in the missing Y_2 values. The algorithm proceeds through the remaining Y_i variables until the entire dataset is now completed. The flexibility of this approach is that we are free to choose the regression models for each Y_i variables, and this works with continuous or discrete data.

Let us examine the case of imputing a continuous variable using a basic example to reduce the notational noise that comes with trying to write out the general case. Given Y_1 has the least number of missing entries, we compute the following regression:

$$\mathbf{Y_1} \approx \mathbf{W}\boldsymbol{\beta} + \boldsymbol{\epsilon}$$

where $\mathbf{W} = [\mathbf{1}, \mathbf{X}]$ is the design matrix of the linear regression and $\boldsymbol{\epsilon}$ is additive Gaussian noise vector with zero mean. Using ordinarily squares, the only random element here is the $\boldsymbol{\epsilon}$, and so we can solve for the expected value of $\boldsymbol{\beta}$ with the usual:

$$\hat{\boldsymbol{\beta}} = (\mathbf{W}^T\mathbf{W})^{-1}\mathbf{W}^T\mathbf{Y_1}$$

We need to account for the fact the $\hat{\boldsymbol{\beta}}$ is a random variable due to $\boldsymbol{\epsilon}$, so we compute the covariance matrix of $\hat{\boldsymbol{\beta}}$ as

$$\mathbf{R}_{\hat{\beta}} = \hat{\sigma}_\epsilon^2(\mathbf{W}^T\mathbf{W})^{-1}$$

where $\hat{\sigma}_\epsilon^2$ is the estimated residual noise variance. We estimate this after the regression by dividing the sum of the squared residuals by the number of degrees of freedom in the regression. In this case, there would be two estimated parameters (two columns in the design matrix), so the number of degrees of freedom k for this particular example would be the count of the number of complete rows in $\mathbf{Y}_1$ minus two.

Next, we need to compute a sample of $\boldsymbol{\beta}$ coefficients by sampling from a two-dimensional standard normal distribution $\mathbf{z}$ and then using the Cholesky factorization of $\mathbf{R}_{\hat{\beta}}$ to correct for the estimated covariance. Note that we have to draw a sample u for $\hat{\sigma}_\epsilon$ from the χ^2 distribution with k degrees of freedom noted above. Then, we compute a sample of the estimated variance of the residual using

$$\sigma_\star^2 = k\hat{\sigma}_\epsilon^2 / u$$

Now, when we impute, we have a sample for the $\mathbf{R}_{\hat{\beta}}$ as the following:

$$\mathbf{R}_{\hat{\beta}} = \sigma_\star^2 (\mathbf{W}^T \mathbf{W})^{-1}$$

Recall that the Cholesky factorization $\mathbf{L}$ is the following:

$$\mathbf{L}\mathbf{L}^T = \mathbf{R}_{\hat{\beta}}$$

Thus for our first imputation, we compute the regression parameters $\boldsymbol{\beta}$ using $\mathbf{z}$ as in the following:

$$\boldsymbol{\beta}_\star = \hat{\boldsymbol{\beta}} + \sigma_\star \mathbf{L}\mathbf{z}$$

With these sampled regression parameters in hand, we draw $\mathbf{v}$ of standard normal random variates and then compute the imputed $Y_{1\star}$ values as

$$Y_{1\star} = \mathbf{W}\boldsymbol{\beta}_\star + \sigma_\star \mathbf{v}$$

Bear in mind all of that was just for a *single* imputation, so for multiple imputation we would repeat the same process with a new set of random variates multiple times to construct the consolidated result. There are a lot of steps to keep track of, so next we will code a short example that should make this clearer.

Example To illustrate MICE, let us generate some 500 samples from jointly Gaussian normal distribution:

```
>>> Rcov = np.array([[1, 0.3, 0.2],
...                  [0.3, 1, 0.3],
...                  [0.2, 0.3, 1]])
>>> df = pd.DataFrame(data =
        multivariate_normal(np.zeros(3),Rcov).rvs(500),
...                  columns=['Y1','Y2','X'])
```

We want to impose MAR missingness, so we compute the following:

```
>>> R1 = expit(df.X+1).round().astype(int) # Y1 column
>>> R2 = expit(df.X+df.Y1+0.5).round().astype(int) # Y2 column
```

Note that the missingness matrix R1 for Y1 is a function of the ever present X1 and R2 is a function of both X1 and Y1. This makes everything MAR. Next, we generate new dataframe with the *R* matrices applied to each of the columns, respectively.

```
>>> dfm = df.copy()
>>> dfm.loc[~R1.astype(bool),'Y1'] = np.nan
>>> dfm.loc[~R2.astype(bool),'Y2'] = np.nan
```

Let us see the fraction of missing data for each column:

```
>>> dfm.Y1.isna().mean(), dfm.Y2.isna().mean()  # fraction
        missing
(0.176, 0.368)
```

Note the difference in the estimated parameters between

```
>>> df.corr()
      Y1     Y2     X
Y1 1.000 0.304 0.244
Y2 0.304 1.000 0.283
X  0.244 0.283 1.000
```

and the complete cases of dfm:

```
>>> dfm.loc[dfm.apply(lambda i:not i.hasnans,axis=1),:].corr()
       Y1     Y2     X
Y1   1.000 0.170 -0.042
Y2   0.170 1.000  0.124
X   -0.042 0.124  1.000
```

Note that the off-diagonal cross-correlation terms are really far from the Rcov matrix that generated this data, which is due to the missing data in the Y1 and Y2 columns.

With all that setup, we are ready to start MICE. Since Y1 has the fewest missing values, we start with the following regression:

```
>>> y1 = smf.ols('Y1 ~ X', data=dfm).fit()
```

which will automatically drop the missing values in Y1. Keep in mind that X has no missing values. The degrees of freedom is the following:

```
>>> dof = y1.nobs - 2 # two params in linear regression
>>> u = stats.chi2(dof).rvs() # random chi2 variate
>>> sigma_star = np.sqrt(y1.ssr/u)
```

where ssr is the sum of the squared residuals. Because statsmodels already provides it, we don't have to compute the *W* matrix and can compute the Cholesky factorization as the following:

```
>>> L = np.linalg.cholesky(y1.normalized_cov_params
        *sigma_star**2)
>>> beta_star = y1.params + sigma_star * L @
        stats.norm([0,0]).rvs()
```

Now we have all the ingredients impute for the missing rows of Y_1. We can shoe-horn this into the existing statsmodels regression object and then use it to predict for the missing Y1 values:

```
>>> y1.params[:] = beta_star # force it
>>> dfm.loc[dfm.Y1.isna(),'Y1'] =
        y1.predict(dfm.loc[dfm.Y1.isna(),'X'])
```

Just to double check if we missed anything,

```
>>> dfm.Y1.hasnans # any more missing values?
False
```

Now that we have a completed Y1, let us impute Y2 using the same pattern. Note the degrees of freedom has changed because we have an extra parameter for Y1.

```
>>> y2 = smf.ols('Y2 ~ Y1+X', data=dfm).fit()
>>> dof = y2.nobs - 3 # now three params in linear regression
>>> u = stats.chi2(dof).rvs() # random chi2 variate
>>> sigma_star = np.sqrt(y2.ssr/u)
```

Again, we compute the Cholesky factorization and then draw a sample from the distribution of $\hat{\beta}$,

```
>>> L =
        np.linalg.cholesky(y2.normalized_cov_params*sigma_star**2)
>>> beta_star =
        y2.params + sigma_star * L @ stats.norm([0,0,0]).rvs()
```

Then, fill the missing values for Y2:

```
>>> y2.params[:] = beta_star
>>> dfm.loc[dfm.Y2.isna(),'Y2'] =
        y2.predict(dfm.loc[dfm.Y2.isna(),['X','Y1']]))
```

Now, we have imputed all the missing values for each of the Y1 and Y2 columns; let us compute the new updated correlation

```
>>> dfm.corr()
      Y1     Y2     X
Y1 1.000  0.166  0.318
Y2 0.166  1.000  0.111
X  0.318  0.111  1.000
```

and compare it to the complete original data df:

```
>>> df.corr()
      Y1     Y2     X
Y1 1.000  0.304  0.244
Y2 0.304  1.000  0.283
X  0.244  0.283  1.000
```

To use the MI estimates, we would have to repeat this entire process a few more times with different random variates to create the multiple imputations that MI requires for its ultimate parameter estimation. Even though this is an artificial example, it shows the individual calculations required for imputing continuous data. There is no restriction on the regression methods used even though we used simple linear regression here..

3.18.5   Diagnostics

There are only suggested guidelines for how many imputations to compute. Equation 3.24 shows the degrees of freedom, and the fraction of missing information varies by the amount of missing data in the target parameter. One strategy is to choose $100 \times r_M$ as the number of imputations. Alternatively, given adequate computational capacity, is to keep increasing the number of imputations until T_M (Eq. 3.23) and $\bar{e}_{MI}$ (see Eq. 3.22) numerically stabilize.

Because bias due to missing data is less of an issue with MCAR, we want to check if the data is MCAR. The imputation model is reasonable if the two conditional probability density functions are equivalent

$$f(U_{obs}|X_1, X_2, \ldots, X_p) = f(U^*_{missing}|X_1, X_2, \ldots, X_p)$$

where $U^*_{missing}$ is the so-completed data vector consisting of both observed and imputed values. This means we want to evaluate the histograms of the residuals of the both of these densities after we have used the X_i covariates (independent variables) to construct the estimated values for the parameters. This equivalence can be checked using the response propensity method as follows:

1. Using the covariates, compute the response propensity

$$p = \mathbb{P}(R = 1|X_1, X_2, \ldots, X_p)$$

   using logistic regression or something using the observed data (i.e., $R = 1$).
2. Evaluate the logistic regression on the rows. This gives a vector of propensity scores.
3. Regress the vector of observed and imputed values against p, and construct a histogram of the residuals for the observed and imputed values, and note the amount of overlap. Significant overlap indicates that the imputations are reasonable for MAR.

Chapter 4
Machine Learning

4.1 Introduction

Machine learning is an exciting and accelerating area. In in this chapter, we provide context and some connections to probability and statistics that should make it easier to think about machine learning and how to apply these methods to real-world problems. Statistics and machine learning both start with data, but machine learning is primarily concerned with prediction, whereas statistics is also concerned with explaining data with models. Keep in mind that the foundational ideas in statistics were developed before computation became cheap and ubiquitous so its foundations are analytical. Machine learning is not particularly interested in explanations but rather in algorithmic predictions, in the context of readily available computing storage and power. This has started to change recently due to the spread of machine learning into sensitive areas like medicine, where the predictions have to be interpreted and justified in terms of emerging privacy and legal standards. Our section on machine learning interpretability introduces some of these ideas.

This chapter starts with the overarching theory of machine learning. Getting a good handle on this theory can help contextualize the many machine learning methods in this chapter. The main Python module for this chapter is Scikit-learn, and we provide a short introduction and include many detailed worked examples that you can try for yourself. The deep learning section covers the fastest growing area of machine learning, especially for image processing, and we use the Tensorflow module for that section with many graphical examples to illustrate the internals of such methods.

© The Author(s), under exclusive license to Springer Nature Switzerland AG 2022   359
J. Unpingco, *Python for Probability, Statistics, and Machine Learning*,
https://doi.org/10.1007/978-3-031-04648-3_4

4.2   Python Machine Learning Modules

Python provides many bindings for machine learning libraries, some specialized for technologies such as neural networks and others geared toward novice users. For our discussion, we focus on the powerful and popular Scikit-learn module. Scikit-learn is distinguished by its consistent and sensible API, its wealth of machine learning algorithms, its clear documentation, and its readily available datasets that make it easy to follow along with the online documentation. Like Pandas, Scikit-learn relies on Numpy for numerical arrays. Since its release in 2007, Scikit-learn has become the most widely used, general-purpose, open-source machine learning module that is popular in both industry and academia. As with all of the Python modules we use, Scikit-learn is available on all the major platforms.

To get started, let us revisit the familiar ground of linear regression using Scikit-learn. First, let us create some data.

```
>>> import numpy as np
>>> from matplotlib.pylab import subplots
>>> from sklearn.linear_model import LinearRegression
>>> X = np.arange(10)          # create some data
>>> Y = X+np.random.randn(10)  # linear with noise
```

We next import and create an instance of the `LinearRegression` class from Scikit-learn.

```
>>> lr = LinearRegression() # create model
```

Scikit-learn has a wonderfully consistent API. All Scikit-learn objects use the `fit` method to compute model parameters and the `predict` method to evaluate the model. For the `LinearRegression` instance, the `fit` method computes the coefficients of the linear fit. This method requires a matrix of inputs where the rows are the samples and the columns are the features. The *target* of the regression are the Y values, which must be correspondingly shaped, as in the following,

```
>>> X,Y = X.reshape((-1,1)), Y.reshape((-1,1))
>>> lr.fit(X,Y)
LinearRegression()
>>> lr.coef_
array([[1.0247141]])
```

Programming Tip

The negative one in the `reshape((-1,1))` call above is for the truly lazy. Using a negative one tells Numpy to figure out what that dimension should be given the other dimension and number of array elements.

The `coef_` property of the linear regression object shows the estimated parameters for the fit. The convention is to denote estimated parameters with a trailing

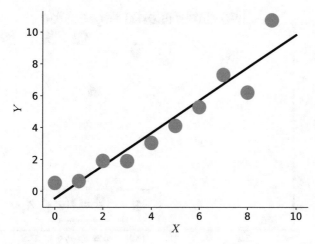

Fig. 4.1 The Scikit-learn module can easily perform basic linear regression. The circles show the *training* data and the fitted line is shown in black

underscore. The model has a `score` method that computes the R^2 value for the regression. Recall from our statistics Sect. 3.8 that the R^2 value is an indicator of the quality of the fit and varies between zero (bad fit) and one (perfect fit).

```
>>> lr.score(X,Y)
0.9056630819239118
```

Now, that we have this fitted, we can evaluate the fit using the `predict` method,

```
>>> xi = np.linspace(0,10,15) # more points to draw
>>> xi = xi.reshape((-1,1)) # reshape as columns
>>> yp = lr.predict(xi)
```

The resulting fit is shown in Fig. 4.1

Multilinear Regression The Scikit-learn module easily extends linear regression to multiple dimensions. For example, for multilinear regression,

$$y = \alpha_0 + \alpha_1 x_1 + \alpha_2 x_2 + \ldots + \alpha_n x_n$$

The problem is to find all of the α terms given the training set $\{x_1, x_2, \ldots, x_n, y\}$. We can create another example dataset and see how this works,

```
>>> X=np.random.randint(20,size=(10,2))
>>> Y=X.dot([1,3])+1 + np.random.randn(X.shape[0])*20
```

Figure 4.2 shows the two-dimensional regression example, where the size of the circles is proportional to the targeted Y value. Note that we salted the output with random noise just to keep things interesting. Nonetheless, the interface with Scikit-learn is the same,

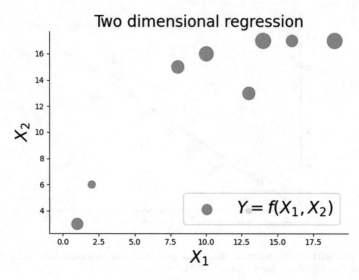

Fig. 4.2 Scikit-learn can easily perform multilinear regression. The size of the circles indicate the value of the two-dimensional function of (X_1, X_2)

```
>>> lr = LinearRegression()
>>> lr.fit(X,Y)
LinearRegression()
>>> print(lr.coef_)
[0.52252609 2.52281536]
```

The coef_ variable now has two terms in it, corresponding to the two input dimensions. Note that the constant offset is already built-in and is an option on the LinearRegression constructor. Figure 4.3 shows how the regression performs.

Polynomial Regression We can extend this to include polynomial regression by using the PolynomialFeatures in the preprocessing submodule. To keep it simple, let us go back to our one-dimensional example. First, let us create some synthetic data,

```
>>> from sklearn.preprocessing import PolynomialFeatures
>>> X = np.arange(10).reshape(-1,1) # create some data
>>> Y = X+X**2+X**3+ np.random.randn(*X.shape)*80
```

Next, we have to create a transformation from X to a polynomial of X,

```
>>> qfit = PolynomialFeatures(degree=2) # quadratic
>>> Xq = qfit.fit_transform(X)
>>> print(Xq)
[[ 1.   0.   0.]
 [ 1.   1.   1.]
 [ 1.   2.   4.]
 [ 1.   3.   9.]
```

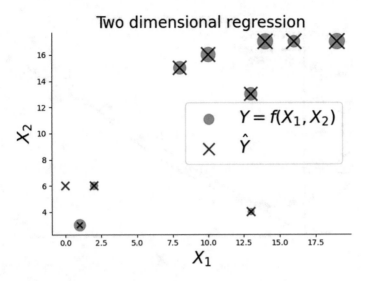

Fig. 4.3 The predicted data is plotted in black. It overlays the training data, indicating a good fit

```
[ 1.  4.  16.]
[ 1.  5.  25.]
[ 1.  6.  36.]
[ 1.  7.  49.]
[ 1.  8.  64.]
[ 1.  9.  81.]]
```

Note there is an automatic constant term in the output 0th column where `fit_transform` has mapped the single-column input into a set of columns representing the individual polynomial terms. The middle column has the linear term, and the last has the quadratic term. With these polynomial features stacked as columns of `Xq`, all we have to do is `fit` and `predict` again. The following draws a comparison between the linear regression and the quadratic regression (see Fig. 4.4),

```
>>> lr=LinearRegression() # create linear model
>>> qr=LinearRegression() # create quadratic model
>>> lr.fit(X,Y)  # fit linear model
LinearRegression()
>>> qr.fit(Xq,Y) # fit quadratic model
LinearRegression()
>>> lp = lr.predict(xi)
>>> qp = qr.predict(qfit.fit_transform(xi))
```

This just scratches the surface of Scikit-learn. We will go through many more examples later, but the main thing is to concentrate on the usage (i.e., `fit`, `predict`) which is standardized across all of the machine learning methods in Scikit-learn.

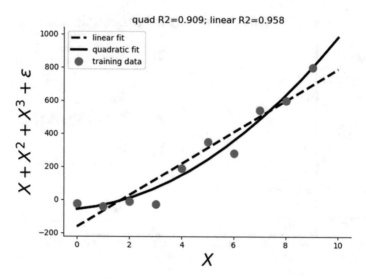

Fig. 4.4 The title shows the R^2 score for the linear and quadratic rogressions

4.3   Theory of Learning

There is nothing so practical as a good theory. In this section, we establish the formal framework for thinking about machine learning. Having a good theory is particularly important for machine learning because of all the parameter-tweaking that is unavoidable, and absent theory, can quickly devolve into superstition. This framework will help us think beyond particular methods for machine learning so we can integrate new methods or combine existing methods intelligently.

Both machine learning and statistics strive to develop understanding from data. Some historical perspective helps. Most of the methods in statistics were derived toward the start of the twentieth century when data were hard to come by. Society was preoccupied with the potential dangers of human overpopulation, and work was focused on studying agriculture and crop yields. At this time, even a dozen data points was considered plenty. Around the same time, the deep foundations of probability were being established by Kolmogorov. Thus, the lack of data meant that the conclusions had to be buttressed by strong assumptions and solid mathematics provided by the emerging theory of probability. Furthermore, inexpensive powerful computers were not yet widely available. The situation today is much different: there are lots of data collected and powerful and easily programmable computers are available. The important problems no longer revolve around a dozen data points on a farm acre, but rather millions of points on a square millimeter of a DNA microarray. Does this mean that statistics will be superseded by machine learning?

In contrast to classical statistics, which is concerned with developing models that characterize, explain, and describe phenomena, machine learning is overwhelmingly concerned with prediction. Areas like exploratory statistics are very closely related

Fig. 4.5 In the classical statistics problem, we observe a sample and model what the urn contains

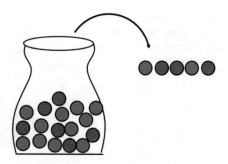

to machine learning, but still not as focused on prediction. In some sense, this is unavoidable because the size of the data machine learning can reduce. In other words, machine learning can help distill a table of a million columns into one hundred columns, but can we still interpret one hundred columns meaningfully? In classical statistics, this was never an issue because data were of a much smaller scale. Whereas mathematical models, usually normal distributions, fitted with observations are common in statistics, machine learning uses data to construct models that sit on complicated data structures and exploit nonlinear optimizations that lack closed-form solutions. A common maxim is that statistics is data plus analytical theory and machine learning is data plus computable structures. This makes it seem like machine learning is completely ad hoc and devoid of underlying theory, but this is not the case, and both machine learning and statistics share many important theoretical results. By way of contrast, let us consider a concrete problem.

Let us consider the classic balls in urns problem (see Fig. 4.5): we have an urn containing red and blue balls, and we draw five balls from the urn, note the color of each ball, and then try to determine the proportion of red and blue balls in the urn. We have already studied many statistical methods for dealing with this problem. Now, let us generalize the problem slightly. Suppose the urn is filled with white balls and there is some target unknown function f that paints each selected ball either red or blue (see Fig. 4.6). The machine learning problem is how to find the f function, given only the observed red/blue balls. So far, this doesn't sound much different from the statistical problem. However, now we want to take our estimated f function, say $\hat{f}$, and use it to predict the next handful of balls from another urn. Now, here's where the story takes a sharp turn. Suppose the next urn *already* has some red and blue balls in it? Then, applying the function f may result in purple balls which were not seen in the *training* data (see Fig. 4.7). What can we do? We have just scraped the surface of the issues machine learning must confront using methods that are not part of the statistics canon.

Fig. 4.6 In the machine learning problem, we want the function that colors the marbles

Fig. 4.7 The problem is further complicated because we may see colored marbles that were not present in the original problem

4.3.1   Introduction to Theory of Machine Learning

Some formality and an example can get us going. We define the unknown target function, $f : \mathcal{X} \mapsto \mathcal{Y}$. The training set is $\{(x, y)\}$ which means that we only see the function's inputs/outputs. The hypothesis set $\mathcal{H}$ is the set of all possible guesses at f. This is the set from which we will ultimately draw our final estimate, $\hat{f}$. The machine learning problem is how to derive the best element from the hypothesis set by using the training set. Let us consider a concrete example in the code below. Suppose $\mathcal{X}$ consists of all three-bit vectors (i.e., $\mathcal{X} = \{000, 001, \dots, 111\}$) as in the code below,

```
>>> df = pd.DataFrame(index=pd.Index(['{0:04b}'.format(i)
...                                    for i in range(2**4)],
...                          dtype='str',
...                          name='x'),
...             columns=['f'])
```

Programming Tip

The string specification above uses Python's advanced string formatting mini-language. In this case, the specification says to convert the integer into a fixed-width, four-character (04b) binary representation.

Next, we define the target function f below which just checks if the number of zeros in the binary representation exceeds the number of ones. If so, then the function outputs 1 and 0 otherwise (i.e., $\mathcal{Y} = \{0, 1\}$).

```
>>> df.f = np.array(df.index.map(lambda i:i.count('0'))
...                     > df.index.map(lambda i:i.count('1')),
                        dtype=int)
>>> df.head(8) # show top half only
        f
x
0000    1
0001    1
0010    1
0011    0
0100    1
0101    0
0110    0
0111    0
```

The hypothesis set for this problem is the set of *all* possible functions of $\mathcal{X}$. The set $\mathcal{D}$ represents all possible input/output pairs. The corresponding hypothesis set $\mathcal{H}$ has 2^{16} elements, one of which matches f. There are 2^{16} elements in the hypothesis set because for each of 16 input elements; there are 2 possible corresponding values (zero or one) for each input. Thus, the size of the hypothesis set is $2 \times 2 \times \ldots \times 2 = 2^{16}$. Now, presented with a training set consisting of the first eight input/output pairs, our goal is to minimize errors over the training set ($E_{in}(\hat{f})$). There are 2^8 elements from the hypothesis set that exactly match f over the training set. But how to pick among these 2^8 elements? It seems that we are stuck here. We need another element from the problem in order to proceed. The extra piece we need is to assume that the training set represents a random sampling (*in-sample* data) from a greater population (*out-of-sample* data) that would be consistent with the population that $\hat{f}$ would ultimately predict upon. In other words, we are assuming a stable probability structure for both the in-sample and out-of-sample data. This is a major assumption!

There is a subtle consequence of this assumption—whatever the machine learning method does once deployed, in order for it to continue to work, it cannot disturb the data environment that it was trained on. Said differently, if the method is not to be trained continuously, then it cannot break this assumption by altering the generative environment that produced the data it was trained on. For example, suppose we develop a model that predicts hospital readmissions based on seasonal weather and patient health. Because the model is so effective, in the next 6 months, the hospital forestalls readmissions by delivering interventions that improve patient health. Clearly using the model cannot change seasonal weather, but because the hospital used the model to change patient health, the training data used to build the model is no longer consistent with the forward-looking health of the patients. Thus, there is little reason to think that the model will continue to work as well going forward.

Returning to our example, let us suppose that the first eight elements from $\mathcal{X}$ are twice as likely as the last eight. The following code is a function that generates elements from $\mathcal{X}$ according to this distribution.

```
>>> np.random.seed(12)
>>> def get_sample(n=1):
...     if n==1:
...         return '{0:04b}'.format(np.random.choice(list(range
                                    (8))*2
                                                  +list(range(8,16)))))
...     else:
...         return [get_sample(1) for _ in range(n)]
...
```

Programming Tip

The function that returns random samples uses the `np.random.choice` function from Numpy which takes samples (with replacement) from the given iterable. Because we want the first eight numbers to be twice as frequent as the rest, we simply repeat them in the iterable using `range(8)*2`. Recall that multiplying a Python list by an integer duplicates the entire list by that integer. It does not do element-wise multiplication as with Numpy arrays. If we wanted the first eight to be 10 times more frequent, then we would use `range(8)*10`, for example. This is a simple but powerful technique that requires very little code. Note that the p keyword argument in `np.random.choice` also provides an explicit way to specify more complicated distributions.

The next block applies the function definition f to the sampled data to generate the training set consisting of eight elements.

```
>>> train = df.loc[get_sample(8),'f'] # 8-element training set
>>> train.index.unique().shape       # how many unique elements?
(6,)
```

Notice that even though there are eight elements, there is redundancy because these are drawn according to an underlying probability. Otherwise, if we just got all 16 different elements, we would have a training set consisting of the complete specification of f, and then we would therefore know what $h \in \mathcal{H}$ to pick! However, this effect gives us a clue as to how this will ultimately work. Given the elements in the training set, consider the set of elements from the hypothesis set that exactly match. How to choose among these? The answer is it does not matter! Why? Because under the assumption that the prediction will be used in an environment that is determined by the same probability, getting something outside of the training set is just as likely as getting something inside the training set. The size of the training set is key here—the bigger the training set, the less likely that there will be real-world data that fall outside of it and the better $\hat{f}$ will perform.[1] The following code shows the elements of the training set in the context of all possible data.

```
>>> df['fhat']=df.loc[train.index.unique(),'f']
>>> df.fhat
x
0000     NaN
0001     NaN
0010   1.000
0011   0.000
0100   1.000
0101     NaN
0110   0.000
0111     NaN
1000   1.000
1001   0.000
1010     NaN
1011     NaN
1100     NaN
1101     NaN
1110     NaN
1111     NaN
Name: fhat, dtype: float64
```

There are NaN symbols where the training set had no values. For definiteness, we fill these in with zeros, although we can fill them with anything we want so long as whatever we do is not determined by the training set.

```
>>> df.fhat.fillna(0,inplace=True) #final specification of fhat
```

Now, let us pretend we have deployed this and generate some test data.

```
>>> test= df.loc[get_sample(50),'f']
>>> (df.loc[test.index,'fhat'] != test).mean()
0.18
```

The result shows the error rate, given the probability mechanism that generates the data. The following Pandas-fu compares the overlap between the training set and the test set in the context of all possible data. The NaN values show the rows where the test data had items absent in the training data. Recall that the method returns zero for these items. As shown, sometimes this works in its favor, and sometimes not.

```
>>> pd.concat([test.groupby(level=0).mean(),
...            train.groupby(level=0).mean()],
...           axis=1,
...           keys=['test','train'])
     test   train
x
```

---

[1] This assumes that the hypothesis set is big enough to capture the entire training set (which it is for this example). We will discuss this trade-off in greater generality shortly.

```
0000 1.000     NaN
0001 1.000     NaN
0010 1.000   1.000
0011 0.000   0.000
0100 1.000   1.000
0101 0.000     NaN
0110 0.000   0.000
0111 0.000     NaN
1000 1.000   1.000
1001 0.000   0.000
1010 0.000     NaN
1011 0.000     NaN
1100 0.000     NaN
1101 0.000     NaN
1110 0.000     NaN
1111 0.000     NaN
```

Where the test data and training data shared elements, the prediction matched; but when the test set produced an unseen element, the prediction may or may not have matched.

Programming Tip

The pd.concat function concatenates the two Series objects in the list. The axis=1 means join the two objects along the columns where each newly created column is named according to the given keys. The level=0 in the groupby for each of the Series objects means group along the index. Because the index corresponds to the 4-bit elements, this accounts for repetition in the elements. The mean aggregation function computes the values of the function for each 4-bit element. Because all functions in each respective group have the same value, the mean just picks out that value because the average of a list of constants is that constant.

Now, we are in position to ask how big the training set should be to achieve a level of performance. For example, on average, how many in-samples do we need for a given error rate? For this problem, we can ask how large (on average) must the training set be in order to capture *all* of the possibilities and achieve perfect out-of-sample error rates? For this problem, this turns out to be 63.[2] Let us start over and retrain with these many in-samples.

```
>>> train=df.loc[get_sample(63),'f']
>>> del df['fhat']
>>> df['fhat']=df.loc[train.index.unique(),'f']
>>> df.fhat.fillna(0,inplace=True) #final specification of fhat
>>> test= df.loc[get_sample(50),'f']
```

---

[2] This is a slight generalization of the classic coupon collector problem.

```
>>> # error rate
>>> (df.loc[test.index,'fhat'] != df.loc[test.index,'f']).mean()
0.0
```

Notice that this bigger training set has a better error rate because it is able to identify the best element from the hypothesis set because the training set captured more of the complexity of the unknown f. This example shows the trade-offs between the size of the training set, the complexity of the target function, the probability structure of the data, and the size of the hypothesis set. Upon exposure to the data, the so-called learning method did nothing besides memorize the data and give any unknown, newly encountered data the zero output. This means that the hypothesis set contains the single hypothesis function that memorizes and defaults to zero output. If the method attempted to change the default zero output based on the particular data, then we could say that meaningful learning took place. What we lack here is *generalization*, which is the topic of the next section.

4.3.2 Theory of Generalization

What we really want to know is how our method will perform once deployed. It would be nice to have some kind of performance guarantee. In other words, we worked hard to minimize the errors in the training set, but what errors can we expect at deployment? In training, we minimized the in-sample error, $E_{\text{in}}(\hat{f})$, but that's not good enough. We want guarantees about the out-of-sample error, $E_{\text{out}}(\hat{f})$. This is what *generalization* means in machine learning. The mathematical statement of this is the following,

$$\mathbb{P}\left(|E_{\text{out}}(\hat{f}) - E_{\text{in}}(\hat{f})| > \epsilon\right) < \delta$$

for ϵ and δ. Informally, this says that the probability of the respective errors differing by more than a given ϵ is less than some quantity, δ. This means that whatever the performance on the training set, it should probably be pretty close to the corresponding performance once deployed. This does not say that the in-sample errors (E_{in}) are any good in an absolute sense. It just says that we would not expect much different after deployment. Thus, *good* generalization means no surprises after deployment, not necessarily good performance. There are two main ways to get at this: cross-validation and probability inequalities. Let us consider the latter first. There are two entangled issues: the complexity of the hypothesis set and the probability of the data. It turns out we can separate these two by deriving a separate notion of complexity free from any particular data probability.

VC Dimension We first need a way to quantify model complexity. Following Wasserman [48], let $\mathcal{A}$ be a class of sets and $F = \{x_1, x_2, \ldots, x_n\}$, a set of n data points. Then, we define

$$N_{\mathcal{A}}(F) = \#\{F \cap A : A \in \mathcal{A}\}$$

This counts the number of subsets of F that can be extracted by the sets of $\mathcal{A}$. The number of items in the set (i.e., cardinality) is noted by the # symbol. For example, suppose $F = \{1\}$ and $\mathcal{A} = \{(x \leq a)\}$. In other words, $\mathcal{A}$ consists of all intervals closed on the right and parameterized by a. In this case, we have $N_{\mathcal{A}}(F) = 1$ because all elements can be extracted from F using $\mathcal{A}$. Specifically, any $a > 1$ means that $\mathcal{A}$ contains F.

The *shatter coefficient* is defined as,

$$s(\mathcal{A}, n) = \max_{F \in \mathcal{F}_n} N_{\mathcal{A}}(F)$$

where $\mathcal{F}$ consists of all finite sets of size n. Note that this sweeps over all finite sets so we don't need to worry about any particular dataset of finitely many points. The definition is concerned with $\mathcal{A}$ and how its sets can pick off elements from the dataset. A set F is *shattered* by $\mathcal{A}$ if it can pick out every element in it. This provides a sense of how the complexity in $\mathcal{A}$ consumes data. In our last example, the set of half-closed intervals shattered every singleton set $\{x_1\}$.

Now, we come to the main definition of the Vapnik-Chervonenkis [47] dimension d_{VC} which is defined as the largest k for which $s(\mathcal{A}, n) = 2^k$, except in the case where $s(\mathcal{A}, n) = 2^n$ for which it is defined as infinity. For our example where $F = \{x_1\}$, we already saw that $\mathcal{A}$ shatters F. How about when $F = \{x_1, x_2\}$? Now, we have two points, and we have to consider whether all subsets can be extracted by $\mathcal{A}$. In this case, there are four subsets, $\{\varnothing, \{x_1\}, \{x_2\}, \{x_1, x_2\}\}$. Note that $\varnothing$ denotes the empty set. The empty set is easily extracted—pick a so that it is smaller than both x_1 and x_2. Assuming that $x_1 < x_2$, we can get the next set by choosing $x_1 < a < x_2$. The last set is likewise do-able by choosing $x_2 < a$. The problem is that we cannot capture the third set, $\{x_2\}$, without capturing x_1 as well. This means that we cannot shatter any finite set with $n = 2$ using $\mathcal{A}$. Thus, $d_{\mathrm{VC}} = 1$.

Here is the climatic result

$$E_{\mathrm{out}}(\hat{f}) \leq E_{\mathrm{in}}(\hat{f}) + \sqrt{\frac{8}{n} \ln\left(\frac{4((2n)^{d_{\mathrm{VC}}} + 1)}{\delta}\right)}$$

with probability at least $1 - \delta$. This basically says that the expected out-of-sample error can be no worse than the in-sample error plus a penalty due to the complexity of the hypothesis set. The expected in-sample error comes from the training set but the complexity penalty comes from just the hypothesis set, thus disentangling these two issues.

A general result like this, for which we do not worry about the probability of the data, is certain to be pretty generous, but nonetheless, it tells us how the complexity penalty enters into the out-of-sample error. In other words, the bound on $E_{\mathrm{out}}(\hat{f})$ gets worse for a more complex hypothesis set. Thus, this generalization bound is a useful guideline but not very practical if we want to get a good estimate of $E_{\mathrm{out}}(\hat{f})$.

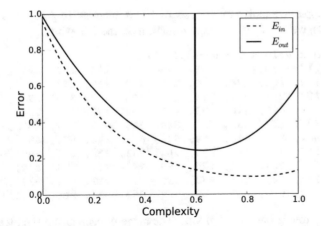

Fig. 4.8 In the ideal situation, there is a *best model* that represents the optimal trade-off between complexity and error. This is shown by the vertical line

4.3.3 Worked Example for Generalization/Approximation Complexity

The stylized curves in Fig. 4.8 illustrate the idea that there is some optimal point of complexity that represents the best generalization given the training set.

To get a firm handle on these curves, let us develop a simple one-dimensional machine learning method and go through the steps to create this graph. Let us suppose we have a training set consisting of x-y pairs $\{(x_i, y_i)\}$. Our method groups the x-data into intervals and then averages the y-data in those intervals. Predicting for new x-data means simply identifying the interval containing the new data and then reporting the corresponding value. In other words, we are building a simple one-dimensional, nearest neighbor classifier. For example, suppose the training set x-data is the following,

```
>>> train = pd.DataFrame(columns=['x','y'])
>>> train['x']=np.sort(np.random.choice(range(2**10),size=90))
>>> train.x.head(10) # first ten elements
0      15
1      30
2      45
3      65
4      76
5      82
6     115
7     145
8     147
9     158
Name: x, dtype: int64
```

In this example, we took a random set of 10-bit integers. To group these into, say, ten intervals, we simply use Numpy reshape as in the following,

```
>>> train.x.values.reshape(10,-1)
array([[ 15,   30,   45,   65,   76,   82,  115,  145,  147],
       [158,  165,  174,  175,  181,  209,  215,  217,  232],
       [233,  261,  271,  276,  284,  296,  318,  350,  376],
       [384,  407,  410,  413,  452,  464,  472,  511,  522],
       [525,  527,  531,  534,  544,  545,  548,  567,  567],
       [584,  588,  610,  610,  641,  645,  648,  659,  667],
       [676,  683,  684,  697,  701,  703,  733,  736,  750],
       [754,  755,  772,  776,  790,  794,  798,  804,  830],
       [831,  834,  861,  883,  910,  910,  911,  911,  937],
       [943,  946,  947,  955,  962,  962,  984,  989,  998]])
```

where every row is one of the groups. The range of each group (i.e., length of the interval) is not preassigned and is learned from the training data. For this example, the y-values correspond to the number of ones in the bit representation of the x-values. The following code defines this target function,

```
>>> f_target=np.vectorize(lambda i:i.count('1'))
```

Programming Tip

The above function uses `np.vectorize` which is a convenience method in Numpy that converts plain Python functions into Numpy versions. This basically saves additional looping semantics and makes it easier to use with other Numpy arrays and functions.

Next, we create the bit representations of all of the x-data below and then complete training set y-values,

```
>>> train['xb']= train.x.map('{0:010b}'.format)
>>> train.y=train.xb.map(f_target)
>>> train.head(5)
    x   y          xb
0  15   4   0000001111
1  30   4   0000011110
2  45   4   0000101101
3  65   2   0001000001
4  76   3   0001001100
```

To train on this data, we just group by the specified amount and then average the y-data over each group.

```
>>> train.y.values.reshape(10,-1).mean(axis=1)
array([3.55555556, 4.88888889, 4.44444444, 4.88888889,
       4.11111111,
       4.        , 6.        , 5.11111111, 6.44444444,
       6.66666667])
```

Note that the `axis=1` keyword argument just means average across the columns. So far, this defines the training. To predict using this method, we have to extract the edges from each of the groups and then fill in with the group-wise mean we just computed for `y`. The following code extracts the edges of each group.

```
>>> le,re=train.x.values.reshape(10,-1)[:,[0,-1]].T
>>> print (le) # left edge of group
[ 15 158 233 384 525 584 676 754 831 943]
>>> print (re) # right edge of group
[147 232 376 522 567 667 750 830 937 998]
```

Next, we compute the group-wise means and assign them to their respective edges.

```
>>> val = train.y.values.reshape(10,-1).mean(axis=1).round()
>>> func = pd.Series(index=range(1024))
>>> func[le]=val     # assign value to left edge
>>> func[re]=val     # assign value to right edge
>>> func.iloc[0]=0   # default 0 if no data
>>> func.iloc[-1]=0  # default 0 if no data
>>> func.head()
0    0.000
1      NaN
2      NaN
3      NaN
4      NaN
dtype: float64
```

The Pandas `Series` object automatically fills in unassigned values with NaN. We have thus far only filled in values at the edges of the groups. Now, we need to fill in the intermediate values.

```
>>> fi=func.interpolate('nearest')
>>> fi.head()
0    0.000
1    0.000
2    0.000
3    0.000
4    0.000
dtype: float64
```

The `interpolate` method of the `Series` object can apply a wide variety of powerful interpolation methods, but we only need the simple nearest neighbor method to create our piecewise approximant. Figure 4.9 shows how this looks for the training data we have created.

Now, with all that established, we can now draw the curves for this machine learning method. Instead of partitioning the training data for cross-validation (which

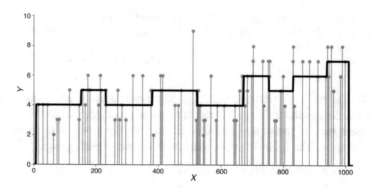

Fig. 4.9 The vertical lines show the training data and the thick black line is the approximant we have learned from the training data

we'll discuss later), we can simulate test data using the same mechanism as for the training data, as shown next,

```
>>> test=pd.DataFrame(columns=['x','xb','y'])
>>> test['x']=np.random.choice(range(2**10),size=500)
>>> test.xb= test.x.map('{0:010b}'.format)
>>> test.y=test.xb.map(f_target)
>>> test.sort_values('x',inplace=True)
```

The curves are the respective errors for the training data and the testing data. For our error measure, we use the mean squared error,

$$E_{\text{out}} = \frac{1}{n} \sum_{i=1}^{n} (\hat{f}(x_i) - y_i)^2$$

where $\{(x_i, y_i)\}_{i=1}^{n}$ come from the test data. The in-sample error (E_{in}) is defined the same except for the in-sample data. In this example, the size of each group is proportional to d_{VC}, so the more groups we choose, the more complexity in the fitting. Now, we have all the ingredients to understand the trade-offs of complexity versus error.

Figure 4.10 shows the curves for our one-dimensional clustering method. The dotted line shows the mean squared error on the training set, and the other line shows the same for the test data. The shaded region is the *complexity penalty* of this method. With enough complexity, the method can exactly memorize the testing data, but that only penalizes the testing error (E_{out}). This effect is exactly what the Vapnik-Chervonenkis theory expresses. The horizontal axis is proportional to the VC dimension. In this case, complexity boils down to the number of intervals used in the sectioning. At the far right, we have as many intervals as there are elements in the dataset, meaning that every element is wrapped in its own interval. The average value of the data in that interval is therefore just the corresponding y value because there are no other elements to average over.

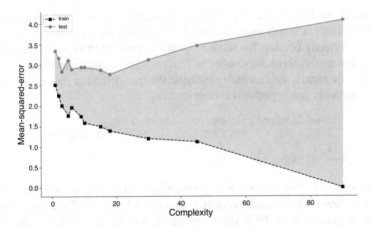

Fig. 4.10 The dotted line shows the mean squared error on the training set and the other line shows the same for the test data. The shaded region is the *complexity penalty* of this method. As the complexity of the model increases, the training error decreases, and the method essentially memorizes the data. However, this improvement in training error comes at the cost of larger testing error

Before we leave this problem, there is another way to visualize the performance of our learning method. This problem can be thought of as a multi-class identification problem. Given a 10-bit integer, the number of ones in its binary representation is in one of the classes $\{0, 1, \ldots, 10\}$. The output of the model tries to put each integer in its respective class. How well this was done can be visualized using a *confusion matrix* as shown in the next code block,

```
>>> from sklearn.metrics import confusion_matrix
>>> cmx = confusion_matrix(test.y.values,fi[test.x].values)
>>> print(cmx)
[[ 1  0  0  0  0  0  0  0  0  0]
 [ 1  0  1  0  1  1  0  0  0  0]
 [ 0  0  3  9  7  4  0  0  0  5]
 [ 1  0  3 23 19  6  6  0  2  0]
 [ 0  0  1 26 27 14 27  2  2  0]
 [ 0  0  3 15 31 28 30  8  1  0]
 [ 0  0  1  8 18 20 25 23  2  2]
 [ 1  0  1 10  5 13  7 19  3  6]
 [ 4  0  1  2  0  2  2  7  4  3]
 [ 2  0  0  0  0  1  0  0  0  0]]
```

The rows of this 10×10 matrix show the true class, and the columns indicate the class that the model predicted. The numbers in the matrix indicate the number of times that association was made. For example, the first row shows that there was one entry in the test set with no ones in its binary representation (i.e., the number zero) and it was correctly classified (namely, it is in the first row, first column of the matrix). The second row shows there were four entries total in the test set with a binary representation containing exactly a single one. This was incorrectly classified as the 0-class (i.e., first column) once, the 2-class (third column) once, the 4-class (fifth column) once, and the 5-class (sixth column) once. It was never classified correctly because the second column is zero for this row. In other words, the diagonal entries show the number of times it was correctly classified.

Using this matrix, we can easily estimate the true-detection probability that we covered earlier in our hypothesis testing section,

```
>>> print(cmx.diagonal()/cmx.sum(axis=1))
[1.          0.          0.10714286 0.38333333 0.27272727
 0.24137931
 0.25252525 0.29230769 0.16        0.         ]
```

In other words, the first element is the probability of detecting 0 when 0 is in force, the second element is the probability of detecting 1 when 1 is in force, and so on. We can likewise compute the false-alarm rate for each of the classes in the following,

```
>>> print((cmx.sum(axis=0)-cmx.diagonal())/(cmx.sum()-cmx.
    sum(axis=1)))
[0.01803607 0.          0.02330508 0.15909091 0.20199501
 0.15885417
 0.17955112 0.09195402 0.02105263 0.03219316]
```

Programming Tip

The Numpy sum function can sum across a particular axis or, if the axis is unspecified, will sum all entries of the array.

In this case, the first element is the probability that 0 is declared when another category is in force, the next element is the probability that 1 is declared when another category is in force, and so on. For a decent classifier, we want a true-detection probability to be greater than the corresponding false-alarm rate; otherwise the classifier is no better than a coin-flip.

The missing feature of this problem, from the learning algorithm standpoint, is that we did not supply the bit representation of every element which was used to derive the target variable, y. Instead, we just used the integer value of each of the 10-bit numbers, which essentially concealed the mechanism for creating the y values. In other words, there was an unknown transformation from the input space $\mathcal{X}$ to $\mathcal{Y}$ that the learning algorithm had to overcome, but that it could not overcome, at least not without memorizing the training data. This lack of knowledge is a key issue in all machine learning problems, although we have made it explicit here with this stylized example. This means that there may be one or more transformations from $\mathcal{X} \rightarrow \mathcal{X}'$ that can help the learning algorithm get traction on the so-transformed space while providing a better trade-off between generalization and approximation than could have been achieved otherwise. Finding such transformations is called *feature engineering*.

4.3.4   Cross-Validation

In the last section, we explored a stylized machine learning example to understand the issues of complexity in machine learning. However, to get an estimate of out-of-sample errors, we simply generated more synthetic data. In practice, this is not an option, so we need to estimate these errors from the training set itself. This is what cross-validation does. The simplest form of cross-validation is k-fold validation. For example, if $K = 3$, then the training data is divided into three sections wherein each of the three sections is used for testing and the remaining two are used for training. This is implemented in Scikit-learn as in the following,

```
>>> import numpy as np
>>> from sklearn.model_selection import KFold
>>> data =np.array(['a',]*3+['b',]*3+['c',]*3) # example
>>> print (data)
['a' 'a' 'a' 'b' 'b' 'b' 'c' 'c' 'c']
>>> kf = KFold(3)
>>> for train_idx,test_idx in kf.split(data):
...     print (train_idx,test_idx)
...
[3 4 5 6 7 8] [0 1 2]
[0 1 2 6 7 8] [3 4 5]
[0 1 2 3 4 5] [6 7 8]
```

In the code above, we construct a sample data array and then see how KFold splits it up into indices for training and testing, respectively. Notice that there are no duplicated elements in each row between training and testing indices. To examine the elements of the dataset in each category, we simply use each of the indices as in the following,

```
>>> for train_idx,test_idx in kf.split(data):
...     print('training', data[ train_idx ])
...     print('testing' , data[ test_idx ])
...
training ['b' 'b' 'b' 'c' 'c' 'c']
testing ['a' 'a' 'a']
training ['a' 'a' 'a' 'c' 'c' 'c']
testing ['b' 'b' 'b']
training ['a' 'a' 'a' 'b' 'b' 'b']
testing ['c' 'c' 'c']
```

This shows how each group is used in turn for training/testing. There is no random shuffling of the data unless the shuffle keyword argument is given. The error over the test set is the *cross-validation error*. The idea is to postulate models of differing complexity and then pick the one with the best cross-validation error. For example, suppose we had the following sine wave data,

```
>>> xi = np.linspace(0,1,30)
>>> yi = np.sin(2*np.pi*xi)
```

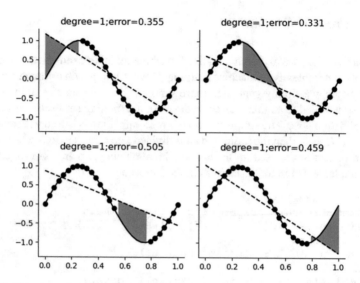

Fig. 4.11 This shows the folds and errors for the linear model. The shaded areas show the errors in each respective test set (i.e., *cross-validation scores*) for the linear model

and we want to fit this with polynomials of increasing order.

Figure 4.11 shows the individual folds in each panel. The circles represent the training data. The diagonal line is the fitted polynomial. The gray shaded areas indicate the regions of errors between the fitted polynomial and the held-out testing data. The larger the gray area, the bigger the cross-validation errors, as are reported in the title of each frame.

After reviewing the last four figures and averaging the cross-validation errors, the one with the least average error is declared the winner. Thus, cross-validation provides a method of using a single dataset to make claims about unseen out-of-sample data insofar as the model with the best complexity can be determined. The entire process to generate the above figures can be captured using `cross_val_score` as shown for the linear regression (compare the output with the values in the titles in each panel of Fig. 4.11),

```
>>> from sklearn.metrics import  make_scorer, mean_squared_error
>>> from sklearn.model_selection import cross_val_score
>>> from sklearn.linear_model import LinearRegression
>>> Xi = xi.reshape(-1,1) # refit column-wise
>>> Yi = yi.reshape(-1,1)
>>> lf = LinearRegression()
>>> scores = cross_val_score(lf,Xi,Yi,cv=4,
...                          scoring=make_scorer(mean_squared
                             _error))
>>> print(scores)
[0.3554451  0.33131438 0.50454257 0.45905672]
```

> **Programming Tip**
>
> The `make_scorer` function is a wrapper that enables `cross_val_score` to compute scores from the given estimator's output.

The process can be further automated by using a pipeline as in the following,

```
>>> from sklearn.pipeline import Pipeline
>>> from sklearn.preprocessing import PolynomialFeatures
>>> polyfitter = Pipeline([('poly', PolynomialFeatures
...    (degree=3)),  ('linear', LinearRegression())])
>>> polyfitter.get_params()
{'memory': None, 'steps': [('poly', PolynomialFeatures
(degree=3)), ('linear', LinearRegression())], 'verbose':
 False, 'poly': PolynomialFeatures(degree=3), 'linear':
LinearRegression(), 'poly__degree': 3, 'poly__include_
bias': True, 'poly__interaction_only': False, 'poly__
order': 'C','linear__copy_X': True, 'linear__fit_
intercept': True, 'linear__n_jobs': None, 'linear__
normalize': 'deprecated', 'linear__positive': False}
```

The `Pipeline` object is a way of stacking standard steps into one big estimator, while respecting the usual `fit` and `predict` interfaces. The output of the `get_params` function contains the polynomial degrees we previously looped over to create Fig. 4.11, etc. We will use these named parameters in the next code block. To do this automatically using this `polyfitter` estimator, we need the Grid Search Cross Validation object, `GridSearchCV`. The next step is to use this to create the grid of parameters we want to loop over as in the following,

```
>>> from sklearn.model_selection import GridSearchCV
>>> gs=GridSearchCV(polyfitter,{'poly__degree':[1,2,3]},
...                   cv=4,return_train_score=True)
```

The `gs` object will loop over the polynomial degrees up to cubic using four-fold cross-validation `cv=4`, like we did manually earlier. The `poly__degree` item comes from the previous `get_params` call. Now, we just apply the usual `fit` method on the training data,

```
>>> _ = gs.fit(Xi,Yi)
>>> gs.cv_results_
{'mean_fit_time': array([0.00063854, 0.00056225, 0.0005722 ]),
 'std_fit_time': array([1.11161160e-04, 4.99674308e-05,
5.23417027e-05]), 'mean_score_time': array([0.00030696,
0.0003503 , 0.00029844]), 'std_score_time': array
([2.13144743e-05, 6.21535275e-05, 2.81502458e-05]),
 'param_poly__degree': masked_array(data=[1, 2, 3],
           mask=[False, False, False],
      fill_value='?',
           dtype=object), 'params': [{'poly__degree': 1},
           {'poly__degree': 2}, {'poly__degree': 3}], 'split0
```

```
_test_score': array([ -2.03118491, -68.54947351,
-1.64899934]), 'split1_test_score': array
([-1.38557769, -3.20386236,  0.81372823]),
'split2_test_score': array([ -7.82417707,
-11.8740862 ,  0.47246476]), 'split3_test_score':
array([ -3.21714294, -60.70054797,  0.14328163]),
 'mean_test_score': array([ -3.61452065,
-36.08199251,  -0.05488118]), 'std_test_score':
array([ 2.51765044, 28.84096377,  0.95040218]),
'rank_test_score': array([2, 3, 1], dtype=int32),
'split0_train_score': array([0.52652515,
0.93434227, 0.99177894]), 'split1_train_score':
array([0.5494882 , 0.60357784, 0.99154288]),
'split2_train_score': array([0.54132528,
0.59737218, 0.99046089]), 'split3_train_score':
array([0.57837263, 0.91061274, 0.99144127]),
 'mean_train_score': array([0.54892781, 0.76147626,
0.99130599]), 'std_train_score': array([0.01888775,
0.16123462, 0.00050307])}
```

The scores shown correspond to the cross-validation scores for each of the parameters (e.g., polynomial degrees) using four-fold cross-validation. The higher scores are better here and the cubic polynomial is best, as we observed earlier. The default R^2 metric is used for the scoring in this case as opposed to mean squared error. The validation results of this pipeline for the quadratic fit are shown in Fig. 4.12 and for the cubic fit, in Fig. 4.13. This can be changed by passing the scoring=make_scorer(mean_squared_error) keyword argument to GridSearchCV. There is also RandomizedSearchCV that does not necessarily evaluate every point on the grid and instead randomly samples the grid according to an input probability distribution. This is very useful for a large number of hyper-parameters.

4.3.5 Bias and Variance

Considering average error in terms of in-samples and out-samples depends on a particular training dataset. What we want is a concept that captures the performance of the estimator for *all* possible training data. For example, our ultimate estimator, $\hat{f}$, is derived from a particular set of training data ($\mathcal{D}$) and is thus denoted, $\hat{f}_D$. This makes the out-of-sample error explicitly, $E_{\text{out}}(\hat{f}_D)$. To eliminate the dependence on a particular set of training dataset, we have to compute the expectation across all training datasets,

$$\mathbb{E}_D E_{\text{out}}(\hat{f}_D) = \text{bias} + \text{var}$$

where

$$\text{bias}(x) = (\overline{\hat{f}}(x) - f(x))^2$$

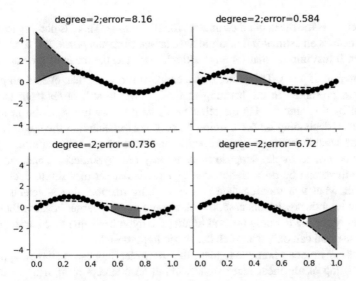

Fig. 4.12 This shows the folds and errors as in Figs. 4.10 and 4.11. The shaded areas show the errors in each respective test set for the quadratic model

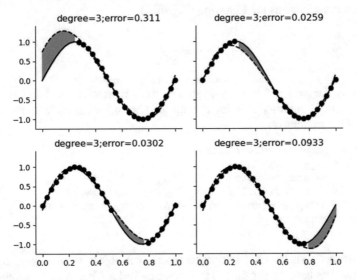

Fig. 4.13 This shows the folds and errors. The shaded areas show the errors in each respective test set for the cubic model

and

$$\text{var}(x) = \mathbb{E}_D(\hat{f}_D(x) - \overline{\hat{f}}(x))^2$$

and where $\bar{\hat{f}}$ is the mean of all estimators for all datasets. There is nothing to say that such a mean is an estimator that could have arisen from any *particular* training data, however. It just implies that for any particular point x, the mean of the values of all the estimators is $\bar{\hat{f}}(x)$. Therefore, bias captures the sense that, even if all possible data were presented to the learning method, it would still differ from the target function by this amount. On the other hand, var shows the variation in the final hypothesis, depending on the training dataset, notwithstanding the target function. Thus, the tension between approximation and generalization is captured by these two terms. For example, suppose there is only one hypothesis. Then, var $= 0$ because there can be no variation due to a particular set of training data because no matter what that training data is, the learning method always selects the one and only hypothesis. In this case, the bias could be very large, because there is no opportunity for the learning method to alter the hypothesis due to the training data, and the method can only ever pick the single hypothesis!

Let us construct an example to make this concrete. Suppose we have a hypothesis set consisting of all linear regressions without an intercept term, $h(x) = ax$. The training data consists of only two points $\{(x_i, \sin(\pi x_i))\}_{i=1}^2$ where x_i is drawn uniformly from the interval $[-1, 1]$. From Sect. 3.8 on linear regression, we know that the solution for a is the following (Fig. 4.14),

$$a = \frac{\mathbf{x}^T \mathbf{y}}{\mathbf{x}^T \mathbf{x}} \tag{4.1}$$

where $\mathbf{x} = [x_1, x_2]$ and $\mathbf{y} = [y_1, y_2]$. The $\bar{\hat{f}}(x)$ represents the solution over all possible sets of training data for a fixed x. The following code shows how to construct the training data,

```
>>> from scipy import stats
>>> def gen_sindata(n=2):
...         x=stats.uniform(-1,2)  # define random variable
...         v = x.rvs((n,1))        # generate sample
...         y = np.sin(np.pi*v)    # use sample for sine
...         return (v,y)
...
```

Again, using Scikit-learn's LinearRegression object, we can compute the a parameter. We have to set fit_intercept=False keyword to suppress the default automatic fitting of the intercept. Figure 4.14 shows the best fitted line.

```
>>> lr = LinearRegression(fit_intercept=False)
>>> lr.fit(*gen_sindata(2))
LinearRegression(fit_intercept=False)
>>> lr.coef_
array([[0.24974914]])
```

Fig. 4.14 For a two-element training set consisting of the points shown, the line is the best fit over the hypothesis set, $h(x) = ax$

Programming Tip

We designed gen_sindata to return a tuple to use the automatic unpacking feature of Python functions in lr.fit(*gen_sindata()). In other words, using the asterisk notation means we don't have to separately assign the outputs of gen_sindata before using them for lr.fit.

In this case, $\overline{f}(x) = \overline{a}x$, where $\overline{a}$ is the expected value of the parameter over *all* possible training datasets. Using our knowledge of probability, we can write this out explicitly as the following,

$$\overline{a} = \mathbb{E}\left(\frac{x_1 \sin(\pi x_1) + x_2 \sin(\pi x_2)}{x_1^2 + x_2^2}\right)$$

where $\mathbf{x} = [x_1, x_2]$ and $\mathbf{y} = [\sin(\pi x_1), \sin(\pi x_2)]$ in Eq. (4.1). However, computing this expectation analytically is hard, but for this specific situation, $\overline{a} \approx 1.43$. To get this value using simulation, we just loop over the process, collect the outputs, and average them as in the following,

```
>>> a_out=[] # output container
>>> for i in range(100):
...      _=lr.fit(*gen_sindata(2))
...      a_out.append(lr.coef_[0,0])
...
>>> np.mean(a_out) # approx 1.43
1.5476180748170176
```

You may have to loop over many more iterations to get close to the purported value. The `var` requires the variance of a,

$$\text{var}(x) = \mathbb{E}((a - \overline{a})x)^2 = x^2 \mathbb{E}(a - \overline{a})^2 \approx 0.71x^2$$

The `bias` is the following,

$$\text{bias}(x) = (\sin(\pi x) - \overline{a}x)^2$$

Fig. 4.15 shows the `bias`, `var`, and mean squared error for this problem. Notice that there is zero bias and zero variance when $x = 0$. This is because the learning method cannot help but get that correct because all the hypotheses happen to match the value of the target function at that point! Likewise, the `var` is zero because all possible pairs, which constitute the training data, are fitted through zero because $h(x) = ax$ has no choice but to go through zero. The errors are worse at the end points. As we discussed in our statistics chapter, those points have the most leverage against the hypothesized models and result in the worst errors. Notice that reducing the edge-errors depends on getting exactly those points near the edges as training data. The sensitivity to a particular dataset is reflected in this behavior.

What if we had more than two points in the training data? What would happen to `var` and `bias`? Certainly, the `var` would decrease because it would be harder and harder to generate training datasets that would be substantially different from each other. The bias would also decrease because more points in the training data means better approximation of the sine function over the interval. What would happen if we changed the hypothesis set to include more complex polynomials? As we have already seen with our polynomial regression earlier in this chapter, we would see the same overall effect as here, but with relatively smaller absolute errors and the same edge effects we noted earlier.

Fig. 4.15 These curves decompose the mean squared error into its constituent bias and variance for this example

4.3.6   Learning Noise

We have thus far not considered the effect of noise in our analysis of learning. The following example should help resolve this. Let us suppose we have the following scalar target function,

$$y(\mathbf{x}) = \mathbf{w}_o^T \mathbf{x} + \eta$$

where $\eta \sim \mathcal{N}(0, \sigma^2)$ is an additive noise term and $\mathbf{w}, \mathbf{x} \in \mathbb{R}^d$. Furthermore, we have n measurements of y. This means the training set consists of $\{(\mathbf{x}_i, y_i)\}_{i=1}^n$. Stacking the measurements together into a vector format,

$$\mathbf{y} = \mathbf{X}\mathbf{w}_o + \boldsymbol{\eta}$$

with $\mathbf{y}, \boldsymbol{\eta} \in \mathbb{R}^n, \mathbf{w}_o \in \mathbb{R}^d$ and $\mathbf{X}$ contain $\mathbf{x}_i$ as columns. The hypothesis set consists of all linear models,

$$h(\mathbf{w}, \mathbf{x}) = \mathbf{w}^T \mathbf{x}$$

We need to the learn the correct $\mathbf{w}$ from the hypothesis set given the training data. So far, this is the usual setup for the problem, but how does the noise factor play into this? In our usual situation, the training set consists of randomly chosen elements from a larger space. In this case, that would be the same as getting random sets of $\mathbf{x}_i$ vectors. That still happens in this case, but the problem is that even if the same $\mathbf{x}_i$ appears twice, it will not be associated with the same y value due the additive noise coming from η. To keep this simple, we assume that there is a fixed set of $\mathbf{x}_i$ vectors and that we get all of them in the training set. For every specific training set, we know how to solve for the MMSE from our earlier statistics work,

$$\mathbf{w} = (\mathbf{X}^T \mathbf{X})^{-1} \mathbf{X}^T \mathbf{y}$$

Given this setup, what is the in-sample mean squared error? Because this is the MMSE solution, we know from our study of the associated orthogonality of such systems that we have,

$$E_{\text{in}} = \|\mathbf{y}\|^2 - \|\mathbf{X}\mathbf{w}\|^2 \tag{4.2}$$

where our best hypothesis, $\mathbf{h} = \mathbf{X}\mathbf{w}$. Now, we want to compute the expectation of this over the distribution of η. For instance, for the first term, we want to compute,

$$\mathbb{E}|\mathbf{y}|^2 = \frac{1}{n}\mathbb{E}(\mathbf{y}^T\mathbf{y}) = \frac{1}{n}\text{Tr}\,\mathbb{E}(\mathbf{y}\mathbf{y}^T)$$

where Tr is the matrix trace operator (i.e., sum of the diagonal elements). Because each η are independent, we have

$$\mathrm{Tr}\,\mathbb{E}(\mathbf{y}\mathbf{y}^T) = \mathrm{Tr}\,\mathbf{X}\mathbf{w}_o\mathbf{w}_o^T\mathbf{X}^T + \sigma^2\mathrm{Tr}\,\mathbf{I} = \mathrm{Tr}\,\mathbf{X}\mathbf{w}_o\mathbf{w}_o^T\mathbf{X}^T + n\sigma^2 \qquad (4.3)$$

where $\mathbf{I}$ is the $n \times n$ identity matrix. For the second term in Eq. (4.2), we have

$$|\mathbf{X}\mathbf{w}|^2 = \mathrm{Tr}\,\mathbf{X}\mathbf{w}\mathbf{w}^T\mathbf{X}^T = \mathrm{Tr}\,\mathbf{X}(\mathbf{X}^T\mathbf{X})^{-1}\mathbf{X}^T\mathbf{y}\mathbf{y}^T\mathbf{X}(\mathbf{X}^T\mathbf{X})^{-1}\mathbf{X}^T$$

The expectation of this is the following,

$$\mathbb{E}|\mathbf{X}\mathbf{w}|^2 = \mathrm{Tr}\,\mathbf{X}(\mathbf{X}^T\mathbf{X})^{-1}\mathbf{X}^T\mathbb{E}(\mathbf{y}\mathbf{y}^T)\mathbf{X}(\mathbf{X}^T\mathbf{X})^{-1}\mathbf{X}^T \qquad (4.4)$$

which, after substituting in Eq. (4.3), yields,

$$\mathbb{E}|\mathbf{X}\mathbf{w}|^2 = \mathrm{Tr}\,\mathbf{X}\mathbf{w}_o\mathbf{w}_o^T\mathbf{X}^T + \sigma^2 d \qquad (4.5)$$

Next, assembling Eq. (4.2) from this and Eq. (4.3) gives,

$$\mathbb{E}(E_{\mathrm{in}}) = \frac{1}{n}E_{\mathrm{in}} = \sigma^2\left(1 - \frac{d}{n}\right) \qquad (4.6)$$

which provides an explicit relationship between the noise power, σ^2, the complexity of the method (d), and the number of training samples (n). This is very illustrative because it reveals the ratio d/n, which is a statement of the trade-off between model complexity and in-sample data size. From our analysis of the VC dimension, we already know that there is a complicated bound that represents the penalty for complexity, but this problem is unusual in that we can actually derive an expression for this without resorting to bounding arguments. Furthermore, this result shows that with a very large number of training examples ($n \rightarrow \infty$), the expected in-sample error approaches σ^2. Informally, this means that the learning method cannot *generalize* from noise and thus can only reduce the expected in-sample error by memorizing the data (i.e., $d \approx n$).

The corresponding analysis for the expected out-of-sample error is similar, but more complicated because we don't have the orthogonality condition. Also, the out-of-sample data has different noise from that used to derive the weights, $\mathbf{w}$. This results in extra cross-terms,

$$E_{\mathrm{out}} = \mathrm{Tr}\left(\mathbf{X}\mathbf{w}_o\mathbf{w}_o^T\mathbf{X}^T + \boldsymbol{\xi}\boldsymbol{\xi}^T + \mathbf{X}\mathbf{w}\mathbf{w}^T\mathbf{X}^T - \mathbf{X}\mathbf{w}\mathbf{w}_o^T\mathbf{X}^T \right.$$
$$\left. -\mathbf{X}\mathbf{w}_o\mathbf{w}^T\mathbf{X}^T\right) \qquad (4.7)$$

where we are using the $\boldsymbol{\xi}$ notation for the noise in the out-of-sample case, which is different from that in the in-sample case. Simplifying this leads to the following,

$$\mathbb{E}(E_{\mathrm{out}}) = \mathrm{Tr}\,\sigma^2\mathbf{I} + \sigma^2\mathbf{X}(\mathbf{X}^T\mathbf{X})^{-1}\mathbf{X}^T \qquad (4.8)$$

Then, assembling all of this gives,

$$\mathbb{E}(E_{\text{out}}) = \sigma^2 \left(1 + \frac{d}{n}\right) \tag{4.9}$$

which shows that even in the limit of large n, the expected out-of-sample error also approaches the noise power limit, σ^2. This shows that memorizing the in-sample data (i.e., $d/n \approx 1$) imposes a proportionate penalty on the out-of-sample performance (i.e., $\mathbb{E}E_{\text{out}} \approx 2\sigma^2$ when $\mathbb{E}E_{\text{in}} \approx 0$).

The following code simulates this important example:

```
>>> def est_errors(d=3,n=10,niter=100):
...      assert n>d
...      wo = np.matrix(arange(d)).T
...      Ein = list()
...      Eout = list()
...      # choose any set of vectors
...      X = np.matrix(np.random.rand(n,d))
...      for ni in range(niter):
...          y = X*wo + np.random.randn(X.shape[0],1)
...          # training weights
...          w = np.linalg.inv(X.T*X)*X.T*y
...          h = X*w
...          Ein.append(np.linalg.norm(h-y)**2)
...          # out of sample error
...          yp = X*wo + np.random.randn(X.shape[0],1)
...          Eout.append(np.linalg.norm(h-yp)**2)
...      return (np.mean(Ein)/n,np.mean(Eout)/n)
...
```

> **Programming Tip**
>
> Python has an `assert` statement to make sure that certain entry conditions for the variables in the function are satisfied. It is a good practice to use reasonable assertions at entry and exit to improve the quality of code.

The following runs the simulation for the given value of d.

```
>>> d=10
>>> xi = arange(d*2,d*10,d//2)
>>> ei,eo=np.array([est_errors(d=d,n=n,niter=100) for n in xi]).T
```

which results in Fig. 4.16. This figure shows the estimated expected in-sample and out-of-sample errors from our simulation compared with our corresponding analytical result. The heavy horizontal line shows the variance of the additive noise $\sigma^2 = 1$. Both these curves approach this asymptote because the noise is the ultimate learning limit for this problem. For a given dimension d, even with an infinite

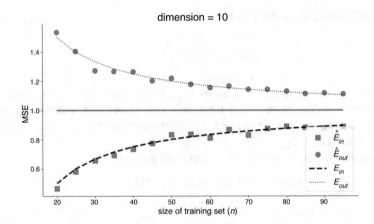

Fig. 4.16 The dots show the learning curves estimated from the simulation and the solid lines show the corresponding terms for our analytical result. The horizontal line shows the variance of the additive noise ($\sigma^2 = 1$ in this case). Both the expected in-sample and out-of-sample errors asymptotically approach this line

amount of training data, the learning method cannot generalize beyond the limit of the noise power. Thus, the expected generalization error is $\mathbb{E}(E_{\text{out}}) - \mathbb{E}(E_{\text{in}}) = 2\sigma^2 \frac{d}{n}$.

4.4 Decision Trees

A decision tree is the easiest classifier to understand, interpret, and explain. A decision tree is constructed by recursively splitting the dataset into a sequence of subsets based on if-then questions. The training set consists of pairs $(\mathbf{x}, y)$ where $\mathbf{x} \in \mathbb{R}^d$ where d is the number of features available and where y is the corresponding label. The learning method splits the training set into groups based on $\mathbf{x}$ while attempting to keep the assignments in each group as uniform as possible. In order to do this, the learning method must pick a feature and an associated threshold for that feature upon which to divide the data. This is tricky to explain in words, but easy to see with an example. First, let us set up the Scikit-learn classifier,

```
>>> from sklearn import tree
>>> clf = tree.DecisionTreeClassifier()
```

Let us also create some example data,

```
>>> import numpy as np
>>> M=np.fromfunction(lambda i,j:j>=2,(4,4)).astype(int)
>>> print(M)
[[0 0 1 1]
```

```
[0 0 1 1]
[0 0 1 1]
[0 0 1 1]]
```

> **Programming Tip**
>
> The fromfunction creates Numpy arrays using the indices as inputs to a function whose value is the corresponding array entry.

We want to classify the elements of the matrix based on their respective positions in the matrix. By just looking at the matrix, the classification is pretty simple—classify as 0 for any positions in the first two columns of the matrix, and classify 1 otherwise. Let us walk through this formally and see if this solution emerges from the decision tree. The values of the array are the labels for the training set and the indices of those values are the elements of **x**. Specifically, the training set has $\mathcal{X} = \{(i, j)\}$ and $\mathcal{Y} = \{0, 1\}$. Now, let us extract those elements and construct the training set.

```
>>> i,j = np.where(M==0)
>>> x=np.vstack([i,j]).T # build nsamp by nfeatures
>>> y = j.reshape(-1,1)*0 # 0 elements
>>> print(x)
[[0 0]
 [0 1]
 [1 0]
 [1 1]
 [2 0]
 [2 1]
 [3 0]
 [3 1]]
>>> print(y)
[[0]
 [0]
 [0]
 [0]
 [0]
 [0]
 [0]
 [0]]
```

Thus, the elements of x are the two-dimensional indices of the values of y. For example, M[x[0,0],x[0,1]]=y[0,0]. Likewise, to complete the training set, we just need to stack the rest of the data to cover all the cases,

```
>>> i,j = np.where(M==1)
>>> x=np.vstack([np.vstack([i,j]).T,x ]) # build nsamp
    nfeatures
>>> y=np.vstack([j.reshape(-1,1)*0+1,y]) # 1 elements
```

With all that established, all we have to do is train the classifier,

```
>>> clf.fit(x,y)
DecisionTreeClassifier()
```

To evaluate how the classifier performed, we can report the score,

```
>>> clf.score(x,y)
1.0
```

For this classifier, the *score* is the accuracy, which is defined as the ratio of the sum of the true-positive (TP) and true-negatives (TN) divided by the sum of all the terms, including the false terms,

$$\text{accuracy} = \frac{TP + TN}{TP + TN + FN + FP}$$

In this case, the classifier gets every point correctly, so $FN = FP = 0$. On a related note, two other common names from information retrieval theory are *recall* (aka sensitivity) and *precision* (aka positive predictive value, $TP/(TP + FP)$). We can visualize this tree in Fig. 4.17. The Gini coefficients (aka categorical variance) in the figure are a measure of the purity of each so-determined class. This coefficient is defined as,

$$\text{Gini}_m = \sum_k p_{m,k}(1 - p_{m,k})$$

where

$$p_{m,k} = \frac{1}{N_m} \sum_{x_i \in R_m} I(y_i = k)$$

which is the proportion of observations labeled k in the mth node and $I(\cdot)$ is the usual indicator function. Note that the maximum value of the Gini coefficient is $\max \text{Gini}_m = 1 - 1/m$. For our simple example, half of the 16 samples are in category 0 and the other half are in the 1 category. Using the notation above, the top box corresponds to the 0th node, so $p_{0,0} = 1/2 = p_{0,1}$. Then, $\text{Gini}_0 = 0.5$. The next layer of nodes in Fig. 4.17 is determined by whether or not the second dimension of the **x** data is greater than 1.5. The Gini coefficients for each of these child nodes is zero because after the prior split, each subsequent category is pure. The value list in each of the nodes shows the distribution of elements in each category at each node.

To make this example more interesting, we can contaminate the data slightly,

```
>>> M[1,0]=1 # put in different class
>>> print(M) # now contaminated
[[0 0 1 1]
 [1 0 1 1]
 [0 0 1 1]
 [0 0 1 1]]
```

Fig. 4.17 Example decision tree. The `Gini` coefficient in each branch measures the purity of the partition in each node. The `samples` item in the box shows the number of items in the corresponding node in the decision tree

Now we have a 1 entry in the previously pure first column's second row. Let us redo the analysis as in the following,

```
>>> i,j = np.where(M==0)
>>> x=np.vstack([i,j]).T
>>> y = j.reshape(-1,1)*0
>>> i,j = np.where(M==1)
>>> x=np.vstack([np.vstack([i,j]).T,x])
>>> y = np.vstack([j.reshape(-1,1)*0+1,y])
>>> clf.fit(x,y)
DecisionTreeClassifier()
```

The result is shown in Fig. 4.18. Note the tree has grown significantly due to this one change! The 0th node has the following parameters, $p_{0,0} = 7/16$ and $p_{0,1} = 9/16$. This makes the Gini coefficient for the 0th node equal to $\frac{7}{16}\left(1 - \frac{7}{16}\right) + \frac{9}{16}(1 - \frac{9}{16}) = 0.492$. As before, the root node splits on $X[1] \leq 1.5$. Let us see if we can reconstruct the succeeding layer of nodes manually, as in the following,

```
>>> y[x[:,1]>1.5] # first node on the right
array([[1],
       [1],
       [1],
       [1],
       [1],
       [1],
       [1],
       [1]])
```

This obviously has a zero Gini coefficient. Likewise, the node on the left contains the following,

```
>>> y[x[:,1]<=1.5] # first node on the left
array([[1],
       [0],
       [0],
       [0],
       [0],
       [0],
       [0],
       [0]])
```

Fig. 4.18 Decision tree for
contaminated data. Note that
just one change in the training
data caused the tree to grow
five times as large as before!

The Gini coefficient in this case is computed as $(1/8)*(1-1/8)+(7/8)*$
$(1-7/8)=0.21875$. This node splits based on $X[1]<0.5$. The child node to
the right derives from the following equivalent logic,

```
>>> np.logical_and(x[:,1]<=1.5,x[:,1]>0.5)
array([False, False, False, False, False, False, False, False,
       False, False,  True,  True, False,  True, False,  True])
```

with corresponding classes,

```
>>> y[np.logical_and(x[:,1]<=1.5,x[:,1]>0.5)]
array([[0],
       [0],
       [0],
       [0]])
```

Programming Tip

The `logical_and` in Numpy provides element-wise logical conjunction.
It is not possible to accomplish this with something like `0.5< x[:,1]`
`<=1.5` because of the way Python parses this syntax.

Notice that for this example as well as for the previous one, the decision tree
was exactly able to memorize (overfit) the data with perfect accuracy. From our
discussion of machine learning theory, this is an indication potential problems in
generalization.

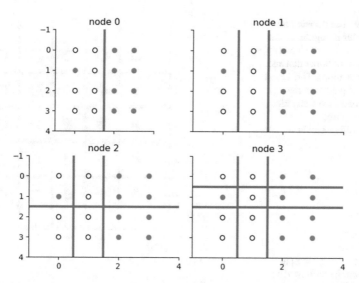

Fig. 4.19 The decision tree divides the training set into regions by splitting successively along each dimension until each region is as pure as possible

The key step in building the decision tree is to come up with the initial split. There are a number of algorithms that can build decision trees based on different criteria, but the general idea is to control the information *entropy* as the tree is developed. In practical terms, this means that the algorithms attempt to build trees that are not excessively deep. It is well-established that this is a very hard problem to solve completely and there are many approaches to it. This is because the algorithms must make global decisions at each node of the tree using the local data available up to that point.

For this example, the decision tree partitions the $\mathcal{X}$ space into different regions corresponding to different $\mathcal{Y}$ labels as shown in Fig. 4.19. The root node at the top of Fig. 4.18 splits the input data based on $X[1] \leq 1.5$. This corresponds to the top left panel in Fig. 4.19 (i.e., node 0) where the vertical line divides the training data shown into two regions, corresponding to the two subsequent child nodes. The next split happens with $X[1] \leq 0.5$ as shown in the next panel of Fig. 4.19 titled node 1. This continues until the last panel on the lower right, where the contaminated element we injected has been isolated into its own sub-region. Thus, the last panel is a representation of Fig. 4.18, where the horizontal/vertical lines correspond to successive splits in the decision tree.

Figure 4.20 shows another example, but now using a simple triangular matrix. As shown by the number of vertical and horizontal partitioning lines, the decision tree that corresponds to this figure is tall and complex. Notice that if we apply a simple rotational transform to the training data, we can obtain Fig. 4.21, which requires a trivial decision tree to fit. Thus, there may be transformations of the training data that simplify the decision tree, but these are very difficult to derive in general. Nonetheless, this highlights a key weakness of decision trees wherein they may be

Fig. 4.20 The decision tree fitted to this triangular matrix is very complex, as shown by the number of horizontal and vertical partitions. Thus, even though the pattern in the training data is visually clear, the decision tree cannot automatically uncover it

Fig. 4.21 Using a simple rotation on the training data in Fig. 4.20, the decision tree can now easily fit the training data with a single partition

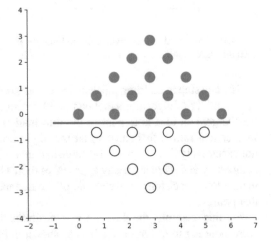

easy to understand, to train, and to deploy but may be completely blind to such time-saving and complexity-saving transformations. Indeed, in higher dimensions, it may be impossible to even visualize the potential of such latent transformations. Thus, the advantages of decision trees can be easily outmatched by other methods that we will study later that *do* have the ability to uncover useful transformations, but which will necessarily be harder to train. Another disadvantage is that because of how decision trees are built, even a single misplaced data point can cause the tree to grow very differently. This is a symptom of high variance.

In all of our examples, the decision tree was able to memorize the training data exactly, as we discussed earlier, this is a sign of potentially high generalization errors. There are pruning algorithms that strategically remove some of the deepest nodes, but these are not yet fully implemented in Scikit-learn, as of this writing. Alternatively, restricting the maximum depth of the decision tree can have a similar effect. The `DecisionTreeClassifier` and `DecisionTreeRegressor` in Scikit-learn both have keyword arguments that specify maximum depth.

4.4.1 Random Forests

It is possible to combine a set of decision trees into a larger composite tree that has better performance than its individual components by using ensemble learning. This is implemented in Scikit-learn as `RandomForestClassifier`. The composite tree helps mitigate the primary weakness of decision trees—high variance. Random forest classifiers help by averaging out the predictions of many constituent trees to minimize this variance by randomly selecting subsets of the training set to train the embedded trees. On the other hand, this randomization can increase bias because there may be a subset of the training set that yields an excellent decision tree, but the averaging effect over randomized training samples washes this out in the same averaging that reduces the variance. This is a key trade-off. The following code implements a simple random forest classifier from our last example.

```
>>> from sklearn.ensemble import RandomForestClassifier
>>> rfc = RandomForestClassifier(n_estimators=4,max_depth=2)
>>> rfc.fit(X_train,y_train.flat)
RandomForestClassifier(max_depth=2, n_estimators=4)
```

Note that we have constrained the maximum depth `max_depth=2` to help with generalization. To keep things simple we have only set up a forest with four individual classifiers.[3] Figure 4.22 shows the individual classifiers in the forest that have been trained above. Even though all the constituent decision trees share the same training data, the random forest algorithm randomly picks feature subsets (with replacement) upon which to train individual trees. This helps avoid the tendency of decision trees to become too deep and lopsided, which hurts both performance and generalization. At the prediction step, the individual outputs of each of the constituent decision trees are put to a majority vote for the final classification. To estimate generalization errors without using cross-validation, the training elements *not* used for a particular constituent tree can be used to test that tree and form a collaborative estimate of generalization errors. This is called the *out-of-bag* estimate.

The main advantage of random forest classifiers is that they require very little tuning and provide a way to trade-off bias and variance via averaging and randomization. Furthermore, they are fast and easy to train in parallel (see the `n_jobs` keyword argument) and fast to predict. On the downside, they are less interpretable than simple decision trees. There are many other powerful tree methods in Scikit-learn like `ExtraTrees` and Gradient Boosted Regression Trees `GradientBoostingRegressor` which are discussed in the online documentation.

---

[3] We have also set the random seed to a fixed value to make the figures reproducible in the Jupyter Notebook corresponding to this section.

Fig. 4.22 The constituent
decision trees of the random
forest and how they
partitioned the training set are
shown in these four panels.
The random forest classifier
uses the individual outputs of
each of the constituent trees
to produce a collaborative
final estimate

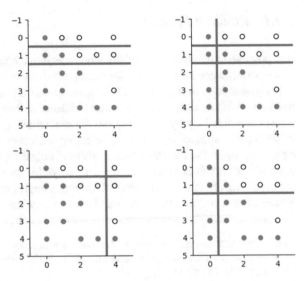

4.4.2   Understanding Boosting Trees

To understand additive modeling using trees, recall the Gram-Schmidt orthogonalization procedure for vectors. The purpose of this orthogonalization procedure is to
create an orthogonal set of vectors starting with a given vector $\mathbf{u}_1$. We have already
discussed the projection operator in Sect. 2.2. The Gram-Schmidt orthogonalization
procedure starts with a vector $\mathbf{v}_1$, which we define as the following:

$$\mathbf{u}_1 = \mathbf{v}_1$$

with the corresponding projection operator $proj_{\mathbf{u}_1}$. The next step in the procedure
is to remove the residual of $\mathbf{u}_1$ from $\mathbf{v}_2$, as in the following:

$$\mathbf{u}_2 = \mathbf{v}_2 - proj_{\mathbf{u}_1}(\mathbf{v}_2)$$

This procedure continues for $\mathbf{v}_3$ as in the following:

$$\mathbf{u}_3 = \mathbf{v}_3 - proj_{\mathbf{u}_1}(\mathbf{v}_3) - proj_{\mathbf{u}_2}(\mathbf{v}_3)$$

and so on. The important aspect of this procedure is that new incoming vectors
(i.e., $\mathbf{v}_k$) are stripped of any pre-existing components already present in the set of
$\{\mathbf{u}_1, \mathbf{u}_2, \dots, \mathbf{u}_M\}$.

Note that this procedure is sequential. That is, the *order* of the incoming
$\mathbf{v}_i$ matters.[4] Thus, any new vector can be expressed using the so-constructed

---

[4] At least up to a rotation of the resulting orthonormal basis.

$\{\mathbf{u}_1, \mathbf{u}_2, \ldots, \mathbf{u}_M\}$ basis set, as in the following:

$$\mathbf{x} = \sum \alpha_i \mathbf{u}_i$$

The idea behind additive trees is to reproduce this procedure for trees instead of vectors. There are many natural topological and algebraic properties that we lack for the general problem, however. For example, we already have well established methods for measuring distances between vectors for the Gram-Schmidt procedure outlined above (namely, the L_2 distance), which we lack here. Thus, we need the concept of *loss function*, which is a way of measuring how well the process is working out at each sequential step. This loss function is parameterized by the training data and by the classification function under consideration: $L_{\mathbf{y}}(f(x))$. For example, if we want a classifier (f) that selects the label y_i based upon the input data $\mathbf{x}_i$ $(f : \mathbf{x}_i \to y_i)$, then the squared error loss function would be the following:

$$L_{\mathbf{y}}(f(x)) = \sum_i (y_i - f(x_i))^2$$

We represent the classifier in terms of a set of basis trees:

$$f(x) = \sum_k \alpha_k u_{\mathbf{x}}(\theta_k)$$

The general algorithm for forward stagewise additive modeling is the following:

- Initialize $f(x) = 0$
- For $m = 1$ to $m = M$, compute the following:

$$(\beta_m, \gamma_m) = \arg\min_{\beta, \gamma} \sum_i L(y_i, f_{m-1}(x_i) + \beta b(x_i; \gamma))$$

- Set $f_m(x) = f_{m-1}(x) + \beta_m b(x; \gamma_m)$

The key point is that the residuals from the prior step are used to fit the basis function for the subsequent iteration. That is, the following equation is being sequentially approximated.

$$f_m(x) - f_{m-1}(x) = \beta_m b(x_i; \gamma_m)$$

Let us see how this works for decision trees and the exponential loss function.

$$L(x, f(x)) = \exp(-yf(x))$$

Recall that for the classification problem, $y \in \{-1, 1\}$. For AdaBoost, the basis functions are the individual classifiers, $G_m(x) \mapsto \{-1, 1\}$. The key step in the algorithm is the minimization step for the objective function

$$J(\beta, G) = \sum_i \exp(y_i(f_{m-1}(x_i) + \beta G(x_i)))$$

$$(\beta_m, G_m) = \arg\min_{\beta, G} \sum_i \exp(y_i(f_{m-1}(x_i) + \beta G(x_i)))$$

Now, because of the exponential, we can factor out the following:

$$w_i^{(m)} = \exp(y_i f_{m-1}(x_i))$$

as a weight on each data element and rewrite the objective function as the following:

$$J(\beta, G) = \sum_i w_i^{(m)} \exp(y_i \beta G(x_i))$$

The important observation here is that $y_i G(x_i) \mapsto 1$ if the tree classifies x_i correctly and $y_i G(x_i) \mapsto -1$ otherwise. Thus, the above sum has terms like the following:

$$J(\beta, G) = \sum_{y_i \neq G(x_i)} w_i^{(m)} \exp(-\beta) + \sum_{y_i = G(x_i)} w_i^{(m)} \exp(\beta)$$

For $\beta > 0$, this means that the best $G(x)$ is the one that incorrectly classifies for the largest weights. Thus, the minimizer is the following:

$$G_m = \arg\min_G \sum_i w_i^{(m)} I(y_i \neq G(x_i))$$

where I is the indicator function (i.e., $I(\text{True}) = 1$, $I(\text{False}) = 0$). For $\beta > 0$, we can rewrite the objective function as the following:

$$J = (\exp(\beta) - \exp(-\beta)) \sum_i w_i^{(m)} I(y_i \neq G(x_i)) + \exp(-\beta) \sum_i w_i^{(m)}$$

and substitute $\theta = \exp(-\beta)$ so that

$$\frac{J}{\sum_i w_i^{(m)}} = \left(\frac{1}{\theta} - \theta\right) \epsilon_m + \theta \tag{4.10}$$

where

$$\epsilon_m = \frac{\sum_i w_i^{(m)} I(y_i \neq G(x_i))}{\sum_i w_i^{(m)}}$$

is the error rate of the classifier with $0 \leq \epsilon_m \leq 1$. Now, finding β is a straightforward calculus minimization exercise on the right side of Eq. (4.10), which gives the following:

$$\beta_m = \frac{1}{2} \log \frac{1 - \epsilon_m}{\epsilon_m}$$

Importantly, β_m can become negative if $\epsilon_m < \frac{1}{2}$, which would violate our assumptions on β. This is captured in the requirement that the base learner be better than just random guessing, which would correspond to $\epsilon_m > \frac{1}{2}$. Practically speaking, this means that boosting cannot fix a base learner that is no better than a random guess. Formally speaking, this is known as the *empirical weak learning assumption* [40].

Now we can move to the iterative weight update. Recall that

$$w_i^{(m+1)} = \exp(y_i f_m(x_i)) = w_i^{(m)} \exp(y_i \beta_m G_m(x_i))$$

which we can rewrite as the following:

$$w_i^{(m+1)} = w_i^{(m)} \exp(\beta_m) \exp(-2\beta_m I(G_m(x_i) = y_i))$$

This means that the data elements that are incorrectly classified have their corresponding weights increased (boosted!) by $\exp(\beta_m)$ and those that are correctly classified have their corresponding weights reduced by $\exp(-\beta_m)$. The reason for the choice of the exponential loss function comes from the following:

$$f^*(x) = \arg\min_{f(x)} \mathbb{E}_{Y|x}(\exp(-Yf(x))) = \frac{1}{2} \log \frac{\mathbb{P}(Y = 1|x)}{\mathbb{P}(Y = -1|x)}$$

This means that boosting is approximating a $f(x)$ that is actually half the log-odds of the conditional class probabilities. This can be rearranged as the following

$$\mathbb{P}(Y = 1|x) = \frac{1}{1 + \exp(-2f^*(x))}$$

The important benefit of this general formulation for boosting, as a sequence of additive approximations, is that it opens the door to other choices of loss function, especially loss functions that are based on robust statistics that can account for errors in the training data (c.f. [14]).

Regression Tree Boosting Example To understand how this works, let us code up a simple one-dimensional regression problem using the squared error loss. Suppose we want to approximate the following cosine function,

```
>>> xi = np.linspace(0,1,20)
>>> yi = np.cos(np.pi*2*1.5*xi)
```

In one dimension, a tree is a rectangular function,

```
>>> h = lambda a,b:np.logical_and(a<=xi,xi <=b).astype(int)
```

Our goal is to compute up with a sequential expansion

$$f_m(x_i) - f_{m-1}(x_i) = \beta_m h_{[a_i,b_i]}(x_i)$$

that minimizes the squared error of the residual, $y_i - F(x_i)$, for each sequential data point,

$$\mathcal{E}(F(x_i), y_i) = ((y_i - F(x_i)) - \beta_m h_{[a,b]}(x_i))^2$$

where $h_{[a,b]}(x)$ is only nonzero in the interval $x \in [a, b]$ and

$$F(x) = \sum_{m=1}^{n} \beta_m h_{[a_m,b_m]}(x)$$

Once we know the $[a_i, b_i]$ parameters of the rectangular functions, then computing the corresponding coefficient is straightforward using inner products since we are approximating using the squared difference,

$$\beta_m = \frac{\langle (y_i - F_0), h_{[a,b]}(x_i) \rangle}{\langle h_{[a,b]}(x_i), h_{[a,b]}(x_i) \rangle}$$

Consider the following example, where we start the sequential iteration with $F_0(x) = 0$. Then

```
>>> # choose a particular interval [a,b]
>>> a,b = [0.0, 0.8]
>>> F0 = 0 # initial condition
>>> beta = (yi-F0) @ h(a,b)/(h(a,b) @ h(a,b))
>>> # residual error
>>> np.linalg.norm((yi-F0)-beta*h(a,b),2)
3.1790811634752045
```

If we want to minimize this over the parameters $[a, b]$, then we can do a grid search using `ParameterGrid` from `sklearn`,

```
>>> from sklearn.model_selection import ParameterGrid
>>> param_grid = {'a': np.linspace(0,1,20), 'b': np.linspace
    (0,1,20)}
>>> minval = [np.inf,{}]
>>> for params in ParameterGrid(param_grid):
...         if params['b'] <= params['a']: continue # skip
        backwards interval
...         a, b = params['a'], params['b']
...         beta = (yi-F0) @ h(a,b)/ (h(a,b) @ h(a,b))
...         # residual error with corresponding params
```

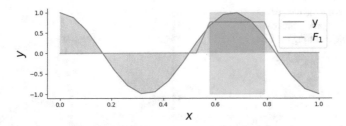

Fig. 4.23 The shaded blue area is the difference between F_0 and the curve y. The orange line is the F_1 iterative fit to y

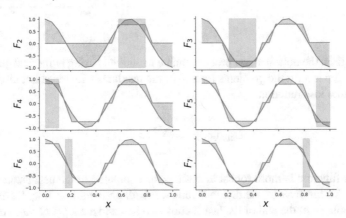

Fig. 4.24 As with Fig. 4.23, this shows F_m iteratively fitting y. Notice that the blue area decreases in the sequence as F_m improves

```
...        err = (np.linalg.norm((yi-F0)-beta*h(a,b),2),
         (a,b,beta))
...        minval = min(minval, err, key=lambda i:i[0])
...
>>> print(minval) # squared err, (a, b, beta)
(2.7589037651370742, (0.5789473684210527, 0.7894736842105263,
   0.7600592101556926))
```

The result is shown in Fig. 4.23 where the shaded blue area is the difference between F_0 and the curve. The shaded gray area shows the interval $[a, b]$ that was found as the minimizer, and the orange line shows F_1, the updated approximation to the curve y. Note that because the squared error is minimized and not the difference, the shaded area is not exactly what is being minimized here. The sequence in Fig. 4.24 shows F_m updating step by step, which is essentially doing the Gram-Schmidt procedure we started with.

Gradient Boosting Gradient boosting is the extension of AdaBoost for more general loss functions, especially those that are robust to outliers. Given a differentiable

Fig. 4.25 Now the data has
certain outlier points that can
cause problems fitting the tree

loss function, the optimization process can be formulated using numerical gradients. The key step is to fit the gradients instead of the residuals with trees. Let us consider the absolute loss function

$$L_{abs}(x, y) = \sum_{i=1}^{n} |x_i - y_i|$$

This loss function is more robust to outliers, as compared to the usual squared loss function. To see how to use this loss function with gradient boosting, let us revisit the example from the end of the last section and just add a couple of outliers,

```
>>> yi[10], yi[4] = 5, -2 # add outliers
```

Figure 4.25 below shows the result of this change,

Proceeding to approximate this using the usual squared loss would ultimately focus on mitigating the outliers, due to their outsized effect on the error function, but what if we want to reduce their influence on the final result? We want to use the absolute loss with gradient boosting, and the key step is to compute the gradient of the absolute loss, L_{abs}, which is the following,

$$\frac{\partial L_{abs}(t)}{\partial t} = \text{sign}(t)$$

The next step is to fit the tree to this gradient instead of the residual, which is what we did previously, Using the same code as before but now we can no longer compute β directly and must instead use a one-dimensional optimization to find it,

```
>>> from scipy.optimize import fmin
>>> F0 = 0 # initial condition
>>> minval = [np.inf,{}]
>>> for params in ParameterGrid(param_grid):
...        if params['b'] <= params['a']: continue # interval
        backward
...        a, b = params['a'], params['b']
```

```
...        obj = lambda beta: np.linalg.norm(np.sign(yi-F0)
       -beta*h(a,b),2)
...        beta, fval, *tmp = fmin(obj, .1, disp=False,
       full_output=True)
...        err =(fval, (a,b,float(beta)))
...        minval = min(minval, err, key=lambda i:i[0])
...
>>> print(minval)
(3.7416573867739413, (0.21052631578947367,
       0.47368421052631576, -0.9999999999999978))
```

Note that the squared norm is used to find the best fitting tree because the absolute loss is used for the F_m function, *not* the individual tree. The above block finds tree that best fits the *gradient* of the absolute loss. With that settled, now we find the next sequential term for our approximation F_m using a line search,

```
>>> a, b, beta = minval[1] # unpack
>>> obj = lambda rho: np.abs(yi - (F0 + rho*h(a,b))).sum()
>>> rho = fmin(obj,0.1, disp=False)
>>> F0 = F0 + rho*h(a,b) # update F_m
```

The `obj` function is the absolute loss that measure the discrepancy between F_m and y. Iterating these steps repeatedly produces the approximation shown in Fig. 4.26 along with the result from `GradientBoostingRegressor` from Scikit-learn, which is obviously better, with many more features than our own crude approximation, but which still matches pretty well.

Recall that from our section on gradient descent for deep learning, we saw that we can proceed in the direction opposite to the gradient in order to improve the differentiable loss function. For AdaBoost, the fitted trees were driven by the incremental re-weighting (i.e., *boosting*) of the residuals. That is, the sequential decision trees fitted the residuals. For gradient boosting, we want to fit the decision trees using the negative gradients. These two approaches turn out to be the same for the sum of squares loss function, but not for more general loss functions.

Fig. 4.26 Absolute loss makes the resulting tree less sensitive to outliers

4.5 Logistic Regression

The Bernoulli distribution we studied earlier answers the question of which of two outcomes ($Y \in \{0, 1\}$) would be selected with probability, p.

$$\mathbb{P}(Y) = p^Y (1 - p)^{1-Y}$$

We also know how to solve the corresponding likelihood function for the maximum likelihood estimate of p given observations of the output, $\{Y_i\}_{i=1}^n$. However, now we want to include other factors in our estimate of p. For example, suppose we observe not just the outcomes, but a corresponding continuous variable, x. That is, the observed data is now $\{(x_i, Y_i)\}_{i=1}^n$. How can we incorporate x into our estimation of p?

The most straightforward idea is to model $p = ax + b$ where a, b are parameters of a fitted line. However, because p is a probability with value bounded between zero and one, we need to wrap this estimate in another function that can map the entire real line into the [0, 1] interval. The logistic (aka sigmoid) function has this property,

$$\theta(s) = \frac{e^s}{1 + e^s}$$

Thus, the new parameterized estimate for p is the following,

$$\hat{p} = \theta(ax + b) = \frac{e^{ax+b}}{1 + e^{ax+b}} \tag{4.11}$$

The *logit* function is defined as the following,

$$\text{logit}(t) = \log \frac{t}{1 - t}$$

It has the important property of extracting the regression components from the probability estimator,

$$\text{logit}(p) = b + ax$$

More continuous variables can be accommodated easily as

$$\text{logit}(p) = b + \sum_k a_k x_k$$

This can be further extended beyond the binary case to multiple target labels. The maximum likelihood estimate of this uses numerical optimization methods that are implemented in Scikit-learn.

Fig. 4.27 This scatterplot shows the binary Y variables and the corresponding x data for each category

Let us construct some data to see how this works. In the following, we assign class labels to a set of randomly scattered points in the two-dimensional plane,

```
>>> v = 0.9
>>> @np.vectorize
... def gen_y(x):
...     if x<5: return np.random.choice([0,1],p=[v,1-v])
...     else:    return np.random.choice([0,1],p=[1-v,v])
...
>>> xi = np.sort(np.random.rand(500)*10)
>>> yi = gen_y(xi)
```

> **Programming Tip**
>
> The `np.vectorize` decorator used in the code above makes it easy to avoid looping in code that uses Numpy arrays by embedding the looping semantics inside of the so-decorated function. Note, however, that this does not necessarily accelerate the wrapped function. It's mainly for convenience.

Figure 4.27 shows a scatterplot of the data we constructed in the above code, $\{(x_i, Y_i)\}$. As constructed, it is more likely that large values of x correspond to $Y = 1$. On the other hand, values of $x \in [4, 6]$ of either category are heavily overlapped. This means that x is not a particularly strong indicator of Y in this region. Figure 4.28 shows the fitted logistic regression curve against the same data. The points along the curve are the probabilities that each point lies in either of the two categories. For large values of x, the curve is near one, meaning that the

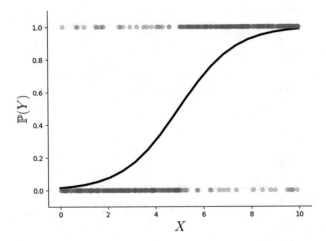

Fig. 4.28 This shows the fitted logistic regression on the data shown in Fig. 4.27. The points along the curve are the probabilities that each point lies in either of the two categories

probability that the associated Y value is equal to one. On the other extreme, small values of x mean that this probability is close to zero. Because there are only two possible categories, this means that the probability of $Y = 0$ is thereby higher. The region in the middle corresponding to the middle probabilities reflect the ambiguity between the two categories because of the overlap in the data for this region. Thus, logistic regression cannot make a strong case for one category here. The following code fits the logistic regression model,

```
>>> from sklearn.linear_model import LogisticRegression
>>> lr = LogisticRegression()
>>> lr.fit(np.c_[xi],yi)
LogisticRegression()
```

For a deeper understanding of logistic regression, we need to alter our notation slightly and once again use our projection methods. More generally we can rewrite Eq. (4.11) as the following,

$$p(\mathbf{x}) = \frac{1}{1 + \exp(-\boldsymbol{\beta}^T \mathbf{x})} \tag{4.12}$$

where $\boldsymbol{\beta}, \mathbf{x} \in \mathbb{R}^n$. From our prior work on projection, we know that the signed perpendicular distance between $\mathbf{x}$ and the linear boundary described by $\boldsymbol{\beta}$ is $\boldsymbol{\beta}^T \mathbf{x}/\|\boldsymbol{\beta}\|$. This means that the probability that is assigned to any point in $\mathbb{R}^n$ is a function of how close that point is to the linear boundary described by the following equation,

$$\boldsymbol{\beta}^T \mathbf{x} = 0$$

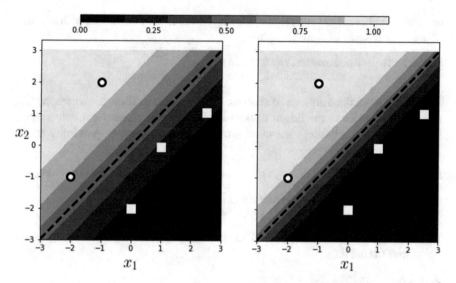

Fig. 4.29 Scaling can arbitrarily increase the probabilities of points near the decision boundary

But there is something subtle hiding here. Note that for any $\alpha \in \mathbb{R}$,

$$\alpha \boldsymbol{\beta}^T \mathbf{x} = 0$$

describes the *same* hyperplane. This means that we can multiply $\boldsymbol{\beta}$ by an arbitrary scalar and still get the same geometry. However, because of $\exp(-\alpha \boldsymbol{\beta}^T \mathbf{x})$ in Eq. (4.12), this scaling determines the intensity of the probability attributed to $\mathbf{x}$. This is illustrated in Fig. 4.29. The panel on the left shows two categories (squares/circles) split by the dotted line that is determined by $\boldsymbol{\beta}^T \mathbf{x} = 0$. The background colors show the probabilities assigned to points in the plane. The right panel shows that by scaling with α, we can increase the probabilities of class membership for the given points, given the exact same geometry. The points near the boundary have lower probabilities because they could easily be on the opposite side. However, by scaling by α, we can raise those probabilities to any desired level at the cost of driving the points further from the boundary closer to one. Why is this a problem? By driving the probabilities arbitrarily using α, we can overemphasize the training set at the cost of out-of-sample data. That is, we may wind up insisting on emphatic class membership of yet unseen points that are close to the boundary that otherwise would have more equivocal probabilities (say, near $1/2$). Once again, this is another manifestation of bias/variance trade-off.

Regularization is a method that controls this effect by penalizing the size of $\boldsymbol{\beta}$ as part of its solution. Algorithmically, logistic regression works by iteratively solving a sequence of weighted least squares problems. Regression adds a $\|\boldsymbol{\beta}\|/C$ term to the least squares error. To see this in action, let us create some data from a logistic regression and see if we can recover it using Scikit-learn. Let us start with a scatter of points in the two-dimensional plane,

```
>>> x0,x1 = np.random.rand(2,20)*6-3
>>> X = np.c_[x0,x1,x1*0+1] # stack as columns
```

Note that X has a third column of all ones. This is a trick to allow the corresponding line to be offset from the origin in the two-dimensional plane. Next, we create a linear boundary and assign the class probabilities according to proximity to the boundary.

```
>>> beta = np.array([1,-1,1]) # last coordinate for affine offset
>>> prd = X.dot(beta)
>>> probs = 1/(1+np.exp(-prd/np.linalg.norm(beta)))
>>> c = (prd>0) # boolean array class labels
```

This establishes the training data. The next block creates the logistic regression object and fits the data.

```
>>> lr = LogisticRegression()
>>> _ = lr.fit(X[:,:-1],c)
```

Note that we have to omit the third dimension because of how Scikit-learn internally breaks down the components of the boundary. The resulting code extracts the corresponding β from the `LogisticRegression` object.

```
>>> betah = np.r_[lr.coef_.flat,lr.intercept_]
```

Programming Tip

The Numpy `np.r_` object provides a quick way to stack Numpy arrays horizontally instead of using `np.hstack`.

The resulting boundary is shown in the left panel in Fig. 4.30. The crosses and triangles represent the two classes we created above, along with the separating gray line. The logistic regression fit produces the dotted black line. The dark circle is the point that logistic regression categorizes incorrectly. The regularization parameter is $C = 1$ by default. Next, we can change the strength of the regularization parameter as in the following,

```
>>> lr = LogisticRegression(C=1000)
```

and re-fit the data to produce the right panel in Fig. 4.30. By increasing the regularization parameter, we essentially nudged the fitting algorithm to *believe* the data more than the general model. That is, by doing this we accepted more variance in exchange for better bias.

Fig. 4.30 The left panel shows the resulting boundary (dashed line) with $C = 1$ as the regularization parameter. The right panel is for $C = 1000$. The gray line is the boundary used to assign the class membership for the synthetic data. The dark circle is the point that logistic regression categorizes incorrectly

Maximum Likelihood Estimation for Logistic Regression Let us again consider the binary classification problem. We define $y_k = \mathbb{P}(C_1|\mathbf{x}_k)$, the conditional probability of the data as a member of given class. Our construction of this problem provides

$$y_k = \theta([\mathbf{w}, w_0] \cdot [\mathbf{x}_k, 1])$$

where θ is the logistic function. Recall that there are only two classes for this problem. The dataset looks like the following,

$$\{(\mathbf{x}_0, r_0), \ldots, (\mathbf{x}_k, r_k), \ldots, (\mathbf{x}_{n-1}, r_{n-1})\}$$

where $r_k \in \{0, 1\}$. For example, we could have the following sequence of observed classes,

$$\{C_0, C_1, C_1, C_0, C_1\}$$

For this case, the likelihood is then the following,

$$\ell = \mathbb{P}(C_0|\mathbf{x}_0)\mathbb{P}(C_1|\mathbf{x}_1)\mathbb{P}(C_1|\mathbf{x}_2)\mathbb{P}(C_0|\mathbf{x}_3)\mathbb{P}(C_1|\mathbf{x}_4)$$

which we can rewrite as the following,

$$\ell(\mathbf{w}, w_0) = (1 - y_0)y_1 y_2(1 - y_3)y_4$$

Recall that there are two mutually exhaustive classes. More generally, this can be written as the following,

$$\ell(\mathbf{w}|\mathcal{X}) = \prod_{k}^{n} y_k^{r_k} (1 - y_k)^{1-r_k}$$

Naturally, we want to compute the logarithm of this as the cross-entropy,

$$E = -\sum_{k} r_k \log(y_k) + (1 - r_k) \log(1 - y_k)$$

and then minimize this to find $\mathbf{w}$ and w_0. This is difficult to do with calculus because the derivatives have nonlinear terms in them that are hard to solve for.

Multi-Class Logistic Regression Using Softmax The logistic regression problem provides a solution for the probability between exactly two alternative classes. To extend to the multi-class problem, we need the *softmax* function. Consider the likelihood ratio between the ith class and the reference class, $\mathcal{C}_k$,

$$\log \frac{p(\mathbf{x}|\mathcal{C}_i)}{p(\mathbf{x}|\mathcal{C}_k)} = \mathbf{w}_i^T \mathbf{x}$$

Note that the bias term is baked into the $\mathbf{w}_i$. Taking the exponential of this and normalizing across all the classes gives the softmax function,

$$y_i(\mathbf{x}) = p(\mathcal{C}_i|\mathbf{x}) = \frac{\exp\left(\mathbf{w}_i^T \mathbf{x}\right)}{\sum_k \exp\left(\mathbf{w}_k^T \mathbf{x}\right)}$$

Note that $\sum_i y_i = 1$. If the $\mathbf{w}_i^T \mathbf{x}$ term is larger than the others, after the exponentiation and normalization, it automatically suppresses the other $y_j \forall j \neq i$, which acts like the maximum function, except this function is differentiable, hence *soft*, as in *softmax*. While that is all straightforward, the trick is deriving the $\mathbf{w}_i$ vectors from the training data $\{\mathbf{x}_i, \mathcal{C}_i\}$ where $r_k(\mathcal{C}_k) = 1$ and $r_k(\mathcal{C}_i) = 0$ is the indicator function for the kth class.

Once again, the launching point is the likelihood function. As with the two-class logistic regression problem, we have the likelihood as the following,

$$\ell = \prod_{k} \prod_{i} y_i(\mathbf{x}_k)^{r_i(\mathcal{C}_k)}$$

where k indexes the data and i indexes the classes, C_i. The log-likelihood of this is the same as the cross-entropy,

$$E = -\sum_k \sum_i r_i(C_k) \log y_i(\mathbf{x}_k)$$

This is the error function we want to minimize. The computation works as before with logistic regression, except there are more derivatives to keep track of in this case.

Understanding Logistic Regression To generalize this technique beyond logistic regression, we need to rethink the problem more abstractly as the dataset $\{x_i, y_i\}$. We have the $y_i \in \{0, 1\}$ data modeled as Bernoulli random variables. We also have the x_i data associated with each y_i, but it is not clear how to exploit this association. What we would like is to construct $\mathbb{E}(Y|X)$ which we already know (see 2.1) is the best MSE estimator. For this problem, we have

$$\mathbb{E}(Y|X) = \mathbb{P}(Y|X)$$

because only $Y = 1$ is nonzero in the summation. Regardless, we don't have the conditional probabilities anyway. One way to look at logistic regression is as a way to build in the functional relationship between y_i and x_i. The simplest thing we could do is approximate,

$$\mathbb{E}(Y|X) \approx \beta_0 + \beta_1 x := \eta(x)$$

If this is the model, then the target would be the y_i data. We can force the output of this linear regression into the interval $[0, 1]$ by composing it with a sigmoidal function,

$$\theta(x) = \frac{1}{1 + \exp(-x)}$$

Then we have a new function $\theta(\eta(x))$ to match against y_i using

$$J(\beta_0, \beta_1) = \sum_i (\theta(\eta(x_i)) - y_i)^2$$

This is a nice setup for an optimization problem. We could certainly solve this numerically using `scipy.optimize`. Unfortunately, this would take us into the black box of the optimization algorithm where we would lose all of our intuitions and experience with linear regression. We can take the opposite approach. Instead of trying to squash the output of the linear estimator into the desired domain, we can map the y_i data into the unbounded space of the linear estimator. Thus, we define the inverse of the above θ function as the *link* function.

$$g(y) = \log\left(\frac{y}{1-y}\right)$$

This means that our approximation to the unknown conditional expectation is the following,

$$g(\mathbb{E}(Y|X)) \approx \beta_0 + \beta_1 x := \eta(x)$$

We cannot apply this directly to the y_i, so we compute the Taylor series expansion centered on $\mathbb{E}(Y|X)$, up to the linear term, to obtain the following,

$$g(Y) \approx g(\mathbb{E}(Y|X)) + (Y - \mathbb{E}(Y|X))g'(\mathbb{E}(Y|X))$$

$$\approx \eta(x) + (Y - \theta(\eta(x)))g'(\theta(\eta(x))) := z$$

Because we do not know the conditional expectation, we replaced these terms with our earlier $\theta(\eta(x))$ function. This new approximation defines our transformed data that we will use to feed the linear model. Note that the β parameters are embedded in this transformation. The $(Y - \theta(\eta(x)))$ term acts as the usual additive noise term. Also,

$$g'(x) = \frac{1}{x(1-x)}$$

The following code applies this transformation to the `xi`, `yi` data

```
>>> b0, b1 = -2,0.5
>>> g = lambda x: np.log(x/(1-x))
>>> theta = lambda x: 1/(1+np.exp(-x))
>>> eta = lambda x: b0 + b1*x
>>> theta_ = theta(eta(xi))
>>> z=eta(xi)+(yi-theta_)/(theta_*(1-theta_))
```

Note the two vertical scales shown in Fig. 4.31. The red scale on the right is the $\{0, 1\}$ domain of the y_i data (red dots) and the left scale is transformed z_i data (black dots). Note that the transformed data is more linear where the original data is less equivocal at the extremes. Also, this transformation is used a specific pair of β_i parameters. The idea is to iterate over this transformation and derive new β_i parameters. With this approach, we have

$$\mathbb{V}(Z|X) = (g')^2 \mathbb{V}(Y|X)$$

Recall that, for this binary variable, we have

Fig. 4.31 The transformation underlying logistic regression

$$\mathbb{P}(Y|X) = \theta(\eta(x)))$$

Thus,

$$\mathbb{V}(Y|X) = \theta(\eta(x))(1 - \theta(\eta(x)))$$

from which we obtain

$$\mathbb{V}(Z|X) = [\theta(\eta(x))(1 - \theta(\eta(x)))]^{-1}$$

The important fact here is that the variance is a function of the X (i.e., heteroskedastic). As we discussed with Gauss-Markov, the appropriate linear regression is weighted least squares where the weights at each data point are inversely proportional to the variance. This ensures that the regression process accounts for this heteroskedasticity. Numpy has a weighted least squares implemented in the `polyfit` function,

```
>>> w = (theta_*(1-theta_))
>>> p = np.polyfit(xi,z,1,w=np.sqrt(w))
```

The output of this fit is shown in Fig. 4.32, along with the raw data and $\mathbb{V}(Z|X)$ for this particular fitted β_i. Iterating a few more times refines the estimated line but it does not take many such iterations to converge. As indicated by the variance line, the fitted line favors the data at either extreme.

Interpreting Logistic Regression Coefficients Logistic regression coefficients are easier to interpret using the *odds ratio*. To get started, let us define the *odds* as the following,

$$\Omega = \frac{\mathbb{P}(y = 1|x_1, x_2, \ldots, x_n)}{\mathbb{P}(y = 0|x_1, x_2, \ldots, x_n)}$$

Fig. 4.32 The output of the weighted least squares fit is shown, along with the raw data and $\mathbb{V}(Z|X)$

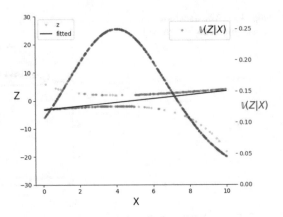

because there are only two outcomes $\mathbb{P}(y = 0) = 1 - \mathbb{P}(y = 1)$. For example, if the $\Omega = 5$, then we can say that the $y = 1$ outcome is five times more likely than the alternative. Under logistic regression, the logarithm of this gives the following,

$$\log\left(\frac{\mathbb{P}(y = 1|x_1, x_2, \ldots, x_n)}{1 - \mathbb{P}(y = 1|x_1, x_2, \ldots, x_n)}\right) = \sum_{i=1}^{n} \beta_i x_i$$

The problem is to interpret the β_i coefficients meaningfully. For example, saying that a unit change in x_i means that the logit changes by $beta_i$, while holding the other variables constant, is not particularly easy to understand because the logarithm is nonintuitive. Let us define the *odds ratio* as the following and consider x_1,

$$\frac{\Omega(x_1 + 1, x_2, \ldots, x_n)}{\Omega(x_1, x_2, \ldots, x_n)} = e^{\beta_1}$$

With this definition, we can say that for a unit increase in x_1, the odds of y increase by e^{β_1}, if everything else stays constant. Thus, defining the odds ratio gets around the nonintuitive nature of the logarithm by focusing on the relative odds.

4.6   Generalized Linear Models

Logistic regression is one example of a wider class of generalized linear models (GLMs). These GLMs have the following three key features

- A target Y variable distributed according to one of the exponential family of distributions (e.g., normal, binomial, Poisson)
- An equation that links the expected value of Y with a linear combination of the observed variables (i.e., $\{x_1, x_2, \ldots, x_n\}$).
- A smooth invertible *link* function $g(x)$ such that $g(\mathbb{E}(Y)) = \sum_k \beta_k x_k$

Exponential Family Here is the one-parameter exponential family,

$$f(y; \lambda) = e^{\lambda y - \gamma(\lambda)}$$

The *natural parameter* is λ and y is the sufficient statistic. For example, for logistic regression, we have $\gamma(\lambda) = \log(1 + e^{\lambda})$ and $\lambda = \log \frac{p}{1-p}$.

An important property of this exponential family is that

$$\mathbb{E}_\lambda(y) = \frac{d\gamma(\lambda)}{d\lambda} = \gamma'(\lambda) \tag{4.13}$$

To see this, we compute the following,

$$1 = \int f(y; \lambda) dy = \int e^{\lambda y - \gamma(\lambda)} dy$$

$$0 = \int \frac{df(y; \lambda)}{d\lambda} dy = \int e^{\lambda y - \gamma(\lambda)} (y - \gamma'(\lambda)) dy$$

$$\int y e^{\lambda y - \gamma(\lambda)} dy = \mathbb{E}_\lambda(y) = \gamma'(\lambda)$$

Using the same technique, we also have,

$$\mathbb{V}_\lambda(Y) = \gamma''(\lambda)$$

which explains the usefulness of this generalized notation for the exponential family.

Deviance The scaled Kullback-Leibler divergence is called the *deviance* as defined below,

$$D(f_1, f_2) = 2 \int f_1(y) \log \frac{f_1(y)}{f_2(y)} dy$$

Hoeffding's Lemma

Using our exponential family notation, we can write out the deviance as the following,

$$\frac{1}{2} D(f(y; \lambda_1), f(y; \lambda_2)) = \int f(y; \lambda_1) \log \frac{f(y; \lambda_1)}{f(y; \lambda_2)} dy$$

$$= \int f(y; \lambda_1)((\lambda_1 - \lambda_2)y - (\gamma(\lambda_1) - \gamma(\lambda_2))) dy$$

$$= \mathbb{E}_{\lambda_1}[(\lambda_1 - \lambda_2)y - (\gamma(\lambda_1) - \gamma(\lambda_2))]$$

$$= (\lambda_1 - \lambda_2)\mathbb{E}_{\lambda_1}(y) - (\gamma(\lambda_1) - \gamma(\lambda_2))$$

$$= (\lambda_1 - \lambda_2)\mu_1 - (\gamma(\lambda_1) - \gamma(\lambda_2))$$

where $\mu_1 := \mathbb{E}_{\lambda_1}(y)$. For the maximum likelihood estimate $\hat{\lambda}_1$, we have $\mu_1 = y$. Plugging this into the above equation gives the following,

$$\frac{1}{2}D(f(y;\hat{\lambda}_1), f(y;\lambda_2)) = (\hat{\lambda}_1 - \lambda_2)y - (\gamma(\hat{\lambda}_1) - \gamma(\lambda_2))$$

$$= \log f(y;\hat{\lambda}_1) - \log f(y;\lambda_2)$$

$$= \log \frac{f(y;\hat{\lambda}_1)}{f(y;\lambda_2)}$$

Taking the negative exponential of both sides gives,

$$f(y;\lambda_2) = f(y;\hat{\lambda}_1)e^{-\frac{1}{2}D(f(y;\hat{\lambda}_1),f(y;\lambda_2))}$$

Because D is always non-negative, the likelihood is maximized when the deviance is zero. In particular, for the scalar case, it means that y itself is the best maximum likelihood estimate for the mean. Also, $f(y;\hat{\lambda}_1)$ is called the *saturated* model. We write Hoeffding's lemma as the following,

$$f(y;\mu) = f(y;y)e^{-\frac{1}{2}D(f(y;y),f(y;\mu))} \tag{4.14}$$

to emphasize that $f(y;y)$ is the likelihood function when the mean is replaced by the sample itself and $f(y;\mu)$ is the likelihood function when the mean is replaced by μ.

Vectorizing Eq. (4.14) using mutual independence gives the following,

$$f(\mathbf{y};\mu) = e^{-\sum_i D(y_i,\mu_i)} \prod f(y_i;y_i)$$

The idea now is to minimize the deviance by deriving,

$$\mu(\beta) = g^{-1}(\mathbf{M}^T\beta)$$

This means the MLE $\hat{\beta}$ is the best $p \times 1$ vector β that minimizes the total deviance where g is the *link* function and $\mathbf{M}$ is the $p \times n$ *structure* matrix. This is the key step with GLM estimation because it reduces the number of parameters from n to p. The structure matrix is where the associated x_i data enters into the problem. Thus, GLM maximum likelihood fitting minimizes the total deviance like plain linear regression minimizes the sum of squares.

With the following,

$$\lambda = \mathbf{M}^T \boldsymbol{\beta}$$

with $2 \times n$ dimensional $\mathbf{M}$. The corresponding joint density function is the following,

$$f(\mathbf{y}; \boldsymbol{\beta}) = e^{\boldsymbol{\beta}^T \boldsymbol{\xi} - \psi(\boldsymbol{\beta})} f_0(\mathbf{y})$$

where

$$\boldsymbol{\xi} = \mathbf{M}\mathbf{y}$$

and

$$\psi(\boldsymbol{\beta}) = \sum \gamma(\mathbf{m}_i^T \boldsymbol{\beta})$$

where now the sufficient statistic is $\boldsymbol{\xi}$ and the parameter vector is $\boldsymbol{\beta}$, which fits into our exponential family format, and $\mathbf{m}_i$ is the ith column of $\mathbf{M}$.

Given this joint density, we can compute the log likelihood as the following,

$$\ell = \boldsymbol{\beta}^T \boldsymbol{\xi} - \psi(\boldsymbol{\beta})$$

To maximize this likelihood, we take the derivative of this with respect to $\boldsymbol{\beta}$ to obtain the following,

$$\frac{d\ell}{d\boldsymbol{\beta}} = \mathbf{M}\mathbf{y} - \mathbf{M}\boldsymbol{\mu}(\mathbf{M}^T \boldsymbol{\beta})$$

since $\gamma'(\mathbf{m}_i^T \boldsymbol{\beta}) = \mathbf{m}_i^T \mu_i(\boldsymbol{\beta})$ and (c.f. Eq. (4.6)), $\gamma' = \mu_\lambda$. Setting this derivative equal to zero gives the conditions for the maximum likelihood solution,

$$\mathbf{M}(\mathbf{y} - \boldsymbol{\mu}(\mathbf{M}^T \boldsymbol{\beta})) = \mathbf{0} \qquad (4.15)$$

where $\boldsymbol{\mu}$ is the element-wise inverse of the link function. This leads us to exactly the same place we started: trying to regress $\mathbf{y}$ against $\boldsymbol{\mu}(\mathbf{M}^T \boldsymbol{\beta})$.

Example The structure matrix $\mathbf{M}$ is where the x_i data associated with the corresponding y_i enters the problem. If we choose

$$\mathbf{M}^T = [\mathbf{1}, \mathbf{x}]$$

where $\mathbf{1}$ is an n-length vector and

$$\beta = [\beta_0, \beta_1]^T$$

with $\mu(x) = 1/(1 + e^{-x})$, we have the original logistic regression problem.

Generally, $\mu(\beta)$ is a nonlinear function, and thus we regress against our transformed variable $\mathbf{z}$

$$\mathbf{z} = \mathbf{M}^T \beta + \text{diag}(g'(\mu))(\mathbf{y} - \mu(\mathbf{M}^T \beta))$$

This fits the format of the Gauss-Markov (see 3.12) problem and has the following solution,

$$\hat{\beta} = (\mathbf{M}\mathbf{R}_z^{-1}\mathbf{M}^T)^{-1}\mathbf{M}\mathbf{R}_z^{-1}\mathbf{z} \tag{4.16}$$

where

$$\mathbf{R}_z := \mathbb{V}(\mathbf{z}) = \text{diag}(g'(\mu))^2 \mathbf{R} = \mathbf{v}(\mu)\,\text{diag}(g'(\mu))^2 \mathbf{I}$$

where g is the link function and $\mathbf{v}$ is the variance function on the designated distribution of the y_i. Thus, $\hat{\beta}$ has the following covariance matrix,

$$\mathbb{V}(\hat{\beta}) = (\mathbf{M}\mathbf{R}_z^{-1}\mathbf{M}^T)^{-1}$$

These results allow inferences about the estimated parameters $\hat{\beta}$. We can easily write Eq. (4.6) as an iteration as follows,

$$\hat{\beta}_{k+1} = (\mathbf{M}\mathbf{R}_{z_k}^{-1}\mathbf{M}^T)^{-1}\mathbf{M}\mathbf{R}_{z_k}^{-1}\mathbf{z}_k$$

Example Consider the data shown in Fig. 4.33. Note that the variance of the data increases for each x and the data increases as a power of x along x. This makes this data a good candidate for a Poisson GLM with $g(\mu) = \log(\mu)$.

We can use our iterative matrix-based approach. The following code initializes the iteration.

```
>>> M    = np.c_[x*0+1,x].T
>>> gi   = np.exp                # inverse g link function
>>> bk   = np.array([.9,0.5])    # initial point
>>> muk  = gi(M.T @ bk).flatten()
>>> Rz   = np.diag(1/muk)
>>> zk   = M.T @ bk + Rz @ (y-muk)
```

and this next block establishes the main iteration

Fig. 4.33 Some data for
Poisson example

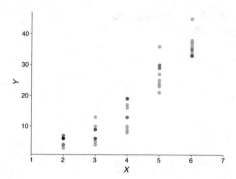

```
>>> while abs(sum(M @ (y-muk))) > .01: # orthogonality
        condition as threshold
...     Rzi = np.linalg.inv(Rz)
...     bk = (np.linalg.inv(M @ Rzi @ M.T)) @ M @ Rzi @ zk
...     muk = gi(M.T @ bk).flatten()
...     Rz =np.diag(1/muk)
...     zk = M.T @ bk + Rz @ (y-muk)
...
```

with corresponding final β computed as the following,

```
>>> print(bk)
[0.72411051 0.48722401]
```

with corresponding estimated $\mathbb{V}(\hat{\beta})$ as

```
>>> print(np.linalg.inv(M @ Rzi @ M.T))
[[ 0.01859313 -0.00357958]
 [-0.00357958  0.00073252]]
```

The orthogonality condition Eq. (4.6) is the following,

```
>>> print(M @ (y-muk))
[-5.83936237e-05 -3.13227667e-04]
```

For comparison, the `statsmodels` module provides the Poisson GLM object.
Note that the reported standard error is the square root of the diagonal elements of
$\mathbb{V}(\hat{\beta})$. A plot of the data and the fitted model is shown below in Fig. 4.34.

```
>>> pm=sm.GLM(y, sm.tools.add_constant(x),
...                       family=sm.families.Poisson())
>>> pm_results=pm.fit()
>>> pm_results.summary()
<class 'statsmodels.iolib.summary.Summary'>
"""
          Generalized Linear Model Regression Results
==============================================================================
Dep. Variable:                    y   No. Observations:                   50
Model:                          GLM   Df Residuals:                       48
Model Family:               Poisson   Df Model:                            1
```

Fig. 4.34 Fitted using the
Poisson GLM

```
Link Function:                   log    Scale:                        1.0000
Method:                         IRLS    Log-Likelihood:              -133.41
Date:             Tue, 08 Feb 2022     Deviance:                     42.659
Time:                       13:39:17    Pearson chi2:                   41.9
No. Iterations:                    5    Pseudo R-squ. (CS):           0.9994
Covariance Type:           nonrobust
=================================================================================
                coef      std err          z       P>|z|      [0.025      0.975]
---------------------------------------------------------------------------------
const         0.7241        0.136      5.310       0.000       0.457       0.991
x1            0.4872        0.027     17.999       0.000       0.434       0.540
=================================================================================
"""
```

4.7   Regularization

We have referred to regularization in Sect. 4.5, but we want to develop this
important idea more fully. Regularization is the mechanism by which we navigate
the bias/variance trade-off. To get started, let us consider a classic constrained least
squares problem,

$$\underset{\mathbf{x}}{\text{minimize}} \quad \|\mathbf{x}\|_2^2$$

$$\text{subject to:} \quad x_0 + 2x_1 = 1$$

where $\|\mathbf{x}\|_2 = \sqrt{x_0^2 + x_1^2}$ is the L_2 norm. Without the constraint, it would be easy to
minimize the objective function—just take $\mathbf{x} = 0$. Otherwise, suppose we somehow
know that $\|\mathbf{x}\|_2 < c$, then the locus of points defined by this inequality is the circle
in Fig. 4.35. The constraint is the line in the same figure. Because every value of c
defines a circle, the constraint is satisfied when the circle touches the line. The circle
can touch the line at many different points, but we are only interested in the smallest
such circle because this is a minimization problem. Intuitively, this means that we

Fig. 4.35 The solution of the constrained L_2 minimization problem is at the point where the constraint (dark line) intersects the L_2 ball (gray circle) centered at the origin. The point of intersection is indicated by the dark circle. The two neighboring squares indicate points on the line that are close to the solution

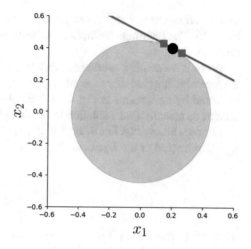

inflate a L_2 ball at the origin and stop when it just touches the constraint. The point of contact is our L_2 minimization solution.

We can obtain the same result using the method of Lagrange multipliers. We can rewrite the entire L_2 minimization problem as one objective function using the Lagrange multiplier, λ,

$$J(x_0, x_1, \lambda) = x_0^2 + x_1^2 + \lambda(1 - x_0 - x_1)$$

and solve this as an ordinary function using calculus. Let us do this using Sympy.

```
>>> import sympy as S
>>> S.var('x:2 l',real=True)
(x0, x1, l)
>>> J=S.Matrix([x0,x1]).norm()**2 + l*(1-x0-2*x1)
>>> sol=S.solve(map(J.diff,[x0,x1,l]))
>>> print(sol)
{l: 2/5, x0: 1/5, x1: 2/5}
```

Programming Tip

Using the `Matrix` object is overkill for this problem, but it does demonstrate how Sympy's matrix machinery works. In this case, we are using the `norm` method to compute the L_2 norm of the given elements. Using `S.var` defines Sympy variables and injects them into the global namespace. It is more Pythonic to do something like `x0 = S.symbols('x0',real=True)` instead but the other way is quicker, especially for variables with many dimensions.

The solution defines the exact point where the line is tangent to the circle in Fig. 4.35. The Lagrange multiplier has incorporated the constraint into the objective function.

There is something subtle and very important about the nature of the solution, however. Notice that there are other points very close to the solution on the circle, indicated by the squares in Fig. 4.35. This closeness could be a good thing, in case it helps us actually find a solution in the first place, but it may be unhelpful in so far as it creates ambiguity. Let us hold that thought and try the same problem using the L_1 norm instead of the L_2 norm. Recall that

$$\|\mathbf{x}\|_1 = \sum_{i=1}^{d} |x_i|$$

where d is the dimension of the vector $\mathbf{x}$. Thus, we can reformulate the same problem in the L_1 norm as in the following,

$$\underset{\mathbf{x}}{\text{minimize}} \quad \|\mathbf{x}\|_1$$

$$\text{subject to:} \quad x_1 + 2x_2 = 1$$

It turns out that this problem is somewhat harder to solve using Sympy, but we have convex optimization modules in Python that can help.

```
>>> from cvxpy import Variable, Problem, Minimize, norm1, norm
>>> x=Variable((2,1),name='x')
>>> constr=[np.matrix([[1,2]]) @ x==1]
>>> obj=Minimize(norm1(x))
>>> p= Problem(obj,constr)
>>> p.solve()
0.5000000003101714
>>> print(x.value)
[[6.2034426e-10]
 [5.0000000e-01]]
```

> **Programming Tip**
>
> The cvxy module provides a unified and accessible interface to the powerful cvxopt convex optimization package, as well as other open-source solver packages.

As shown in Fig. 4.36, the constant-norm contour in the L_1 norm is shaped like a diamond instead of a circle. Furthermore, the solutions found in each case are different. Geometrically, this is because inflating the circular L_2 reaches out in all directions whereas the L_1 ball creeps out along the principal axes. This effect is much more pronounced in higher dimensional spaces where L_1 balls get more

Fig. 4.36 The diamond is the L_1 ball in two dimensions and the line is the constraint. The point of intersection is the solution to the optimization problem. Note that for L_1 optimization, the two nearby points on the constraint (squares) do not touch the L_1 ball. Compare this with Fig. 4.35

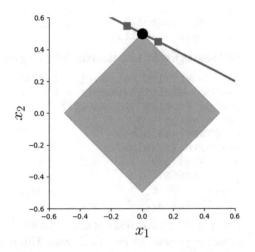

spikey.[5] Like the L_2 case, there are also neighboring points on the constraint line, but notice that these are not close to the boundary of the corresponding L_1 ball, as they were in the L_2 case. This means that these would be harder to confuse with the optimal solution because they correspond to a substantially different L_1 ball.

To double-check our earlier L_2 result, we can also use the cvxpy module to find the L_2 solution as in the following code,

```
>>> constr=[np.matrix([[1,2]])@x==1]
>>> obj=Minimize(norm(x,2)) #L2 norm
>>> p= Problem(obj,constr)
>>> p.solve()
0.44721359549995604
>>> print(x.value)
[[0.2]
 [0.4]]
```

The only change to the code is the L_2 norm and we get the same solution as before.

Let us see what happens in higher dimensions for both L_2 and L_1 as we move from two dimensions to four dimensions.

```
>>> x=Variable((4,1),name='x')
>>> constr=[np.matrix([[1,2,3,4]])@x==1]
>>> obj=Minimize(norm1(x))
>>> p= Problem(obj,constr)
>>> p.solve()
0.2500000009072798
>>> print(x.value)
[[3.88487210e-10]
 [8.33295420e-10]
```

---

[5] We discussed the geometry of high-dimensional space when we covered the curse of dimensionality in the statistics chapter.

```
[7.97158511e-10]
[2.49999999e-01]]
```

And also in the L_2 case with the following code,

```
>>> constr=[np.matrix([[1,2,3,4]])@x==1]
>>> obj=Minimize(norm(x,2))
>>> p= Problem(obj,constr)
>>> p.solve()
0.1825741858350547
>>> print(x.value)
[[0.03333333]
 [0.06666667]
 [0.1       ]
 [0.13333333]]
```

Note that the L_1 solution has selected out only one dimension for the solution, as the other components are effectively zero. This is not so with the L_2 solution, which has meaningful elements in multiple coordinates. This is because the L_1 problem has many pointy corners in the four-dimensional space that poke at the hyperplane that is defined by the constraint. This essentially means the subsets (namely, the points at the corners) are found as solutions because these touch the hyperplane. This effect becomes more pronounced in higher dimensions, which is the main benefit of using the L_1 norm as we will see in the next section.

4.7.1 Ridge Regression

Now that we have a sense of the geometry of the situation, let us revisit our classic linear regression problem. To recap, we want to solve the following problem,

$$\min_{\beta \in \mathbb{R}^p} \|y - X\beta\|$$

where $X = [x_1, x_2, \ldots, x_p]$ and $x_i \in \mathbb{R}^n$. Furthermore, we assume that the p column vectors are linearly independent (i.e., $\text{rank}(X) = p$). Linear regression produces the β that minimizes the mean squared error above. In the case where $p = n$, there is a unique solution to this problem. However, when $p < n$, then there are infinitely many solutions.

To make this concrete, let us work this out using Sympy. First, let us define an example X and y matrix,

```
>>> import sympy as S
>>> from sympy import Matrix
>>> X = Matrix([[1,2,3],
...             [3,4,5]])
>>> y = Matrix([[1,2]]).T
```

Now, we can define our coefficient vector β using the following code,

```
>>> b0,b1,b2=S.symbols('b:3',real=True)
>>> beta = Matrix([[b0,b1,b2]]).T # transpose
```

Next, we define the objective function we are trying to minimize

```
>>> obj=(X*beta -y).norm(ord=2)**2
```

> **Programming Tip**
>
> The Sympy `Matrix` class has useful methods like the `norm` function used above to define the objective function. The `ord=2` means we want to use the L_2 norm. The expression in parenthesis evaluates to a `Matrix` object.

Note that it is helpful to define real variables using the keyword argument whenever applicable because it relieves Sympy's internal machinery of dealing with complex numbers. Finally, we can use calculus to solve this by setting the derivatives of the objective function to zero.

```
>>> sol=S.solve([obj.diff(i) for i in beta])
>>> beta.subs(sol)
Matrix([
[        b2],
[1/2 - 2*b2],
[        b2]])
```

Notice that the solution does not uniquely specify all the components of the `beta` variable. This is a consequence of the $p < n$ nature of this problem where $p = 2$ and $n = 3$. While the existence of this ambiguity does not alter the solution,

```
>>> obj.subs(sol)
0
```

But it does change the length of the solution vector `beta`,

```
>>> beta.subs(sol).norm(2)
sqrt(2*b2**2 + (2*b2 - 1/2)**2)
```

If we want to minimize this length, we can easily use the same calculus as before,

```
>>> S.solve((beta.subs(sol).norm()**2).diff())
[1/6]
```

This provides the solution of minimum length in the L_2 sense,

```
>>> betaL2=beta.subs(sol).subs(b2,S.Rational(1,6))
>>> betaL2
Matrix([
[1/6],
[1/6],
[1/6]])
```

But what is so special about solutions of minimum length? For machine learning, driving the objective function to zero is symptomatic of overfitting the data. Usually, at the zero bound, the machine learning method has essentially memorized the training data, which is bad for generalization. Thus, we can effectively stall this problem by defining a region for the solution that is away from the zero-bound.

$$\underset{\beta}{\text{minimize}} \quad \|y - \mathbf{X}\beta\|_2^2$$

$$\text{subject to:} \quad \|\beta\|_2 < c$$

where c is the tuning parameter. Using the same process as before, we can rewrite this as the following,

$$\min_{\beta \in \mathbb{R}^p} \|y - \mathbf{X}\beta\|_2^2 + \alpha \|\beta\|_2^2$$

where α is the tuning parameter. These are the *penalized* or Lagrange forms of these problems derived from the constrained versions. The objective function is penalized by the $\|\beta\|_2$ term. For L_2 penalization, this is called *ridge* regression. This is implemented in Scikit-learn as `Ridge`. The following code sets this up for our example,

```
>>> from sklearn.linear_model import Ridge
>>> clf = Ridge(alpha=100.0,fit_intercept=False)
>>> clf.fit(np.array(X).astype(float),np.array(y).astype(float))
Ridge(alpha=100.0, fit_intercept=False)
```

Note that the `alpha` scales of the penalty for the $\|\beta\|_2$. We set the `fit_intercept=False` argument to omit the extra offset term from our example. The corresponding solution is the following,

```
>>> print(clf.coef_)
[[0.0428641   0.06113005 0.07939601]]
```

To double-check the solution, we can use some optimization tools from Scipy and our previous Sympy analysis, as in the following,

```
>>> from scipy.optimize import minimize
>>> f   = S.lambdify((b0,b1,b2),obj+beta.norm()**2*100.)
>>> g   = lambda x:f(x[0],x[1],x[2])
>>> out = minimize(g,[.1,.2,.3]) # initial guess
>>> out.x
array([0.0428641 , 0.06113005, 0.07939601])
```

Programming Tip

We had to define the additional g function from the lambda function we created from the Sympy expression in f because the `minimize` function expects a single object vector as input instead of a three separate arguments.

---

which produces the same answer as the `Ridge` object. To better understand the meaning of this result, we can re-compute the mean squared error solution to this problem in one step using matrix algebra instead of calculus,

```
>>> betaLS=X.T*(X*X.T).inv()*y
>>> betaLS
Matrix([
[1/6],
[1/6],
[1/6]])
```

Notice that this solves the posited problem exactly,

```
>>> X*betaLS-y
Matrix([
[0],
[0]])
```

This means that the first term in the objective function goes to zero,

$$\| y - \mathbf{X}\boldsymbol{\beta}_{LS} \| = 0$$

But, let us examine the L_2 length of this solution versus the ridge regression solution,

```
>>> print(betaLS.norm().evalf(), np.linalg.norm(clf.coef_))
0.288675134594813 0.1089859641257551
```

Thus, the ridge regression solution is shorter in the L_2 sense, but the first term in the objective function is not zero for ridge regression,

```
>>> print((y-X*clf.coef_.T).norm()**2)
1.86870864136429
```

Ridge regression solution trades fitting error ($\| y - \mathbf{X}\boldsymbol{\beta} \|_2$) for solution length ($\| \boldsymbol{\beta} \|_2$).

Let us see this in action with a familiar example from 3.13.4. Consider Fig. 4.37. For this example, we created our usual chirp signal and attempted to fit it with a high-dimensional polynomial, as we did in Sect. 4.3.4. The lower panel is the same except with ridge regression. The shaded gray area is the space between the true signal and the approximant in both cases. The horizontal hash marks indicate the subset of x_i values that each regressor was trained on. Thus, the training set represents a non-uniform sample of the underlying chirp waveform. The top panel shows the usual polynomial regression. Note that the regressor fits the given points

Fig. 4.37 The top figure shows polynomial regression and the lower panel shows polynomial ridge regression. The ridge regression does not match as well throughout most of the domain, but it does not flare as violently at the ends. This is because the ridge constraint holds the coefficient vector down at the expense of poorer performance along the middle of the domain

extremely well, but fails at the endpoint. The ridge regressor misses many of the points in the middle, as indicated by the gray area, but does not overshoot at the ends as much as the plain polynomial regression. This is the basic trade-off for ridge regression. The main steps are shown in the following,

```
# create chirp signal
xi = np.linspace(0,1,100)[:,None]
# sample chirp randomly
xin= np.sort(np.random.choice(xi.flatten(),20,replace=False))
    [:,None]
# create sampled waveform
y = np.cos(2*pi*(xin+xin**2))
# create full waveform for reference
yi = np.cos(2*pi*(xi+xi**2))

# create polynomial features
from sklearn.preprocessing import PolynomialFeatures
qfit = PolynomialFeatures(degree=8) # quadratic
Xq = qfit.fit_transform(xin)
```

```
# reformat input as polynomial
Xiq = qfit.fit_transform(xi)

from sklearn.linear_model import LinearRegression
lr=LinearRegression() # create linear model
lr.fit(Xq,y) # fit linear model

# create ridge regression model and fit
clf = Ridge(alpha=1e-9,fit_intercept=False)
clf.fit(Xq,y)
```

4.7.2 Lasso Regression

Lasso regression follows the same basic pattern as ridge regression, except with the L_1 norm in the objective function.

$$\min_{\beta \in \mathbb{R}^p} \|y - \mathbf{X}\beta\|^2 + \alpha\|\beta\|_1$$

The interface in Scikit-learn is likewise the same. The following is the same problem as before using lasso instead of ridge regression,

```
>>> X = np.array([[1,2,3],
...               [3,4,5]])
>>> y = np.array([[1,2]]).T
>>> from sklearn.linear_model import Lasso
>>> lr = Lasso(alpha=1.0,fit_intercept=False)
>>> _=lr.fit(X,y)
>>> print(lr.coef_)
[0.          0.          0.32352941]
```

As before, we can use the optimization tools in Scipy to solve this also,

```
>>> from scipy.optimize import fmin
>>> obj = 1/4.*(X*beta-y).norm(2)**2 + beta.norm(1)*1
>>> f = S.lambdify((b0,b1,b2),obj.subs(l,1.0))
>>> g = lambda x:f(x[0],x[1],x[2])
>>> fmin(g,[0.1,0.2,0.3])
Optimization terminated successfully.
         Current function value: 0.360297
         Iterations: 121
         Function evaluations: 221
array([2.27469304e-06, 4.02831864e-06, 3.23134859e-01])
```

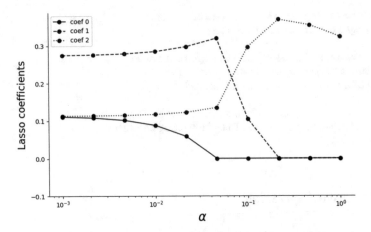

Fig. 4.38 As α increases, more of the model coefficients are driven to zero for lasso regression

---

Programming Tip

The `fmin` function from Scipy's optimization module uses an algorithm that does not depend upon derivatives. This is useful because, unlike the L_2 norm, the L_1 norm has sharp corners that make it harder to estimate derivatives.

This result matches the previous one from the Scikit-learn `Lasso` object. Solving it using Scipy is motivating and provides a good sanity check, but specialized algorithms are required in practice. The following code block re-runs the lasso with varying α and plots the coefficients in Fig. 4.38. Notice that as α increases, all but one of the coefficients are driven to zero. Increasing α makes the trade-off between fitting the data in the L_1 sense and wanting to reduce the number of nonzero coefficients (equivalently, the number of features used) in the model. For a given problem, it may be more practical to focus on reducing the number of features in the model (i.e., large α) than the quality of the data fit in the training data. The lasso provides a clean way to navigate this trade-off.

The following code loops over a set of α values and collects the corresponding lasso coefficients to be plotted in Fig. 4.38

```
>>> o=[]
>>> alphas= np.logspace(-3,0,10)
>>> for a in alphas:
...         clf = Lasso(alpha=a,fit_intercept=False)
...         _=clf.fit(X,y)
...         o.append(clf.coef_)
...
```

Fig. 4.39 In the
two-dimensional plane, the
two classes (gray and white
circles) are easily separated
by any one of the lines shown

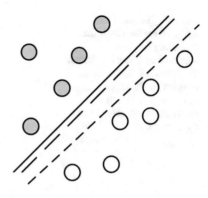

4.8 Support Vector Machines

Support vector machines (SVM) originated from the statistical learning theory
developed by Vapnik-Chervonenkis. As such, it represents a deep application of
statistical theory that incorporates the VC dimension concepts we discussed in the
first section. Let us start by looking at some pictures. Consider the two-dimensional
classification problem shown in Fig. 4.39. Figure 4.39 shows two classes (gray and
white circles) that can be separated by any of the lines shown. Specifically, any such
separating line can be written as the locus of points ($\mathbf{x}$) in the two-dimensional plane
that satisfy the following,

$$\beta_0 + \boldsymbol{\beta}^T \mathbf{x} = 0$$

To classify an arbitrary $\mathbf{x}$ using this line, we just compute the sign of $\beta_0 + \boldsymbol{\beta}^T \mathbf{x}$
and assign one class to the positive sign and the other class to the negative sign.
To uniquely specify such a separating line (or hyperplane in a higher-dimensional
space), we need additional criteria.

Figure 4.40 shows the data with two bordering parallel lines that form a margin
around the central separating line. The *maximal margin algorithm* finds the widest
margin and the unique separating line. As a consequence, the algorithm uncovers
the elements in the data that touch the margins. These are the *support* elements. The
other elements away from the border are not relevant to the solution. This reduces
model variance because the solution is insensitive to the removal of elements other
than these supporting elements (usually a small minority).

To see how this works for linearly separable classes, consider a training set
consisting of $\{(\mathbf{x}, y)\}$ where $y \in \{-1, 1\}$. For any point $\mathbf{x}_i$, we compute the
functional margin as $\hat{\gamma}_i = y_i(\beta_0 + \boldsymbol{\beta}^T \mathbf{x}_i)$. Thus, $\hat{\gamma}_i > 0$ when $\mathbf{x}_i$ is correctly
classified. The geometrical margin is $\gamma = \hat{\gamma}/\|\boldsymbol{\beta}\|$. When $\mathbf{x}_i$ is correctly classified,
the geometrical margin is equal to the perpendicular distance from $\mathbf{x}_i$ to the line. Let
us see how the maximal margin algorithm works.

Fig. 4.40 The maximal
margin algorithm finds the
separating line that
maximizes the margin shown.
The elements that touch the
margins are the support
elements. The dotted
elements are not relevant to
the solution

Let M be the width of the margin. The maximal margin algorithm is can
be formulated as a quadratic programming problem. We want to simultaneously
maximize the margin M while ensuring that all of the data points are correctly
classified.

$$\underset{\beta_0, \beta, \|\beta\|=1}{\text{maximize}} \quad M$$
$$\text{subject to:} \quad y_i(\beta_0 + \beta^T \mathbf{x}_i) \geq M, \; i = 1, \ldots, N.$$

The first line says we want to generate a maximum value for M by adjusting β_0 and
β while keeping $\|\beta\| = 1$. The functional margins for each ith data element are
the constraints to the problem and must be satisfied for every proposed solution. In
words, the constraints enforce that the elements have to be correctly classified and
outside of the margin around the separating line. With some reformulation, it turns
out that $M = 1/\|\beta\|$ and this can be put into the following standard format,

$$\underset{\beta_0, \beta}{\text{minimize}} \quad \|\beta\|$$
$$\text{subject to:} \quad y_i(\beta_0 + \beta^T \mathbf{x}_i) \geq 1, \; i = 1, \ldots, N.$$

This is a convex optimization problem and can be solved using powerful methods
in that area.

The situation becomes more complex when the two classes are not separable and
we have to allow some unavoidable mixing between the two classes in the solution.
This means that the constraints have to modified as in the following,

$$y_i(\beta_0 + \beta^T \mathbf{x}_i) \geq M(1 - \xi_i)$$

where the ξ_i are the slack variables and represent the proportional amount that the
prediction is on the wrong side of the margin. Thus, elements are misclassified when
$\xi_i > 1$. With these additional variables, we have a more general formulation of the

convex optimization problem,

$$\underset{\beta_0,\beta}{\text{minimize}} \quad \|\beta\|$$

$$\text{subject to:} \quad y_i(\beta_0 + \beta^T x_i) \geq 1 - \xi_i,$$

$$\xi_i \geq 0, \sum \xi_i \leq \text{constant}, \ i = 1, \ldots, N.$$

which can be rewritten in the following equivalent form,

$$\underset{\beta_0,\beta}{\text{minimize}} \quad \frac{1}{2}\|\beta\| + C\sum \xi_i$$

$$\text{subject to:} \quad y_i(\beta_0 + \beta^T x_i) \geq 1 - \xi_i, \xi_i \geq 0 \ i = 1, \ldots, N.$$

(4.17)

Because the ξ_i terms are all positive, the objective is to maximize the margin (i.e., minimize $\|\beta\|$) while minimizing the proportional drift of the predictions to the wrong side of the margin (i.e., $C\sum \xi_i$). Thus, large values of C shunt algorithmic focus toward the correctly classified points near the decision boundary and small values focus on further data. The value C is a hyperparameter for the SVM.

The good news is that all of these complicated pieces are handled neatly inside of Scikit-learn. The following sets up the linear *kernel* for the SVM (more on kernels soon),

```
>>> from sklearn.datasets import make_blobs
>>> from sklearn.svm import SVC
>>> sv = SVC(kernel='linear')
```

We can create some synthetic data using make_blobs and then fit it to the SVM,

```
>>> X,y=make_blobs(n_samples=200, centers=2, n_features=2,
...                 random_state=0,cluster_std=.5)
>>> sv.fit(X,y)
SVC(kernel='linear')
```

After fitting, the SVM now has the estimated support vectors and the coefficients of the β in the sv.support_vectors_ and sv.coef_ attributes, respectively. Figure 4.41 shows the two sample classes (white and gray circles) and the line separating them that was found by the maximal margin algorithm. The two parallel dotted lines show the margin. The large circles enclose the support vectors, which are the data elements that are relevant to the solution. Notice that only these elements can touch the edges of the margins.

Figure 4.42 shows what happens when the value of C changes. Increasing this value emphasizes the ξ part of the objective function in Eq. (4.8). As shown in the top left panel, a small value for C means that the algorithm is willing to accept many support vectors at the expense of maximizing the margin. That is, the proportional amount that predictions are on the wrong side of the margin is more acceptable with smaller C. As the value of C increases, there are fewer support vectors because the

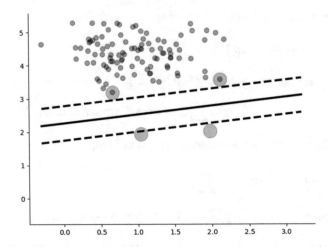

Fig. 4.41 The two class shown (white and gray circles) are linearly separable. The maximal margin solution is shown by the dark black line in the middle. The dotted lines show the extent of the margin. The large circles indicate the support vectors for the maximal margin solution

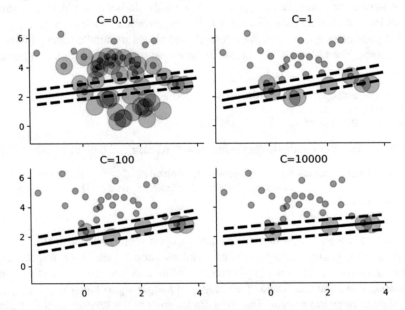

Fig. 4.42 The maximal margin algorithm finds the separating line that maximizes the margin shown. The elements that touch the margins are the support elements. The dotted elements are not relevant to the solution

optimization process prefers to eliminate support vectors that are far away from the margins and accept fewer of these that encroach into the margin. Note that as the value of C progresses through this figure, the separating line tilts slightly.

4.8.1 Kernel Tricks

Support vector machines provide a powerful method to deal with linear separations, but they can also apply to nonlinear boundaries by exploiting the so-called *kernel trick*. The convex optimization formulation of the SVM includes a *dual* formulation that leads to a solution that requires only the inner products of the features. The kernel trick is to substitute inner products by nonlinear kernel functions. This can be thought of as mapping the original features onto a possibly infinite dimensional space of new features. That is, if the data are not linearly separable in two-dimensional space (for example) maybe they are separable in three-dimensional space (or higher)?

To make this concrete, suppose the original input space is $\mathbb{R}^n$, and we want to use a nonlinear mapping $\psi : \mathbf{x} \mapsto \mathcal{F}$ where $\mathcal{F}$ is an inner product space of higher dimension. The kernel trick is to calculate the inner product in $\mathcal{F}$ using a kernel function, $K(\mathbf{x}_i, \mathbf{x}_j) = \langle \psi(\mathbf{x}_i), \psi(\mathbf{x}_j) \rangle$. The long way to compute this is to first compute $\psi(\mathbf{x})$ and then do the inner product. The kernel trick way to do it is to use the kernel function and avoid computing ψ. In other words, the kernel function returns what the inner product in $\mathcal{F}$ would have returned if ψ had been applied. For example, to achieve an nth polynomial mapping of the input space, we can use $\kappa(\mathbf{x}_i, \mathbf{x}_j) = (\mathbf{x}_i^T \mathbf{x}_j + \theta)^n$. For example, suppose the input space is $\mathbb{R}^2$ and $\mathcal{F} = \mathbb{R}^4$ and we have the following mapping,

$$\psi(\mathbf{x}) : (x_0, x_1) \mapsto (x_0^2, x_1^2, x_0 x_1, x_1 x_0)$$

The inner product in $\mathcal{F}$ is then,

$$\langle \psi(\mathbf{x}), \psi(\mathbf{y}) \rangle = \langle \mathbf{x}, \mathbf{y} \rangle^2$$

In other words, the kernel is the square of the inner product in input space. The advantage of using the kernel instead of simply enlarging the feature space is computational because you only need to compute the kernel on all distinct pairs of the input space. The following example should help make this concrete. First we create some Sympy variables,

```
>>> import sympy as S
>>> x0,x1=S.symbols('x:2',real=True)
>>> y0,y1=S.symbols('y:2',real=True)
```

Next, we create the ψ function that maps into $\mathbb{R}^4$ and the corresponding kernel function,

```
>>> psi = lambda x,y: (x**2,y**2,x*y,x*y)
>>> kern = lambda x,y: S.Matrix(x).dot(y)**2
```

Notice that the inner product in $\mathbb{R}^4$ is equal to the kernel function, which only uses the $\mathbb{R}^2$ variables.

```
>>> print(S.Matrix(psi(x0,x1)).dot(psi(y0,y1)))
x0**2*y0**2 + 2*x0*x1*y0*y1 + x1**2*y1**2
>>> print(S.expand(kern((x0,x1),(y0,y1)))) # same as above
x0**2*y0**2 + 2*x0*x1*y0*y1 + x1**2*y1**2
```

Polynomial Regression Using Kernels Recall our favorite linear regression problem from the regularization chapter,

$$\min_{\beta} \| y - \mathbf{X}\beta \|^2$$

where $\mathbf{X}$ is a $n \times m$ matrix with $m > n$. As we discussed, there are multiple solutions to this problem. The least squares solution is the following:

$$\beta_{LS} = \mathbf{X}^T (\mathbf{X}\mathbf{X}^T)^{-1}\mathbf{y}$$

Given a new feature vector $\mathbf{x}$, the corresponding estimator for $\mathbf{y}$ is the following,

$$\hat{\mathbf{y}} = \mathbf{x}^T \beta_{LS} = \mathbf{x}^T \mathbf{X}^T (\mathbf{X}\mathbf{X}^T)^{-1}\mathbf{y}$$

Using the kernel trick, the solution can be written more generally as the following,

$$\hat{\mathbf{y}} = \mathbf{k}(\mathbf{x})^T \mathbf{K}^{-1}\mathbf{y}$$

where the $n \times n$ kernel matrix $\mathbf{K}$ replaces $\mathbf{X}\mathbf{X}^T$ and where $\mathbf{k}(\mathbf{x})$ is a n-vector of components $\mathbf{k}(\mathbf{x}) = [\kappa(\mathbf{x}_i, \mathbf{x})]$ and where $\mathbf{K}_{i,j} = \kappa(\mathbf{x}_i, \mathbf{x}_j)$ for the kernel function κ. With this more general setup, we can substitute $\kappa(\mathbf{x}_i, \mathbf{x}_j) = (\mathbf{x}_i^T \mathbf{x}_j + \theta)^n$ for nth-order polynomial regression [5]. Note that ridge regression can also be incorporated by inverting $(\mathbf{K} + \alpha\mathbf{I})$, which can help stabilize poorly conditioned $\mathbf{K}$ matrices with a tunable α hyper-parameter [5].

For some kernels, the enlarged $\mathcal{F}$ space is infinite-dimensional. Mercer's conditions provide technical restrictions on the kernel functions. Powerful and well-studied kernels have been implemented in Scikit-learn. The advantage of kernel functions may evaporate for when $n \to m$ in which case using the ψ functions instead can be more practicable.

4.9   Dimensionality Reduction

The features from a particular dataset that will ultimately prove important for machine learning can be difficult to know ahead of time. This is especially true for problems that do not have a strong physical underpinning. The row dimension of the input matrix (X) for fitting data in Scikit-learn is the number of samples and the column dimension is the number of features. There may be a large number of column dimensions in this matrix, and the purpose of dimensionality reduction is to somehow reduce these to only those columns that are important for the machine learning task.

Fortunately, Scikit-learn provides some powerful tools to help uncover the most relevant features. Principal component analysis (PCA) consists of taking the input X matrix and (1) subtracting the mean, (2) computing the covariance matrix, and (3) computing the eigenvalue decomposition of the covariance matrix. For example, if X has more columns than is practicable for a particular learning method, then PCA can reduce the number of columns to a more manageable number. PCA is widely used in statistics and other areas beyond machine learning, so it is worth examining what it does in some detail. First, we need the decomposition module from Scikit-learn.

```
>>> from sklearn import decomposition
>>> import numpy as np
>>> pca = decomposition.PCA()
```

Let us create some very simple data and apply PCA.

```
>>> x = np.linspace(-1,1,30)
>>> X = np.c_[x,x+1,x+2] # stack as columns
>>> pca.fit(X)
PCA()
>>> print(pca.explained_variance_ratio_)
[1.00000000e+00 4.44789028e-32 6.56191018e-33]
```

> **Programming Tip**
>
> The np.c_ is a shorcut method for creating stacked column-wise arrays.

In this example, the columns are just constant offsets of the first column. The *explained variance ratio* is the percentage of the variance attributable to the transformed columns of X. You can think of this as the information that is relatively concentrated in each column of the transformed matrix X. Figure 4.43 shows the graph of this dominant transformed column in the bottom panel. Note that a constant offset in each of the columns does not change its respective variance and thus, as far as PCA is concerned, the three columns are identical from an information standpoint.

To make this more interesting, let us change the slope of each of the columns as in the following,

```
>>> X = np.c_[x,2*x+1,3*x+2,x] # change slopes of columns
>>> pca.fit(X)
PCA()
>>> print(pca.explained_variance_ratio_)
[1.00000000e+00 1.36443189e-32 1.71524927e-33 3.68262901e-34]
```

Fig. 4.43 The top panel shows the columns of the feature matrix and the bottom panel shows the dominant component that PCA has extracted

However, changing the slope did not impact the explained variance ratio. Again, there is still only one dominant column. This means that PCA is invariant to both constant offsets and scale changes. This works for functions as well as simple lines,

```
>>> x = np.linspace(-1,1,30)
>>> X = np.c_[np.sin(2*np.pi*x),
...           2*np.sin(2*np.pi*x)+1,
...           3*np.sin(2*np.pi*x)+2]
>>> pca.fit(X)
PCA()
>>> print(pca.explained_variance_ratio_)
[1.00000000e+00 3.63254364e-32 3.47777102e-33]
```

Once again, there is only one dominant column, which is shown in the bottom panel of Fig. 4.44. The top panel shows the individual columns of the feature matrix. To sum up, PCA is able to identify and eliminate features that are merely linear transformations of existing features. This also works when there is additive noise in the features, although more samples are needed to separate the uncorrelated noise from between features.

To see how PCA can simplify machine learning tasks, consider Fig. 4.45 wherein the two classes are separated along the diagonal. After PCA, the transformed data lie along a single axis where the two classes can be split using a one-dimensional interval, which greatly simplifies the classification task. The class identities are preserved under PCA because the principal component is along the same direction that the classes are separated. On the other hand, if the classes are separated along the direction *orthogonal* to the principal component, then the two classes become mixed under PCA and the classification task becomes much harder (see Fig. 4.46). Note that in both cases, the explained_variance_ratio_ is the

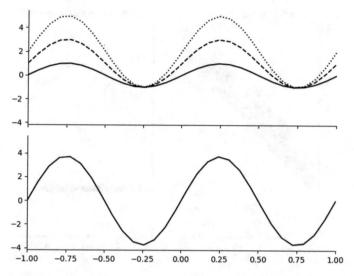

Fig. 4.44 The top panel shows the columns of the feature matrix and the bottom panel shows the dominant component that PCA has computed

Fig. 4.45 The left panel shows the original two-dimensional data space of two easily distinguishable classes, and the right panel shows the reduced data space transformed using PCA. Because the two classes are separated along the principal component discovered by PCA, the classes are preserved under the transformation

same because the explained variance ratio does not account for class membership (Fig. 4.46).

Fig. 4.46 As compared with Fig. 4.45, the two classes differ along the coordinate direction that is orthogonal to the principal component. As a result, the two classes are no longer distinguishable after transformation

PCA works by decomposing the covariance matrix of the data using the singular value decomposition (SVD). This decomposition exists for all matrices and returns the following factorization for an arbitrary matrix $\mathbf{A}$,

$$\mathbf{A} = \mathbf{U}\mathbf{S}\mathbf{V}^T$$

Because of the symmetry of the covariance matrix, $\mathbf{U} = \mathbf{V}$. The elements of the diagonal matrix $\mathbf{S}$ are the singular values of $\mathbf{A}$ whose squares are the eigenvalues of $\mathbf{A}^T\mathbf{A}$. The eigenvector matrix $\mathbf{U}$ is orthogonal: $\mathbf{U}^T\mathbf{U} = \mathbf{I}$. The singular values are in decreasing order so that the first column of $\mathbf{U}$ is the axis corresponding to the largest singular value. This is the first dominant column that PCA identifies. The entries of the covariance matrix are of the form $\mathbb{E}(x_i x_j)$ where x_i and x_j are different features.[6] This means that the covariance matrix is filled with entries that attempt to uncover mutually correlated relationships between all pairs of columns of the feature matrix. Once these have been tabulated in the covariance matrix, the SVD finds optimal orthogonal transformations to align the components along the directions most strongly associated with these correlated relationships. Simultaneously, because orthogonal matrices have columns of unit-length, the SVD collects the absolute squared lengths of these components into the $\mathbf{S}$ matrix. In our example above in Fig. 4.45, the two feature vectors were obviously correlated along

---

[6] Note that these entries are constructed from the data using an estimator of the covariance matrix because we do not have the full probability densities at hand.

the diagonal, meaning that PCA selected that diagonal direction as the principal component.

We have seen that PCA is a powerful dimension reduction method that is invariant to linear transformations of the original feature space. However, this method performs poorly with transformations that are nonlinear. In that case, there are a wide range of extensions to PCA, such as kernel PCA, that are available in Scikit-learn, which allow for embedding parameterized nonlinearities into the PCA at the risk of overfitting.

4.9.1 Generalized PCA

PCA is equivalent to the following optimization problem

$$\underset{\mathbf{Z}}{\text{minimize}} \quad \|A - Z\|_F^2$$

$$\text{subject to:} \quad Rank(Z) < k$$

where $A, Z \in \mathbb{R}^{m \times n}$. The rank constraint can be expressed by setting the non-unique factorization

$$Z = XY$$

where $X \in \mathbb{R}^{m \times k}$ and $Y \in \mathbb{R}^{k \times n}$. Let $x_i \in \mathbb{R}^k$ by the ith row of X and $y_j \in \mathbb{R}^k$ the jth column of Y. Thus, any entry of XY can be expressed as

$$(XY)_{ij} = x_i y_j$$

where the inner product is compactly expressed as $x_i y_j$. With all this notation, the objective function in the optimization problem can be written as,

$$\sum_i^n \sum_j^m (A_{ij} - x_i y_j)^2$$

You can think of y_j as the coefficients mapping the k features in x_i into the original feature j.

Singular Value Decomposition (SVD) The standard way to solve this problem is with the singular value decomposition,

$$A = U \Sigma V^T$$

where $U \in \mathbb{R}^{m \times r}$ and $V \in \mathbb{R}^{n \times r}$ have orthonormal columns and $\Sigma = diag(\sigma_1, \ldots, \sigma_r)$ is the diagonal matrix of ordered singular values (largest to smallest) and where $r = Rank(A)$. With this decomposition, the Frobenius norm can be rewritten as the following,

$$\|A - XY\|_F^2 = \|\Sigma - U^T XYV\|_F^2$$

which means that

$$X = U_k \Sigma_k^{1/2}, Y = \Sigma_k^{1/2} V_k^T$$

where k indicates taking the top k singular values and their associated columns from U and V.

Alternating Minimization The alternating minimization method switches between minimizing the objective for X while Y is fixed and vice versa. This algorithm is easier to extend for versions of PCA than SVD. The stable stationary points of the iteration are viable solutions, but there may be certain unstable stationary points that will not converge to the global solution of the problem, but as a practical matter, alternating minimization converges to the optimal value. The minimization proceeds using the iterative update, given some starting conditions on X and Y. Here is the basic iteration,

$$X = AY^T (YY^T)^{-1}, \ Y = (X^T X)^{-1} X^T A$$

```
>>> A = np.arange(1,17).reshape((4,4)) # rank 2 matrix
>>> X, Y = np.ones((4,1)), np.ones((1,4))
>>> err = [np.linalg.norm(A-X*Y,'fro')]
>>> for i in range(4):
...      X = A@Y.T @ np.linalg.inv((Y@Y.T))
...      Y = np.linalg.inv((X.T @X)) @ X.T @ A
...      err.append(np.linalg.norm(A-X*Y,'fro'))
...
>>> err
[35.21363372331802, 2.0823168251814144, 2.0713231580658165,
    2.0713230668794935, 2.071323066878739]
```

Let us compute the usual SVD solution,

```
>>> u,s,v = np.linalg.svd(A)
>>> u = np.matrix(u) # convert to Numpy matrix
>>> v = np.matrix(v.T)
```

Here is the result of the alternating minimization solution,

```
>>> X @ Y
array([[ 2.22916951,  2.46831501,  2.70746051,  2.94660601],
       [ 5.63832164,  6.24320129,  6.84808093,  7.45296057],
       [ 9.04747378, 10.01808757, 10.98870135, 11.95931514],
       [12.45662592, 13.79297384, 15.12932177, 16.4656697 ]])
```

and compared to the SVD solution,

```
>>> u[:,0]*s[0]*v[:,0].T
matrix([[ 2.22916951,  2.46831501,  2.70746051,  2.94660601],
        [ 5.63832164,  6.24320129,  6.84808093,  7.45296058],
        [ 9.04747378, 10.01808757, 10.98870135, 11.95931514],
        [12.45662591, 13.79297384, 15.12932177, 16.4656697 ]])
```

Quadratically Regularized PCA We can extend PCA to limit the norm of the X and Y matrices by adding a regularization parameter, γ.

$$\underset{X,Y}{\text{minimize}} \quad \|A - XY\|_F^2 + \gamma \|X\|_F^2 + \gamma \|Y\|_F^2$$

when $\gamma = 0$, this reduces to regular PCA. Also, there is no reason that γ has to be the same for both X and Y. The alternating minimization solution is the same except for including the γ term as a diagonal loading on the inverse.

$$X = AY^T(YY^T + \gamma I)^{-1}, \ Y = (X^TX + \gamma I)^{-1}X^T A$$

Note the lengths of the previous solution,

```
>>> np.linalg.norm(X), np.linalg.norm(Y)
(19.209665893798814, 2.010584517471534)
```

Using alternating minimization, we can compute the regularized solution,

```
>>> X = np.ones((4,1))
>>> Y = np.ones((1,4))
>>> gamma = 15
>>> I = np.eye(1)   # seeking rank-1 solution
>>> err = [np.linalg.norm(A-X*Y,'fro')]
>>> for i in range(6):
...     X = A@Y.T @ np.linalg.inv(Y@Y.T+I*gamma)
...     Y = np.linalg.inv(X.T @X+I*gamma) @ X.T @ A
...     err.append(np.linalg.norm(A-X*Y,'fro'))
...
>>> err
[35.21363372331802, 18.593730968257972, 15.24757497172155,
 15.147525413246921, 15.14259575959744, 15.142350192552591,
 15.142337953273561]
```

Because this solution is penalized for size, the norms here are smaller than before,

```
>>> # norms are smaller but approximation is worse
>>> np.linalg.norm(X), np.linalg.norm(Y)
(4.860314307022855, 4.860314517029094)
```

SVD with soft-thresholding of the singular values gives the identical result,

$$\tilde{\Sigma}_k = diag((\sigma_1 - \gamma)_+, \cdots, (\sigma_n - \gamma)_+)$$

and then truncating up to k terms, where $(a)_+ = \max(0, a)$ Then,

$$X = U_k \tilde{\Sigma}_k^{1/2}, Y = \tilde{\Sigma}_k^{1/2} V_k^T$$

Here is the alternating minimization solution,

```
>>> X @ Y # alternating minimization
array([[ 1.36342005,   1.50968788,   1.6559557 ,   1.80222353],
       [ 3.44854923,   3.8185099 ,   4.18847057,   4.55843123],
       [ 5.53367841,   6.12733192,   6.72098543,   7.31463894],
       [ 7.61880759,   8.43615395,   9.2535003 ,  10.07084665]])
```

which is identical to the following soft-threshold SVD solution,

```
>>> u[:,0]*max(s[0]-gamma,0)*v[:,0].T # soft-threshold
matrix([[ 1.36342009,   1.50968792,   1.65595575,   1.80222358],
        [ 3.44854933,   3.81851001,   4.18847068,   4.55843136],
        [ 5.53367857,   6.12733209,   6.72098562,   7.31463914],
        [ 7.6188078 ,   8.43615418,   9.25350055,  10.07084692]])
```

Interpreting PCA   Even though we can compute PCA easily, to understand it, we can think of PCA in a maximum variance problem. Suppose we have a set of N data vectors $\{x_i\}$ where $x_i \in \mathbb{R}^n$. We want to project the data onto a smaller subspace ($k < n$) while maximizing the variance of the projected data. To keep it simple, let us suppose $k = 1$ and we want to find the unit vector u that leads to maximum variance. To estimate the variance, we compute the mean

$$\bar{x} = \frac{1}{N} \sum_i^N u^T x_i$$

and the variance as

$$\mathbb{V} = u^T \mathbf{R} u$$

where $\mathbf{R}$ is the covariance matrix

$$\mathbf{R} = \frac{1}{N} \sum_i^N (x_i - \bar{x})(x_i - \bar{x})^T$$

With all that said, the problem is to maximize the variance, keeping in mind that u is a unit vector; otherwise we could increase the variance by simply scaling up u. Using Lagrange multipliers, we can write this as an unconstrained maximization problem,

$$u^T \mathbf{R} u + \lambda(1 - u^T u)$$

with the following stationary points satisfying,

$$\mathbf{R}u = \lambda u$$

which shows the u is an eigenvector of $\mathbf{R}$, and since

$$u^T \mathbf{R}u = \lambda$$

we see that λ is the eigenvalue. This means that the maximum variance is obtained when u is the eigenvector corresponding to the maximum eigenvalue. The general case for multiple u vectors follows the same pattern, just using the other orthogonal eigenvectors. Let us compute an example using multivariate Gaussian data,

```
>>> from scipy.stats import multivariate_normal as mvn
>>> rxy = 0.6
>>> R =[[1,rxy],
...     [rxy,1]]
>>> x,y=mvn([0,0], R).rvs(5000).T
>>> s,u=np.linalg.eig(R)
>>> u
array([[ 0.70710678, -0.70710678],
       [ 0.70710678,  0.70710678]])
```

Figure 4.47 shows the scatter of the two-dimensional Gaussian random variates. The arrows indicate the eigenvectors drawn in proportion to their eigenvalues. Note the variance of the data along each axis is maximized.

4.9.2 Independent Component Analysis

Independent component analysis (ICA) via the FastICA algorithm is also available in Scikit-learn. This method is fundamentally different from PCA in that it is the

Fig. 4.47 The scatter of two-dimensional Gaussian random variates where the arrows show the eigenvectors with length proportional to their corresponding eigenvalues

small differences between components that are emphasized, not the large principal components. This method is adopted from signal processing. Consider a matrix of signals (**X**) where the rows are the samples and the columns are the different signals. For example, these could be EKG signals from multiple leads on a single patient. The analysis starts with the following model,

$$\mathbf{X} = \mathbf{S}\mathbf{A}^T \qquad\qquad (4.18)$$

In other words, the observed signal matrix is an unknown mixture (**A**) of some set of conformable, independent random sources **S**,

$$\mathbf{S} = [\mathbf{s}_1(t), \mathbf{s}_2(t), \dots, \mathbf{s}_n(t)]$$

The distribution on the random sources is otherwise unknown, except there can be at most one Gaussian source; otherwise, the mixing matrix **A** cannot be identified because of technical reasons. The problem in ICA is to find **A** in Eq. (4.18) and thereby un-mix the $s_i(t)$ signals, but this cannot be solved without a strategy to reduce the inherent arbitrariness in this formulation.

To make this concrete, let us simulate the situation with the following code,

```
>>> from numpy import matrix, c_, sin, cos, pi
>>> t = np.linspace(0,1,250)
>>> s1 = sin(2*pi*t*6)
>>> s2 =np.maximum(cos(2*pi*t*3),0.3)
>>> s2 = s2 - s2.mean()
>>> s3 = np.random.randn(len(t))*.1
>>> # normalize columns
>>> s1=s1/np.linalg.norm(s1)
>>> s2=s2/np.linalg.norm(s2)
>>> s3=s3/np.linalg.norm(s3)
>>> S =c_[s1,s2,s3] # stack as columns
>>> # mixing matrix
>>> A = matrix([[  1,   1,1],
...             [0.5, -1,3],
...             [0.1, -2,8]])
>>> X= S*A.T # do mixing
```

The individual signals ($s_i(t)$) and their mixtures ($X_i(t)$) are shown in Fig. 4.48. To recover the individual signals using ICA, we use the FastICA object and fit the parameters on the X matrix,

```
>>> from sklearn.decomposition import FastICA
>>> ica = FastICA()
>>> # estimate unknown S matrix
>>> S_=ica.fit_transform(X)
```

The results of this estimation are shown in Fig. 4.49, showing that ICA is able to recover the original signals from the observed mixture. Note that ICA is unable

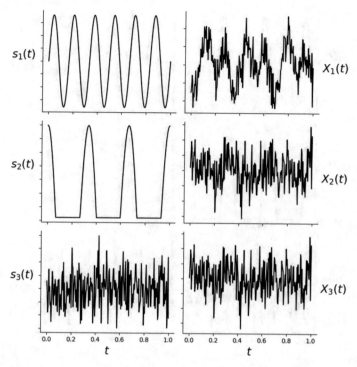

Fig. 4.48 The left column shows the original signals and the right column shows the mixed signals. The object of ICA is to recover the left column from the right

to distinguish the signs of the recovered signals or preserve the order of the input signals.

To develop some intuition as to how ICA accomplishes this feat, consider the following two-dimensional situation with two uniformly distributed independent variables, $u_x, u_y \sim \mathcal{U}[0, 1]$. Suppose we apply the following orthogonal rotation matrix to these variables,

$$\begin{bmatrix} u'_x \\ u'_y \end{bmatrix} = \begin{bmatrix} \cos(\phi) & -\sin(\phi) \\ \sin(\phi) & \cos(\phi) \end{bmatrix} \begin{bmatrix} u_x \\ u_y \end{bmatrix}$$

The so-rotated variables u'_x, u'_y are no longer independent, as shown in Fig. 4.50. Thus, one way to think about ICA is as a search through orthogonal matrices so that the independence is restored. This is where the prohibition against Gaussian distributions arises. The two dimensional Gaussian distribution of independent variables is proportional the following,

$$f(\mathbf{x}) \propto \exp(-\frac{1}{2}\mathbf{x}^T\mathbf{x})$$

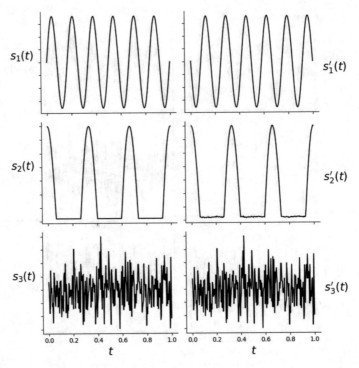

Fig. 4.49 The left column shows the original signals and the right column shows the signals that ICA was able to recover. They match exactly, outside of a possible sign change

Now, if we similarly rotated the **x** vector as,

$$\mathbf{y} = \mathbf{Q}\mathbf{x}$$

the resulting density for **y** is obtained by plugging in the following,

$$\mathbf{x} = \mathbf{Q}^T \mathbf{y}$$

because the inverse of an orthogonal matrix is its transpose, we obtain

$$f(\mathbf{y}) \propto \exp(-\frac{1}{2}\mathbf{y}^T \mathbf{Q}\mathbf{Q}^T \mathbf{y}) = \exp(-\frac{1}{2}\mathbf{y}^T \mathbf{y})$$

In other words, the transformation is lost on the **y** variable. This means that ICA cannot search over orthogonal transformations if it is blind to them, which explains the restriction of Gaussian random variables. Thus, ICA is a method that seeks to maximize the non-Gaussian-ness of the transformed random variables. There are many methods to doing this, some of which involve cumulants and others that use the *negentropy*,

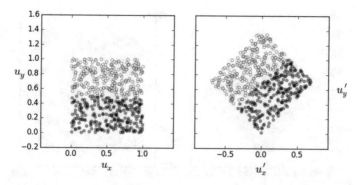

Fig. 4.50 The left panel shows two classes labeled on the u_x, u_y uniformly independent random variables. The right panel shows these random variables after a rotation, which removes their mutual independence and makes it hard to separate the two classes along the coordinate directions

$$\mathcal{J}(Y) = \mathcal{H}(Z) - \mathcal{H}(Y)$$

where $\mathcal{H}(Z)$ is the information entropy of the Gaussian random variable Z that has the same variance as Y. Further details would take us beyond our scope, but that is the outline of how the FastICA algorithm works.

The implementation of this method in Scikit-learn includes two different ways of extracting more than one independent source component. The *deflation* method iteratively extracts one component at a time using an incremental normalization step. The *parallel* method also uses the single-component method but carries out normalization of all the components simultaneously, instead of for just the newly computed component. Because ICA extracts independent components, a whitening step is used beforehand to balance the correlated components from the data matrix. Whereas PCA returns uncorrelated components along dimensions optimal for Gaussian random variables, ICA returns components that are as far from the Gaussian density as possible.

The left panel on Fig. 4.50 shows the original uniform random sources. The white and black colors distinguish between two classes. The right panel shows the mixture of these sources, which is what we observe as input features. The top row of Fig. 4.51 shows the PCA (left) and ICA (right) transformed data spaces. Notice that ICA is able to un-mix the two random sources whereas PCA transforms along the dominant diagonal. Because ICA is able to preserve the class membership, the data space can be reduced to two non-overlapping sections, as shown. However, PCA cannot achieve a similar separation because the classes are mixed along the dominant diagonal that PCA favors as the main component in the decomposition.

For a good principal component analysis treatment, see [3, 11, 17, 37]. Independent component analysis is discussed in more detail in [16].

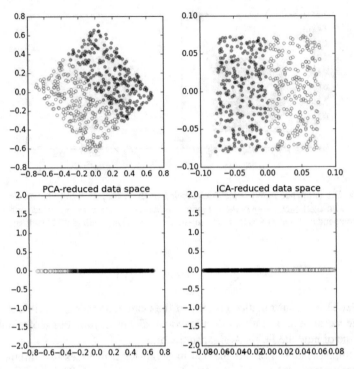

Fig. 4.51 The panel on the top left shows two classes in a plane after a rotation. The bottom left panel shows the result of dimensionality reduction using PCA, which causes mixing between the two classes. The top right panel shows the ICA transformed output and the lower right panel shows that, because ICA was able to un-rotate the data, the lower dimensional data maintains the separation between the classes

4.10   Clustering

Clustering is the simplest member of a family of machine learning methods that do not require supervision to learn from data. Unsupervised methods have training sets that do not have a target variable. These unsupervised learning methods rely upon a meaningful metric to group data into clusters. This makes it an excellent exploratory data analysis method because there are very few assumptions built into the method itself. In this section, we focus on the popular K-means clustering method that is available in Scikit-learn.

Let us manufacture some data to get going with `make_blobs` from Scikit-learn. Figure 4.52 shows some example clusters in two dimensions. Clustering methods work by minimizing the following objective function,

$$J = \sum_k \sum_i \|\mathbf{x}_i - \boldsymbol{\mu}_k\|^2$$

Fig. 4.52 The four clusters are pretty easy to see in this example and we want clustering methods to determine the extent and number of such clusters automatically

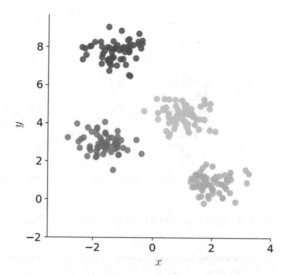

The *distortion* for the kth cluster is the summand,

$$\sum_i \|\mathbf{x}_i - \mu_k\|^2$$

Thus, clustering algorithms work to minimize this by adjusting the centers of the individual clusters, μ_k. Intuitively, each μ_k is the *center of mass* of the points in the cloud. The Euclidean distance is the typical metric used for this,

$$\|\mathbf{x}\|^2 = \sum_i x_i^2$$

There are many clever algorithms that can solve this problem for the best μ_k cluster-centers. The K-means algorithm starts with a user-specified number of K clusters to optimize over. This is implemented in Scikit-learn with the KMeans object that follows the usual fitting conventions in Scikit-learn,

```
>>> from sklearn.cluster import KMeans
>>> kmeans = KMeans(n_clusters=4)
>>> kmeans.fit(X)
KMeans(n_clusters=4)
```

where we have chosen $K = 4$. How do we choose the value of K? This is the eternal question of generalization versus approximation—too many clusters provide great approximation but bad generalization. One way to approach this problem is to compute the mean distortion for increasingly larger values of K until it no longer makes sense. To do this, we want to take every data point and compare it to the centers of all the clusters. Then, take the smallest value of this across all clusters and average those. This gives us an idea of the overall mean performance for the K clusters. The following code computes this explicitly.

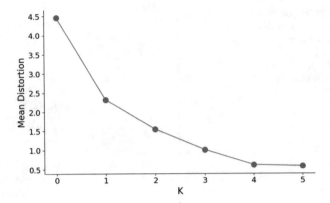

Fig. 4.53 The mean distortion shows that there is a diminishing value in using more clusters

---

Programming Tip

The `cdist` function from Scipy computes all the pairwise differences between the two input collections according to the specified metric.

```
>>> from scipy.spatial.distance import cdist
>>> m_distortions=[]
>>> for k in range(1,7):
...     kmeans = KMeans(n_clusters=k)
...     _=kmeans.fit(X)
...     tmp=cdist(X,kmeans.cluster_centers_,'euclidean')
...     m_distortions.append(sum(np.min(tmp,axis=1))/X.shape[0])
...
```

Note that code above uses the `cluster_centers_`, which are estimated from K-means algorithm. The resulting Fig. 4.53 shows the point of diminishing returns for added additional clusters.

Another figure-of-merit is the silhouette coefficient, which measures how compact and separated the individual clusters are. To compute the silhouette coefficient, we need to compute the mean intra-cluster distance for each sample (a_i) and the mean distance to the next nearest cluster (b_i). Then, the silhouette coefficient for the ith sample is

$$sc_i = \frac{b_i - a_i}{\max(a_i, b_i)}$$

The mean silhouette coefficient is just the mean of all these values over all the samples. The best value is one and the worst is negative one, with values near zero indicating overlapping clusters and negative values showing that samples have been incorrectly assigned to the wrong cluster. This figure-of-merit is implemented in Scikit-learn as in the following,

Fig. 4.54 The figure shows how the silhouette coefficient varies as the clusters move closer and become more compact

```
>>> from sklearn.metrics import silhouette_score
```

Figure 4.54 shows how the silhouette coefficient varies as the clusters become more dispersed and/or closer together.

K-means is easy to understand and to implement but can be sensitive to the initial choice of cluster-centers. The default initialization method in Scikit-learn uses a very effective and clever randomization to come up with the initial cluster-centers. Nonetheless, to see why initialization can cause instability with K-means, consider the following Fig. 4.55. In Fig. 4.55, there are two large clusters on the left and a very sparse cluster on the far right. The large circles at the centers are the cluster-centers that K-means found. Given $K = 2$, how should the cluster-centers be chosen? Intuitively, the first two clusters should have their own cluster-center somewhere between them and the sparse cluster on the right should have its own cluster-center.[7] Why isn't this happening?

The problem is that the objective function for K-means is trading the distance of the far-off sparse cluster with its small size. If we keep increasing the number of samples in the sparse cluster on the right, then K-means will move the cluster-centers out to meet them, as shown in Fig. 4.55. That is, if one of the initial cluster-centers was right in the middle of the sparse cluster, the algorithm would have immediately captured it and then moved the next cluster-center to the middle of the other two clusters (bottom panel of Fig. 4.55). Without some thoughtful

---

[7] Note that we are using the `init=random` keyword argument for this example in order to illustrate this.

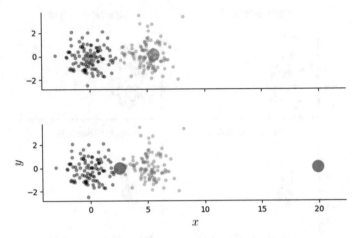

Fig. 4.55 The large circles indicate the cluster-centers found by the K-means algorithm

initialization, this may not happen and the sparse cluster would have been merged into the middle cluster (top panel of Fig. 4.55). Furthermore, such problems are hard to visualize with high-dimensional clusters. Nonetheless, K-means is generally very fast, easy-to-interpret, and easy to understand. It is straightforward to parallelize using the n_jobs keyword argument so that many initial cluster-centers can be easily evaluated. Many extensions of K-means use different metrics beyond Euclidean and incorporate adaptive weighting of features. This enables the clusters to have ellipsoidal instead of spherical shapes.

4.11   Ensemble Methods

With the exception of the random forest, we have so far considered machine learning models as stand-alone entities. Combinations of models that jointly produce a classification are known as *ensembles*. There are two main methodologies that create ensembles: *bagging* and *boosting*.

4.11.1   Bagging

Bagging refers to bootstrap aggregating, where bootstrap here is the same as we discussed in Sect. 3.11. Basically, we resample the data with replacement and then train a classifier on the newly sampled data. Then, we combine the outputs of each of the individual classifiers using a majority-voting scheme (for discrete outputs) or a weighted average (for continuous outputs). This combination is particularly

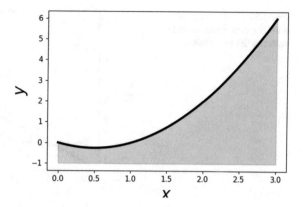

Fig. 4.56 Two regions in the plane are separated by a nonlinear boundary. The training data is sampled from this plane. The objective is to correctly classify the so-sampled data

effective for models that are easily influenced by a single data element. The resampling process means that these elements cannot appear in every bootstrapped training set so that some of the models will not suffer these effects. This makes the so-computed combination of outputs less volatile. Thus, bagging helps reduce the collective variance of individual high-variance models.

To get a sense of bagging, let us suppose we have a two-dimensional plane that is partitioned into two regions with the following boundary: $y = -x + x^2$. Pairs of (x_i, y_i) points above this boundary are labeled one and points below are labeled zero. Figure 4.56 shows the two regions with the nonlinear separating boundary as the black curved line.

The problem is to take samples from each of these regions and classify them correctly using a perceptron (see Sect. 4.12). A perceptron is the simplest possible linear classifier that finds a line in the plane to separate two purported categories. Because the separating boundary is nonlinear, there is no way that the perceptron can completely solve this problem. The following code sets up the perceptron available in Scikit-learn.

```
>>> from sklearn.linear_model import Perceptron
>>> p=Perceptron()
>>> p
Perceptron()
```

The training data and the resulting perceptron separating boundary are shown in Fig. 4.57. The circles and crosses are the sampled training data and the gray separating line is the perceptron's separating boundary between the two categories. The black squares are those elements in the training data that the perceptron misclassified. Because the perceptron can only produce linear separating boundaries, and the boundary in this case is nonlinear, the perceptron makes mistakes near where the boundary curves. The next step is to see how bagging can improve upon this by using multiple perceptrons.

The following code sets up the bagging classifier in Scikit-learn. Here we select only three perceptrons. Figure 4.58 shows each of the three individual classifiers

Fig. 4.57 The perceptron
finds the best linear boundary
between the two classes

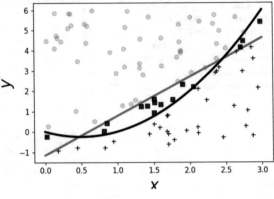

Fig. 4.58 Each panel with
the single gray line is one of
the perceptrons used for the
ensemble bagging classifier
on the lower right

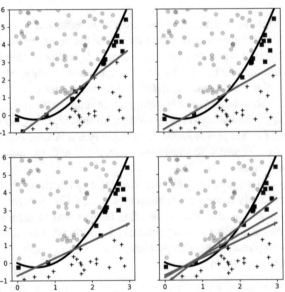

and the final bagged classifier in the panel on the bottom right. As before, the
black circles indicate misclassifications in the training data. Joint classifications are
determined by majority voting.

```
>>> from sklearn.ensemble import BaggingClassifier
>>> bp = BaggingClassifier(Perceptron(),max_samples=0.50,
    n_estimators=3)
>>> bp
BaggingClassifier(base_estimator=Perceptron(),max_samples=0.5,
                  n_estimators=3)
```

The `BaggingClassifier` can estimate its own out-of-sample error if passed
the `oob_score=True` flag upon construction. This keeps track of which samples

were used for training and which were not and then estimates the out-of-sample error using those samples that were unused in training. The `max_samples` keyword argument specifies the number of items from the training set to use for the base classifier. The smaller the `max_samples` used in the bagging classifier, the better the out-of-sample error estimate, but at the cost of worse in-sample performance. Of course, this depends on the overall number of samples and the degrees-of-freedom in each individual classifier. The VC dimension surfaces again!

4.11.2 Boosting

As we discussed, bagging is particularly effective for individual high-variance classifiers because the final majority-vote tends to smooth out the individual classifiers and produce a more stable collaborative solution. On the other hand, boosting is particularly effective for high-bias classifiers that are slow to adjust to new data. On the one hand, boosting is similar to bagging in that it uses a majority-voting (or averaging for numeric prediction) process at the end; and it also combines individual classifiers of the same type. On the other hand, boosting is serially iterative, whereas the individual classifiers in bagging can be trained in parallel. Boosting uses the misclassifications of prior iterations to influence the training of the next iterative classifier by weighting those misclassifications more heavily in subsequent steps. This means that, at every step, boosting focuses more and more on specific misclassifications up to that point, letting the prior classifications be carried by earlier iterations.

The primary implementation for boosting in Scikit-learn is the Adaptive Boosting (*AdaBoost*) algorithm, which does classification (`AdaBoostClassifier`) and regression (`AdaBoostRegressor`). The first step in the basic AdaBoost algorithm is to initialize the weights over each of the training set indices, $D_0(i) = 1/n$ where there are n elements in the training set. Note that this creates a discrete uniform distribution over the *indices*, not over the training data $\{(x_i, y_i)\}$ itself. In other words, if there are repeated elements in the training data, then each gets its own weight. The next step is to train the base classifier h_k and record the classification error at the kth iteration, ϵ_k. Two factors can next be calculated using ϵ_k,

$$\alpha_k = \frac{1}{2} \log \frac{1 - \epsilon_k}{\epsilon_k}$$

and the normalization factor,

$$Z_k = 2\sqrt{\epsilon_k(1 - \epsilon_k)}$$

Fig. 4.59 The individual perceptron classifiers embedded in the AdaBoost classifier are shown along with the misclassified points (in black). Compare this to the lower right panel of Fig. 4.58

For the next step, the weights over the training data are updated as in the following,

$$D_{k+1}(i) = \frac{1}{Z_k} D_k(i) \exp\left(-\alpha_k y_i h_k(x_i)\right)$$

The final classification result is assembled using the α_k factors, $g = \text{sgn}(\sum_k \alpha_k h_k)$.

To redo the problem above using boosting with perceptrons, we set up the AdaBoost classifier in the following,

```
>>> from sklearn.ensemble import AdaBoostClassifier
>>> clf=AdaBoostClassifier(Perceptron(),n_estimators=3,
...                         algorithm='SAMME',
...                         learning_rate=0.5)
>>> clf
AdaBoostClassifier(algorithm='SAMME',base_estimator=Perceptron(),
                   learning_rate=0.5, n_estimators=3)
```

The `learning_rate` above controls how aggressively the weights are updated. The resulting classification boundaries for the embedded perceptrons are shown in Fig. 4.59. Compare this to the lower right panel in Fig. 4.58. The performance for both cases is about the same.

4.12 Deep Learning

Neural networks have a long history going back to the 1960s, but the recent availability of large-scale, high-quality data and new parallel computing infrastructures have reinvigorated neural networks in terms of size and complexity. This new reinvigoration, with many new and complex topologies, is called *deep learning*. There have been exciting developments in image and video processing, speech recognition, and automated video captioning based on deep learning systems. However, this is still a very active area of research. Fortunately, big companies with major investments in this area have made much of their research software open source (e.g., Tensorflow, PyTorch), with corresponding Python-bindings. To build up our understanding of neural networks, we begin with Rosenblatt's 1960 perceptron.

Perceptron Learning The perceptron is the primary ancestor of the most popular deep learning technologies (i.e., multilayer perceptron), and it is the best place to start as it will reveal the basic mechanics and themes of more complicated neural networks. The job of the perceptron is to create a linear classifier that can separate points in $\mathbb{R}^n$ between two classes. The basic idea is that given a set of associations:

$$\{(\mathbf{x}_0, y_0), \ldots, (\mathbf{x}_m, y_m)\}$$

where each $\mathbf{x} \in \mathbb{R}^{n-1}$ is augmented with a unit-entry to account for an offset term, and a set of weights $\mathbf{w} \in \mathbb{R}^n$, compute the following as an estimate of the label $y \in \{-1, 1\}$.

$$\hat{y} = \mathbf{w}^T \mathbf{x}$$

Concisely, this means that we want $\mathbf{w}$ such that

$$\mathbf{w}^T \mathbf{x}_i \underset{C_1}{\overset{C_2}{\gtrless}} 0$$

where $\mathbf{x}_i$ is in class C_2 if $\mathbf{x}_i^T \mathbf{w} > 0$ and class C_1 otherwise. To determine these weights, we apply the following learning rule:

$$\mathbf{w}^{(k+1)} = \mathbf{w}^{(k)} - (y - \hat{y})\mathbf{x}_i$$

The output of the perceptron can be summarized as

$$\hat{y} = \mathrm{sgn}(\mathbf{x}_i^T \mathbf{w})$$

The sign is the *activation* function of the perceptron. With this setup, we can write out the perceptron's output as the following:

```
>>> def yhat(x,w):
...         return np.sign(np.dot(x,w))
...
```

Let us create some fake data to play with:

```
>>> npts = 100
>>> X=np.random.rand(npts,2)*6-3 # random scatter in 2-d plane
>>> labels=np.ones(X.shape[0],dtype=np.int)# labels are 0 or 1
>>> labels[(X[:,1]<X[:,0])]=-1
>>> X = np.c_[X,np.ones(X.shape[0])]# augment with offset term
```

Note that we added a column of ones to account for the offset term. Certainly, by our construction, this problem is linearly separable, so let us see if the perceptron can find the boundary between the two classes. Let us start by initializing the weights,

```
>>> winit = np.random.randn(3)
```

and then apply the learning rule,

```
>>> w= winit
>>> for i,j in zip(X,labels):
...      w = w - (yhat(i,w)-j)*i
...
```

Note that we are taking a single ordered pass through the data. In practice, we would have randomly shuffled the input data to ensure that there is no incidental structure in the order of the data that would influence training. Now, let us examine the accuracy of the perceptron,

```
>>> from sklearn.metrics import accuracy_score
>>> print(accuracy_score(labels, [yhat(i,w) for i in X]))
0.96
```

We can re-run the training rule over the data to try to improve the accuracy. A pass through the data is called an *epoch*.

```
>>> for i,j in zip(X,labels):
...      w = w - (yhat(i,w)-j)*i
...
>>> print(accuracy_score(labels, [yhat(i,w) for i in X]))
0.98
```

Note that our initial weight for this epoch is the last weight from the previous pass. It is common to randomly shuffle the data between epochs. More epochs will result in better accuracy in this case.

We can redo this entire example with Keras. First, we define the model,

```
>>> from tensorflow.keras.optimizers import SGD
>>> from keras.models import Sequential
>>> from keras.layers import Dense
>>> model = Sequential()
>>> model.add(Dense(1,input_shape=(2,),activation='softsign'))
>>> model.compile(SGD(), 'hinge')
```

Note that we use the softsign activation instead of the sgn that we used earlier because we need a differentiable activation function. Given the form of the weight update in perceptron learning, it is equivalent to the hinge loss function. Stochastic gradient descent (SGD) is chosen for updating the weights. The softsign function is defined as the following:

$$s(t) = \frac{x}{1 + |x|}$$

Fig. 4.60 The `softsign` function is a smooth approximation to the `sign` function. This makes it easier to differentiate for backpropagation

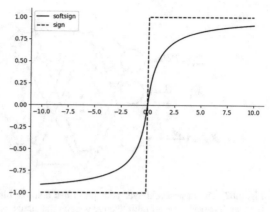

Fig. 4.61 Trajectory of the loss function

We can pull it out from the `tensorflow` backend that `keras` uses as in the following, plotted in Fig. 4.60

```
>>> import tensorflow as tf
>>> xi = np.linspace(-10,10,100)
>>> y_ = tf.math.softsign(xi)
```

Next, all we have to do is `fit` the model on data,

```
>>> h=model.fit(X[:,:2], labels, epochs=300, verbose=0)
```

The h variable is the *history* that contains the internal metrics and parameters involved in the `fit` training phase. We can extract the trajectory of the loss function from this history and draw the `loss` in Fig. 4.61.

Multilayer Perceptron The multilayer perceptron (MLP) generalizes the perceptron by stacking them as fully connected individual layers. The basic topology is shown in Fig. 4.62. In the previous section, we saw that the basic perceptron could generate a linear boundary for data that is linearly separable. The MLP can create

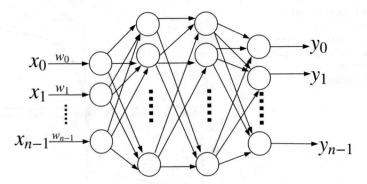

Fig. 4.62 The basic multilayer perceptron has a single hidden layer between input and output. Each of the arrows has a multiplicative weight associated with it

Fig. 4.63 Data from
`make_moons`

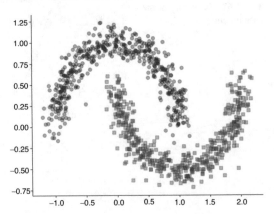

more complex nonlinear boundaries. Let us examine the *moons* dataset from Scikit-learn,

```
>>> from sklearn.datasets import make_moons
>>> X, y =make_moons(n_samples=1000,noise=0.1,random_state=1234)
```

The purpose of the `noise` term is to make data for each of the categories harder to disambiguate. These data are shown in Fig. 4.63.

The challenge for the MLP is to derive a nonlinear boundary between these two classes. We construct our MLP using `keras`,

```
>>> from tensorflow.keras.optimizers import Adam
>>> model = Sequential()
>>> model.add(Dense(4,input_shape=(2,),activation='sigmoid'))
>>> model.add(Dense(2,activation='sigmoid'))
>>> model.add(Dense(1,activation='sigmoid'))
>>> model.compile(Adam(learning_rate=0.05),'binary_crossentropy'
```

This MLP has three layers. The input layer has four units and the next layer has two units and the output layer has one unit to distinguish between the two available classes. Instead of plain stochastic gradient descent, we use the more advanced Adam optimizer. A quick summary of the model elements and parameters comes from the `model.summary()` method,

```
>>> model.summary()
Model: "sequential_1"
```

Layer (type)	Output Shape	Param #
dense_1 (Dense)	(None, 4)	12
dense_2 (Dense)	(None, 2)	10
dense_3 (Dense)	(None, 1)	3

```
Total params: 25
Trainable params: 25
Non-trainable params: 0
```

As usual, we split the input data into train and test sets,

```
>>> from sklearn.model_selection import train_test_split
>>> X_train,X_test,y_train,y_test=train_test_split(X,y,
...                                    test_size=0.3,
...                                    random_state=1234)
```

Thus, we reserve 30% of the data for testing. Next, we train the MLP,

```
>>> h=model.fit(X_train, y_train, epochs=100, verbose=0)
```

To compute the accuracy metric using the test set, we need to compute the model prediction on the this set.

```
>>> y_train_ = (model.predict(X_train) > 0.5).astype("int32")
>>> y_test_ = (model.predict(X_test) > 0.5).astype("int32")
>>> print(accuracy_score(y_train,y_train_))
1.0
>>> print(accuracy_score(y_test,y_test_))
0.9966666666666667
```

To visualize the so-derived boundary between these two classes, we use the `contourf` function from matplotlib which generates a filled contour plot shown in Fig. 4.64.

Instead of computing the accuracy separately, we can assign it as a *metric* for keras to track by supplying it on the `compile` step, as in the following,

```
>>> model.compile(Adam(learning_rate=0.05), 'binary_
                   crossentropy',
...                metrics=['accuracy'])
```

Then, we can train again,

```
>>> h=model.fit(X_train, y_train, epochs=100, verbose=0)
```

Fig. 4.64 The derived boundary separates the two classes

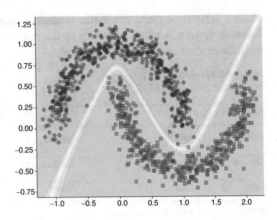

Now, we can evaluate the model on the test data,

```
>>> loss,acc=model.evaluate(X_test,y_test,verbose=0)
>>> print(acc)
0.996666669845581
```

where `loss` is the loss function and `acc` is the corresponding accuracy. The Keras documentation has other metrics that can be specified during the `compile` step.

Backpropagation We have seen that the MLP can generate complicated nonlinear boundaries for classification problems. The key algorithm underpinning MLP is backpropagation. The idea is that when we stack layers into the MLP, we are applying function composition, which basically means we take the output of one function and then feed it into the input of another.

$$h = (f \circ g)(x) = f(g(x))$$

For example, for the simple perceptron, we have $g(\mathbf{x}) = \mathbf{w}^T \mathbf{x}$ and $f(x) = \text{sgn}(x)$. They key property of this composition is that derivatives use the chain rule from calculus.

$$h'(x) = f'(g(x))g'(x)$$

Notice this has turned the differentiation operation into a multiplication operation. Explaining backpropagation in general is a notational nightmare, so let us see if we can get the main idea from a specific example. Consider the following two layer MLP with one input and one output.

There is only one input (x_1). The output of the first layer is

$$z_1 = f(x_1 w_1 + b_1) = f(p_1)$$

where f is the sigmoid function and b_1 is the bias term. The output of the second layer is

$$z_2 = f(z_1 w_2 + b_2) = f(p_2)$$

To keep it simple, let us suppose that the loss function for this MLP is the squared error,

$$J = \frac{1}{2}(z_2 - y)^2$$

where y is the target label. Backpropagation has two phases. The forward phase computes the MLP loss function given the values of the inputs and corresponding weights. The backward phase applies the incremental weight updates to each weight based on the forward phase. To implement gradient descent, we have to calculate the derivative of the loss function with respect to each of the weights.

$$\frac{\partial J}{\partial w_2} = \frac{\partial J}{\partial z_2} \frac{\partial z_2}{\partial p_2} \frac{\partial p_2}{\partial w_2}$$

The first term is the following,

$$\frac{\partial J}{\partial z_2} = z_2 - y$$

The second term is the following,

$$\frac{\partial z_2}{\partial p_2} = f'(p_2) = f(p_2)(1 - f(p_2))$$

Note that by property of the sigmoid function, we have $f'(x) = (1 - f(x))f(x)$. The third term is the following,

$$\frac{\partial p_2}{\partial w_2} = z_1$$

Thus, the update for w_2 is the following,

$$\Delta w_2 \propto (z_2 - y)z_1(1 - z_2)z_2$$

The corresponding analysis for b_2 gives the following,

$$\Delta b_2 = (z_2 - y)z_2(1 - z_2)$$

Let us keep going backward to w_1,

$$\frac{\partial J}{\partial w_1} = \frac{\partial J}{\partial z_2} \frac{\partial z_2}{\partial p_2} \frac{\partial p_2}{\partial z_1} \frac{\partial z_1}{\partial p_1} \frac{\partial p_1}{\partial w_1}$$

The first new term is the following,

$$\frac{\partial p_2}{\partial z_1} = w_2$$

and then the next two terms,

$$\frac{\partial z_1}{\partial p_1} = f(p_1)(1 - f(p_1)) = z_1(1 - z_1)$$

$$\frac{\partial p_1}{\partial w_1} = x_1$$

This makes the update for w_1,

$$\Delta w_1 \propto (z_2 - y)z_2(1 - z_2)w_2 z_1(1 - z_1)x_1$$

To understand why this is called backpropagation, we can define

$$\delta_2 := (z_2 - y)z_2(1 - z_2)$$

This makes the weight update for w_2,

$$\Delta w_2 \propto \delta_2 z_1$$

This means that the weight update for w_2 is proportional to the output of the prior layer (z_1) and a factor that accounts steepness of the activation function. Likewise, the weight update for w_1 can be written as the following,

$$\Delta w_1 \propto \delta_1 x_1$$

where

$$\delta_1 := \delta_2 w_2 z_1(1 - z_1)$$

Note that this weight update is proportional to the input (prior layer's output) just as the weight update for w_2 was proportional to the prior layer output z_1. Also, the δ factors propagate recursively backward to the input layer. These characteristics permit efficient numerical implementations for large networks because the subsequent computations are based on prior calculations. This also means that each individual unit's calculations are localized upon the output of the prior layer. This helps segregate the individual processing behavior of each unit within each layer.

Functional Deep Learning Keras has an alternative API that makes it possible to understand the performance of neural networks using the composition of functions ideas we discussed. The key objects for this functional interpretation are the `Input` object and the `Model` object.

```
>>> from keras.layers import Input
>>> from keras.models import Model
>>> import keras.backend as K
```

We can re-create the data from our earlier classification example

```
>>> X, y = make_moons(n_samples=1000, noise=0.1, random_state=1234)
```

The first step is to construct a placeholder for the input using the `Input` object,

```
>>> inputs = Input(shape=(2,))
```

Next, we can stack the `Dense` layers as before but now tie their inputs to the previous layer's outputs by calling `Dense` as a function.

```
>>> l1=Dense(3,input_shape=(2,),activation='sigmoid')(inputs)
>>> l2=Dense(2,input_shape=(3,),activation='sigmoid')(l1)
>>> outputs=Dense(1,input_shape=(3,),activation='sigmoid')(l1)
```

This means that output $= (\ell_2 \circ \ell_1)(\text{input})$ where ℓ_1 and ℓ_2 are the middle layers. With that established, we collect the individual pieces in the `Model` object and then `fit` and `train` as usual.

```
>>> model = Model(inputs=inputs,outputs=outputs)
>>> model.compile(Adam(learning_rate=0.05),
...                     'binary_crossentropy',
...                     metrics=['accuracy'])
>>> h=model.fit(X_train, y_train, epochs=500, verbose=0)
```

This gives the same result as before. The advantage of the functional perspective is that now we can think of the individual layers as mappings between multi-dimensional $\mathbb{R}^n$ spaces. For example, $\ell_1 : \mathbb{R}^2 \mapsto \mathbb{R}^3$ and $\ell_2 : \mathbb{R}^3 \mapsto \mathbb{R}^2$. Now, we can investigate the performance of the network from the inputs just up until the final mapping to $\mathbb{R}$ at the output by defining the functional mapping $(\ell_2 \circ \ell_1)(\text{inputs}) : \mathbb{R}^2 \mapsto \mathbb{R}^2$, as shown in Fig. 4.65.

To get this result, we have to define a Keras `function` using the `inputs`.

```
>>> l2_function = K.function([inputs], [l2])
>>> # functional mapping just before output layer
>>> l2o=l2_function([X_train])
```

the `l2o` list contains the output of the `l2` layer that is shown in Fig. 4.65.

Fig. 4.65 The embedded representation of the input just before the final output that shows the internal divergence of the two target classes

4.12.1   Understanding Gradient Descent

As gradient descent is fundamental to training neural networks, let us examine how it works in detail. Consider a smooth function f over $\mathbb{R}^n$ suppose we want to find the minimum value of $f(\mathbf{x})$ over this domain, as in the following,

$$\mathbf{x}^* = \arg\min_{\mathbf{x}} f(\mathbf{x})$$

The idea with gradient descent is to choose an initial point $\mathbf{x}^{(0)} \in \mathbb{R}^n$

$$\mathbf{x}^{(k+1)} = \mathbf{x}^{(k)} - \alpha \nabla f(\mathbf{x}^{(k)})$$

where α is the step size (learning rate). The intuition here is that ∇f is the direction of increase and so that moving in the opposite direction scaled by α moves toward a lower function value. This approach turns out to be very fast for well-conditioned, strongly convex f but in general there are practical issues to manage.

Figure 4.66 shows the function $f(x) = 2 - 3x^3 + x^4$ and its first-order Taylor series approximation at a selected points along the curve for a given width parameter. That is, the Taylor approximation approximates the function at a specific point with a corresponding interval around that point for which the approximation is assumed valid. The size of this width is determined by the α step parameter. Crucially, the quality of the approximation varies along the curve. In particular, there are sections where two nearby approximations overlap given the width, as indicated by the dark shaded regions. This is key because gradient descent works by using such first-order approximations to estimate the next step in the minimization. That is, the gradient descent algorithm never actually *sees* $f(x)$, but rather only the given first-order approximant. It judges the direction of the next iterative step by sliding down the slope of the approximant to the edge of a region (determined by α) and then using that next point for the next calculated approximant. As shown by the

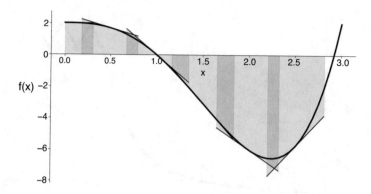

Fig. 4.66 The piecewise linear approximant to $f(x)$

shaded regions, it is possible that the algorithm will overshoot the minimum because the step size (α) is too large. This can cause oscillations as shown in Fig. 4.67.

Let us consider the following Python implementation of gradient descent, using Sympy.

```
>>> import sympy as sm
>>> x = sm.symbols('x',real=True)
>>> fx = 2 - 3*x**3 + x**4
>>> df = fx.diff(x) # compute derivative
>>> x0 =.1 # initial guess
>>> xlist = [(x0,fx.subs(x,x0))]
>>> alpha = 0.1 # step size
>>> for i in range(20):
...         x0 = x0 - alpha*df.subs(x,x0)
...         xlist.append((x0,fx.subs(x,x0)))
...
```

Figure 4.67 shows the sequential steps. Note that the algorithm oscillates at the end because the step size is too large. Practically speaking, it is not possible to know the optimal step size for general functions without strong assumptions on $f(x)$.

Figure 4.68 shows how the algorithm moves along the function as well as the approximant ($\hat{f}(x)$) that the algorithm sees along the way. Note that initial steps are crowded around the initial point because the corresponding gradient is small there. Toward the middle, the algorithm makes a big jump because the gradient is steep, before finally oscillating toward the end. Sections of the underlying function that are relatively flat can cause the algorithm to converge very slowly. Furthermore, if there are multiple local minima, then the algorithm cannot guarantee finding the global minimum.

As we have seen, the step size is key to both performance and convergence. Indeed, a step size that is too big can cause divergence and one that is too small can take a very long time to converge.

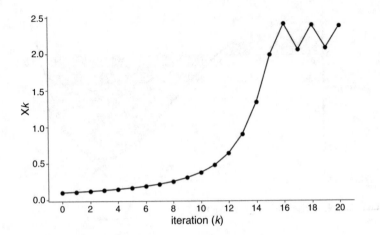

Fig. 4.67 The step size may cause oscillations

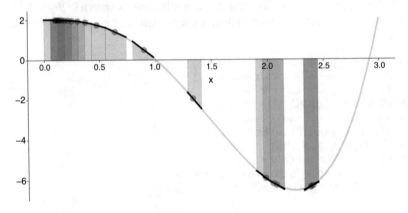

Fig. 4.68 Gradient descent produces a sequence of approximants

Newton's Method Consider the following second-order Taylor series expansion

$$J(\mathbf{x}) = f(\mathbf{x}_0) + \nabla f(\mathbf{x}_0)^T (\mathbf{x} - \mathbf{x}_0) + \frac{1}{2}(\mathbf{x} - \mathbf{x}_0)^T \nabla^2 f(\mathbf{x}_0)(\mathbf{x} - \mathbf{x}_0)$$

where $\mathbf{H}(\mathbf{x}) := \nabla^2 f(\mathbf{x})$ is the Hessian matrix of second derivatives. The (i, j)th entry of this matrix is the following:

$$\frac{\partial^2 f(\mathbf{x})}{\partial x_i \partial x_j}$$

We can use basic matrix calculus to find the minimum by computing:

$$\nabla_{\mathbf{x}} J(\mathbf{x}) = \nabla f(\mathbf{x}_0) + \mathbf{H}(\mathbf{x})(\mathbf{x} - \mathbf{x}_0) = 0$$

Solving this for **x** gives the following:

$$\mathbf{x} = \mathbf{x}_0 - \mathbf{H}(\mathbf{x})^{-1}\nabla f(\mathbf{x}_0)$$

Thus, after renaming some terms, the descent algorithm works by the following update equation:

$$\mathbf{x}^{(k+1)} = \mathbf{x}^{(k)} - \mathbf{H}(\mathbf{x}^{(k)})^{-1}\nabla f(\mathbf{x}^{(k)})$$

There are a number of practical problems with this update equation. First, it requires computing the Hessian matrix at every step. For a significant problem, this means managing a potentially very large matrix. For example, given 1000 dimensions the corresponding Hessian has 1000×1000 elements. Some other issues are that the Hessian may not be numerically stable enough to invert, the functional form of the partial derivatives may have to be separately approximated, and the initial guess has to be in a region where the convexity of the function matches the derived assumption. Otherwise, just based on these equations, the algorithm will converge on the local *maximum* and not the local *minimum*. Consider a slight change of the previous code to implement Newton's method:

```
>>> x0 = 2.0 # init guess is near to solution
>>> xlist = [(x0,fx.subs(x,x0))]
>>> df2 = fx.diff(x,2) # 2nd derivative
>>> for i in range(5):
...     x0 = x0 - df.subs(x,x0)/df2.subs(x,x0)
...     xlist.append((x0,fx.subs(x,x0)))
...
>>> xlist = np.array(xlist).astype(float)
>>> print(xlist)
[[ 2.          -6.         ]
 [ 2.33333333 -6.4691358 ]
 [ 2.25555556 -6.54265522]
 [ 2.25002723 -6.54296874]
 [ 2.25        -6.54296875]
 [ 2.25        -6.54296875]]
```

Note that it took very few iterations to get to the minimum (as compared to our prior method), but if the initial guess is too far away from the actual minimum, the algorithm may not find the local minimum at all and instead find the local maximum. Naturally, there are many extensions to this method to account for these effects, but the main thrust of this section is to illustrate how higher-order derivatives (when available) in a computationally feasible context can greatly accelerate convergence of descent algorithms.

Managing Step Size The problem of determining a good step size (learning rate) can be approached with an *exact line search*. That is, along the ray that extends along $\mathbf{x} + q\nabla f(x)$, find

$$q_{min} = \arg\min_{q \geq 0} f(\mathbf{x} + q \nabla f(\mathbf{x}))$$

In words, this means that given a direction from a point $\mathbf{x}$ along the direction $\nabla f(\mathbf{x})$, find the minimum for this one-dimensional problem. Thus, the minimization procedure alternates at each iteration between moving to a new $\mathbf{x}$ position in $\mathbb{R}^n$ and finding a new step size by solving the one-dimensional minimization.

While this is conceptually clean, the problem is that solving the one-dimensional line search at every step means evaluating the objective function $f(\mathbf{x})$ at many points along the one-dimensional slice. This can be very time-consuming for an objective function that is computationally expensive to evaluate. With Newton's method, we have seen that higher-order derivatives can accelerate convergence, and we can apply those ideas to the one-dimensional line search, as with the *backtracking* algorithm.

- Fix parameters $\beta \in [0, 1)$ an $\alpha > 0$.
- If $f(x - \alpha \nabla f(x)) > f(x) - \alpha \|\nabla f(x)\|_2^2$, then reduce $\alpha \rightarrow \beta\alpha$. Otherwise, do the usual gradient descent update: $x^{(k+1)} = x^{(k)} - \alpha \nabla f(x^{(k)})$.

To gain some intuition about how this works, return to our second-order Taylor series expansion of the function f about $\mathbf{x}_0$,

$$f(\mathbf{x}_0) + \nabla f(\mathbf{x}_0)^T (\mathbf{x} - \mathbf{x}_0) + \frac{1}{2}(\mathbf{x} - \mathbf{x}_0)^T \nabla^2 f(\mathbf{x}_0)(\mathbf{x} - \mathbf{x}_0)$$

We have already discussed the numerical issues with the Hessian term, so one approach is to simply replace that term with an $n \times n$ identity matrix $\mathbf{I}$ to obtain the following:

$$h_\alpha(\mathbf{x}) = f(\mathbf{x}_0) + \nabla f(\mathbf{x}_0)^T (\mathbf{x} - \mathbf{x}_0) + \frac{1}{2\alpha}\|\mathbf{x} - \mathbf{x}_0\|^2$$

This is our more tractable *surrogate* function. But what is the relationship between this surrogate and what we are actually trying to minimize? The key difference is that the curvature information that is contained in the Hessian term has now been reduced to a single $1/\alpha$ factor. Intuitively, this means that local complicated curvature of f about a given point $\mathbf{x}_0$ has been replaced with a uniform bowl-shaped structure, the steepness of which is determined by scaling $1/\alpha$. Given a specific α, we already know how to step directly to the minimum of $h_\alpha(\mathbf{x})$; namely, using the following gradient descent update equation:

$$\mathbf{x}^{(k+1)} = \mathbf{x}^{(k)} - \alpha \nabla f(\mathbf{x}^{(k)})$$

That is the immediate solution to the surrogate problem, but it does not directly supply the next iteration for the function we really want: f. Let us suppose that our minimization of the surrogate has taken us to a new point $\mathbf{x}^{(k)}$ that satisfies the following inequality,

$$f(\mathbf{x}^{(k+1)}) \le h_\alpha(\mathbf{x}^{(k+1)})$$

or, more explicitly,

$$f(\mathbf{x}^{(k+1)}) \le f(\mathbf{x}^{(k)}) + \nabla f(\mathbf{x}^{(k)})^T(\mathbf{x}^{(k+1)} - \mathbf{x}^{(k)}) + \frac{1}{2\alpha}\|\mathbf{x}^{(k+1)} - \mathbf{x}^{(k)}\|^2$$

We can substitute the update equation into this and simplify as,

$$f(\mathbf{x}^{(k+1)}) \le f(\mathbf{x}^{(k)}) - \alpha\nabla f(\mathbf{x}^{(k)})^T(\nabla f(\mathbf{x}^{(k)})) + \frac{\alpha}{2}\|\nabla f(\mathbf{x}^{(k)})\|^2$$

which ultimately boils down to the following,

$$f(\mathbf{x}^{(k+1)}) \le f(\mathbf{x}^{(k)}) - \frac{\alpha}{2}\|\nabla f(\mathbf{x}^{(k)})\|^2 \qquad (4.19)$$

The important observation here is that if we have not reached the minimum of f, then the last term is always positive and we have moved downward,

$$f(\mathbf{x}^{(k+1)}) < f(\mathbf{x}^{(k)})$$

which is what we were after. Conversely, if the inequality in Eq. (4.19) holds for some $\alpha > 0$, then we know that $h_\alpha > f$. This is the key observation behind the backtracking algorithm. That is, we can test a sequence of values for α until we find one that satisfies Eq. (4.19). For example, we can start with some initial α and then scale it up or down until the inequality is satisfied which means that we have found the correct step size and then can proceed with the descent step. This is what the backtracking algorithm is doing as shown in Fig. 4.69. The dotted line is the $h_\alpha(x)$ and the gray line is $f(x)$. The algorithm hops to the quadratic minimum of the $h_\alpha(x)$ function which is close to the actual minimum of $f(x)$.

The basic implementation of backtracking is shown below:

```
>>> x0 = 1
>>> alpha = 0.5
>>> xnew = x0 - alpha*df.subs(x,x0)
>>> while fx.subs(x,xnew) > (fx.subs(x,x0) - (alpha/2.)*(fx.subs(x,x0))**2):
...         alpha = alpha * 0.8
...         xnew = x0 - alpha*df.subs(x,x0)
...
>>> print(alpha,xnew)
0.32000000000000006 2.60000000000000
```

Stochastic Gradient Descent A common variation on gradient descent is to alter how the weights are updated. Specifically, suppose we want to minimize an objective function of the following form:

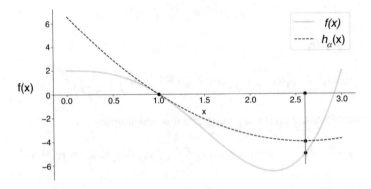

Fig. 4.69 The approximation $h_\alpha(x)$ (dotted line) moves the next iteration from $x = 1$ to the indicated point that is near the minimum of $f(x)$ by finding an appropriate step size (α)

$$\min_x \sum_{i=1}^{m} f_i(x)$$

where i indexes the ith data element for an error function. Equivalently, each summand is parameterized by a data element.

For the usual gradient descent algorithm, we would compute the incremental weights, component-wise as in the following

$$x^{(k+1)} = x^{(k)} - \alpha_k \sum_{i=1}^{m} \partial f_i(x^{(k)})$$

by summing over all of the data. The key idea for stochastic gradient descent is to *not* sum over all of the data but rather to update the weights for each randomized ith data element:

$$x^{(k+1)} = x^{(k)} - \alpha_k \partial f_i(x^{(k)})$$

A compromise between batch and this jump-every-time stochastic gradient descent is *mini-batch* gradient descent in which a randomized subset (σ_r, $|\sigma_r| = M_b$) of the data is summed over at each step as in the following:

$$x^{(k+1)} = x^{(k)} - \alpha_k \sum_{i \in \sigma_r} \partial f_i(x^{(k)})$$

Each step update for the standard gradient descent algorithm processes m data points for each of the p dimensions, $\mathcal{O}(mp)$, whereas for stochastic gradient descent, we have $\mathcal{O}(p)$. Mini-batch gradient descent is somewhere in-between these estimates. For very large, high-dimensional data, the computational costs of gradient descent can become prohibitive, thus favoring stochastic gradient descent. Outside of the computational advantages, stochastic gradient descent has other favorable attributes. For example, the noisy jumping around helps the algorithm avoid getting stalled in local minima, and this helps the algorithm when the starting point is far away from the actual minimum. The obverse is that stochastic gradient descent can struggle to clinch the minimum when it is close to it. Another advantage is robustness to a minority of *bad* data elements. Because only random subsets of the data are actually used in the update, the few individual outlier data points (perhaps due to poor data integrity) do not necessarily contaminate every step update.

Momentum The gradient descent algorithm can be considered as a particle moving along a high-dimensional landscape in search of a minimum. Using a physical analogy, we can add the concept of momentum to the particle's motion. Consider the position of the particle ($x^{(k)}$) at any time k under a net force proportional to $-\nabla J$. This setup induces an estimated velocity term for the particle motion proportional to $\eta(x^{(k+1)} - x^{(k)})$. That is, the particle's velocity is estimated proportional to the difference in two successive positions. The simplest version of stochastic gradient descent update that incorporates this momentum is the following:

$$x^{(k+1)} = x^{(k)} - \alpha \nabla f(x^{(k)}) + \eta(x^{(k+1)} - x^{(k)})$$

Momentum is particularly useful when gradient descent sloshes up and down a steep ravine in the error surface instead of pursuing the descending ravine directly to the local minimum. This oscillatory behavior can cause slow convergence. There are many extensions to this basic idea such as Nesterov momentum.

Advanced Stochastic Gradient Descent Methods that aggregate histories of the step updates can provide superior performance to the basic stochastic gradient descent algorithm. For example, Adam (Adaptive Moment Estimator) implements an adaptive step size for each parameter. It also keeps track of an exponentially decaying mean and variance of past gradients using the exponentially weighted moving average (EWMA). This smoothing technique computes the following recursion,

$$y_n = ax_n + (1-a)y_{n-1}$$

with $y_0 = x_0$ as the initial condition. The $0 < a < 1$ factor controls the amount of mixing between the previous moving average and the new data point at n. For example, if $a = 0.9$, the EWMA favors the new data x_n over the prior value y_{n-1} ($1 - a = 0.1$) of the EWMA. This calculation is common in a wide variety of time-series applications (e.g., signal processing, quantitative finance). The impulse response of the EWMA ($x = \delta_n$) is $(1 - a)^n$. You can think of this as the weighted window function that is applied to x_n. As opposed to the standard moving average that considers a fixed window of data to average over, this exponential window retains prior memory of the entire sequence, albeit weighted by powers of $(1 - a)$.

Fig. 4.70 The exponential moving average versus the plain moving average

To see this, we can generate the response to an impulse data series using `pandas` (Fig. 4.70),

```
>>> import pandas as pd
>>> x = pd.Series([1]+[0]*20)
>>> ma =x.rolling(window=3, center=False).mean()
>>> ewma  = x.ewm(1).mean()
```

As shown in Fig. 4.71, the single nonzero data point thereafter influences the EWMA whereas for the fixed-width moving window average, the effect terminates after the window passes. Note that mini-batch smoothes out data at each iteration by averaging over training data and EWMA smoothes out the descent motion across iterations of the algorithm.

Advanced stochastic gradient descent algorithms are themselves an area of intense interest and development. Each method has its strengths and weaknesses pursuant to the data at hand (i.e., sparse vs. dense data), and there is no clear favorite appropriate to all circumstances. As a practical matter, some variants have parallel implementations that accelerate performance (i.e., Nui's Hogwild update scheme).

Python Example Using Sympy Each of these methods will make more sense with some Python. We emphasize that this implementation is strictly expository and unsuitable for a large-scale application. Let us reconsider the classification problem in the section on logistic regression with the target $y_i \in \{0, 1\}$. The logistic regression seeks to minimize the cross-entropy:

$$ J(\boldsymbol{\beta}) = \sum_i^m \log(1 + \exp(\mathbf{x}_i^T \boldsymbol{\beta})) - y_i \mathbf{x}_i^T \boldsymbol{\beta} $$

Fig. 4.71 Different variations of gradient descent

with the corresponding gradient,

$$\nabla_\beta J(\beta) = \sum_i^m \frac{1}{1 + \exp(-\mathbf{x}_i^T \beta)} \mathbf{x}_i - y_i \mathbf{x}_i$$

To get started, let us create some sample data for logistic regression

```
>>> import numpy as np
>>> import sympy as sm
>>> npts = 100
>>> X=np.random.rand(npts,2)*6-3 # random scatter in 2-d plane
>>> labels=np.ones(X.shape[0],dtype=np.int)# labels are 0 or 1
>>> labels[(X[:,1]<X[:,0])]=0
```

This provides the data in the X Numpy array and the target labels in the `labels` array. Next, we want to develop the objective function with Sympy,

```
>>> x0,x1 = sm.symbols('x:2',real=True) # data placeholders
>>> b0,b1 = sm.symbols('b:2',real=True) # parameters
>>> bias = sm.symbols('bias',real=True) # bias term
>>> y = sm.symbols('y',real=True) # label placeholders
>>> summand = sm.log(1+sm.exp(x0*b0+x1*b1+bias))-y*(x0*b0+x1*b1+bias)
>>> J = sum([summand.subs({x0:i,x1:j,y:y_i})
...             for (i,j),y_i in zip(X,labels)])
```

We can use Sympy to compute the gradient as in the following:

```
>>> from sympy.tensor.array import derive_by_array
>>> grad = derive_by_array(summand,(b0,b1,bias))
```

Using the `sm.latex` function renders `grad` as the following:

$$\left[-x_0 y + \frac{x_0 e^{b_0 x_0 + b_1 x_1 + bias}}{e^{b_0 x_0 + b_1 x_1 + bias} + 1} \quad -x_1 y + \frac{x_1 e^{b_0 x_0 + b_1 x_1 + bias}}{e^{b_0 x_0 + b_1 x_1 + bias} + 1} \quad -y + \frac{e^{b_0 x_0 + b_1 x_1 + bias}}{e^{b_0 x_0 + b_1 x_1 + bias} + 1} \right]$$

which matches our previous computation of the gradient. For standard gradient descent, the gradient is computed by summing over all of the data,

```
>>> grads=np.array([grad.subs({x0:i,x1:j,y:y_i})
...                      for (i,j),y_i in zip(X,labels)]).sum(axis=0)
```

Now, to implement gradient descent, we set up the following loop:

```
>>> # convert expression into function
>>> Jf = sm.lambdify((b0,b1,bias),J)
>>> gradsf = sm.lambdify((b0,b1,bias),grads)
>>> niter = 200
>>> winit = np.random.randn(3)*20
>>> alpha = 0.1 # learning rate (step-size)
>>> WK = winit  # initialize
>>> Jout=[] # container for output
>>> for i in range(niter):
...     WK = WK - alpha * np.array(gradsf(*WK))
...     Jout.append(Jf(*WK))
...
```

For stochastic gradient descent, the above code changes to the following:

```
>>> import random
>>> sgdWK = winit  # initialize
>>> Jout=[] # container for output
>>> # don't sum along all data as before
>>> grads=np.array([grad.subs({x0:i,x1:j,y:y_i})
...                      for (i,j),y_i in zip(X,labels)])
>>> for i in range(niter):
...     gradsf = sm.lambdify((b0,b1,bias),random.choice(grads))
...     sgdWK = sgdWK - alpha * np.array(gradsf(*sgdWK))
...     Jout.append(Jf(*sgdWK))
...
```

The main difference here is that the gradient calculation no longer sums across all of the input data (i.e., `grads` list) and is instead randomly chosen by the `random.choice` function the above body of the loop. The extension to batch gradient descent from this code just requires averaging over a sub-selection of the data for the gradients in the `batch` variable.

```
>>> mbsgdWK = winit  # initialize
>>> Jout=[] # container for output
>>> mb = 10 # number of elements in batch
>>> for i in range(niter):
...     batch = np.vstack([random.choice(grads)
...                            for i in range(mb)]).mean(axis=0)
```

```
...        gradsf = sm.lambdify((b0,b1,bias),batch)
...        mbsgdWK = mbsgdWK-alpha*np.array(gradsf(*mbsgdWK))
...        Jout.append(Jf(*mbsgdWK))
...
```

It is straightforward to incorporate momentum into this loop using a Python deque, as in the following,

```
>>> from collections import deque
>>> momentum = deque([winit,winit],2)
>>> mbsgdWK = winit  # initialize
>>> Jout=[] # container for output
>>> mb = 10 # number of elements in batch
>>> for i in range(niter):
...        batch=np.vstack([random.choice(grads)
...                            for i in range(mb)]).mean(axis=0)
...        gradsf=sm.lambdify((b0,b1,bias),batch)
...        mbsgdWK=mbsgdWK-alpha*np.array(gradsf(*mbsgdWK))+0.5*
           (momentum[1]-momentum[0])
...        Jout.append(Jf(*mbsgdWK))
...
```

Figure 4.71 shows the three variants of the gradient descent algorithm. Notice that the stochastic gradient descent algorithm is the most erratic, as it is characterized by taking a new direction for every randomly selected data element. Mini-batch gradient descent smoothes these out by averaging across multiple data elements. The momentum variant is somewhere in-between the two as the effect of the momentum term is not pronounced in this example.

Python Example Using Theano The code shown makes each step of the gradient descent algorithms explicit using Sympy, but this implementation is far too slow. The theano module provides thoughtful and powerful high-level abstractions for algorithm implementation that relies upon underlying C/C++ and GPU execution models. This means that calculations that are prototyped with theano can be executed downstream outside of the Python interpreter which makes them much faster. The downside of this approach is that calculations can become much harder to debug because of the multiple levels of abstraction. Nonetheless, theano is a powerful tool for algorithm development and execution.

To get started we need some basics from theano.

```
>>> import theano.tensor as T
>>> from theano import function, shared
```

The next step is to define variables, which are essentially placeholders for values that will be computed downstream later. The next block defines two named variables as a double-sized float matrix and vector. Note that we did not have to specify the dimensions of each at this point.

```
>>> x = T.dmatrix("x") # double matrix
>>> y = T.dvector("y") # double vector
```

The parameters of our implementation of gradient descent come next, as the following:

```
>>> w = shared(np.random.randn(2), name="w") # parameters to fit
>>> b = shared(0.0, name="b") # bias term
```

Variables that are shared are ones whose values can be set separately via other computations or directly via the set_value() method. These values can also be retrieved using the get_value() method. Now, we need to define the probability of obtaining a 1 from the given data as p. The cross-entropy function and the T.dot function are already present (along with a wide range of other related functions) in theano. The conformability of the constituent arguments is the responsibility of the user.

```
>>> p=1/(1+T.exp(-T.dot(x,w)-b)) # probability of 1
>>> error = T.nnet.binary_crossentropy(p,y)
>>> loss = error.mean()
>>> gw, gb = T.grad(loss, [w, b])
```

The error variable is TensorVariable type which has many built-in methods such as mean. The so-derived loss function is therefore also a TensorVariable. The last T.grad line is the best part of Theano because it can compute these gradients automatically.

```
>>> train = function(inputs=[x,y],
...                   outputs=[error],
...                   updates=((w, w - alpha * gw),
...                            (b, b - alpha * gb)))
```

The last step is to set up training by defining the training function in theano. The user will supply the previously defined and named input variables (x and y), and theano will return the previously defined error variable. Recall that the w and b variables were defined as shared variables. This means that the function train can update their values between calls using the update formula specified in the updates keyword variable. In this case, the update is just plain gradient descent with the previously defined alpha step-size variable.

We can execute the training plain using the train function in the following loop:

```
>>> training_steps=1000
>>> for i in range(training_steps):
...         error = train(X, labels)
...
```

The train(X,labels) call is where the X and labels arrays we defined earlier replace the placeholder variables. The update step refreshes all of the shared variables at each iterative step. At the end of the iteration, the so-computed parameters are in the w and b variables with values available via get_value(). The implementation for stochastic gradient descent requires just a little modification to this loop, as in the following:

```
>>> for i in range(training_steps):
...         idx = np.random.randint(0,X.shape[0])
...         error = train([X[idx,:]], [labels[idx]])
...
```

where the `idx` variable selects a random data element from the set and uses that for the update step at every iteration. Likewise, batch stochastic gradient descent follows with the following modification,

```
>>> batch_size = 50
>>> indices = np.arange(X.shape[0])
>>> for i in range(training_steps):
...         idx = np.random.permutation(indices)[:batch_size]
...         error = train(X[idx,:], labels[idx])
...
>>> print (w.get_value())
[-4.74033204  4.74110223]
>>> print (b.get_value()) # bias term
0.09935801044334443
```

Here, we set up an `indices` variable that is used for randomly selecting subsets in the `idx` variable that are passed to the `train` function. All of these implementations parallel the corresponding previous implementations in Sympy, but these are many orders of magnitude faster due to `theano`.

4.12.2 Image Processing Using Convolutional Neural Networks

In this section, we develop the convolutional neural network (CNN) which is the fundamental deep learning image processing application. We deconstruct every layer of this network to develop insight into the purpose of the individual operations. CNNs take image as inputs and images can be represented as Numpy arrays, which make them fast and easy to use with any of the scientific Python tools. The individual entries of the Numpy array are the pixels and the row/column dimensions are the height/width of the image, respectively. The array values are between 0 through 255 and correspond to the intensity of the pixel at that location. Three-dimensional images have a third third depth-dimension as the color channel (e.g., red, green, blue). Two-dimensional image arrays are grayscale.

Programming Tip

Matplotlib makes it easy to draw images using the underlying Numpy arrays. For instance, we can draw Fig. 4.72 using the following MNIST image from `sklearn.datasets`, which represents grayscale hand-drawn digits (the number zero in this case).

Fig. 4.72 Image of a hand
drawn number zero from the
MNIST dataset

Fig. 4.73 Samples of the
other hand-drawn digits from
MNIST

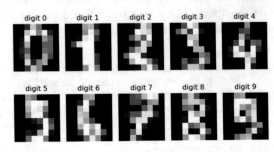

```
>>> from matplotlib.pylab import subplots, cm
>>> from sklearn import datasets
>>> mnist = datasets.load_digits()
>>> fig, ax = subplots()
>>> ax.imshow(mnist.images[0],
...           interpolation='nearest',
...           cmap=cm.gray)
<matplotlib.image.AxesImage object at 0x7fcee01db8b0>
```

The cmap keyword argument specifies the colormap as gray. The
interpolation keyword means that the resulting image from imshow
does not try to visually smooth out the data, which can be confusing when
working at the pixel level. The other hand-drawn digits are shown below in
Fig. 4.73.

Convolution Convolution is an intensive calculation, and it is the core of convolutional neural networks. The purpose of convolution is to create alternative
representations of the input image that emphasize or deemphasize certain features
represented by the *kernel*. The convolution operation consists of a kernel and an
input matrix. The convolution operation is a way of aligning and comparing image
data with the corresponding data in an image kernel. You can think of an image
kernel as a template for a canonical feature that the convolution operation will
uncover. To keep it simple, suppose we have the following 3x3 kernel matrix,

```
>>> kern = np.eye(3,dtype=np.int)
>>> kern
array([[1, 0, 0],
       [0, 1, 0],
       [0, 0, 1]])
```

Using this kernel, we want to find anything in an input image that looks like a diagonal line. Let us suppose we have the following input Numpy image

```
>>> tmp = np.hstack([kern,kern*0])
>>> x = np.vstack([tmp,tmp])
>>> x
array([[1, 0, 0, 0, 0, 0],
       [0, 1, 0, 0, 0, 0],
       [0, 0, 1, 0, 0, 0],
       [1, 0, 0, 0, 0, 0],
       [0, 1, 0, 0, 0, 0],
       [0, 0, 1, 0, 0, 0]])
```

Note that this image is just the kernel stacked into a larger Numpy array. We want to see if the convolution can pull out the kernel that is embedded in the image. Of course, in a real application, we would not know whether or not the kernel is present in the image, but this example helps us understand the convolution operation step-by-step. There is a convolution function available in the `scipy` module.

```
>>> from scipy.ndimage.filters import convolve
>>> res = convolve(x,kern,mode='constant',cval=0)
>>> res
array([[2, 0, 0, 0, 0, 0],
       [0, 3, 0, 0, 0, 0],
       [0, 0, 2, 0, 0, 0],
       [2, 0, 0, 1, 0, 0],
       [0, 3, 0, 0, 0, 0],
       [0, 0, 2, 0, 0, 0]])
```

Each step of the convolution operation is represented in Fig. 4.74. The `kern` matrix (light blue square) is overlaid upon the `x` matrix and the element-wise product is computed and summed. Thus, the `0,0` array output corresponds to this operation applied to the top-left 3×3 slice of the input, which results in 3. The convolution operation is sensitive to boundary conditions. For this example, we have chosen `mode=constant` and `cval=0` which means that the input image is bordered by zeros when the kernel sweeps outside of the input image boundary. This is the simplest option for managing the edge conditions and `scipy.ndimage.filters.convolve` provides other practical alternatives. It is also common to normalize the output of the convolution operation by dividing the number of pixels in the kernel (i.e., 3 in this example). Another way to think about the convolution operation is as a matched filter that peaks when it finds a compatible sub-feature. The final output of the convolution operation is shown in Fig. 4.75. The values of the individual pixels are shown in color. Notice where the maximum values of the output image are located on the diagonals.

However, the convolution operation is not a perfect detector and results in nonzero values for other cases. For example, suppose the input image is a forward slash diagonal line. The step-by-step convolution with the kernel is shown in Fig. 4.76 with corresponding output in Fig. 4.77 that looks nothing like the kernel or the input image.

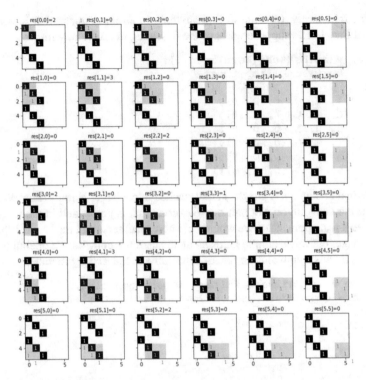

Fig. 4.74 The convolution process that produces the `res` array. As shown in the sequence, the light blue `kern` array is slid around, overlaid, multiplied, and summed upon the `x` array to generate the values of shown in the title. The output of the convolution is shown in Fig. 4.75

Fig. 4.75 The `res` array output of the convolution is shown in Fig. 4.74. The values (in red) shown are the individual outputs of the convolution operation. The grayscale indicates relative magnitude of the shown values (darker is greater)

We can use multiple kernels to explore an input image. For example, suppose we have the input image shown on the left in Fig. 4.78. The two kernels are shown in the upper row, with corresponding outputs on the bottom row. Each kernel is able to emphasize its particular feature but extraneous features appear in both outputs. We can have as many outputs as we have kernels but because each output image is as large as the input image, we need a way to reduce the size of this data.

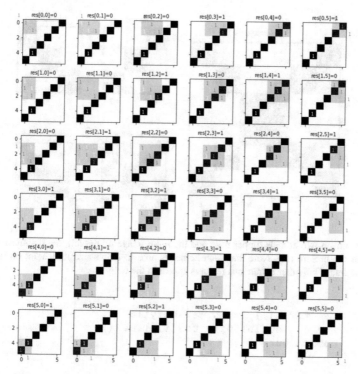

Fig. 4.76 The input array is a forward slash diagonal. This sequence shows the step-by-step convolution operation. The output of this convolution is shown in Fig. 4.77

Fig. 4.77 The output of the convolution operation shown in Fig. 4.76. Note that the output has nonzero elements where there is no match between the input image and the kernel

Maximum Pooling To reduce the size of the output images, we can apply *maximum pooling* to replace a tiled subset of the image with the maximum pixel value in that particular subset. The following Python code illustrates maximum pooling,

```
>>> def max_pool(res,width=2,height=2):
...     m,n = res.shape
...     xi = [slice(i,i+width) for i in range(0,m,width)]
...     yi = [slice(i,i+height) for i in range(0,n,height)]
...     out = np.zeros((len(xi),len(yi)),dtype=res.dtype)
...     for ni,i in enumerate(xi):
```

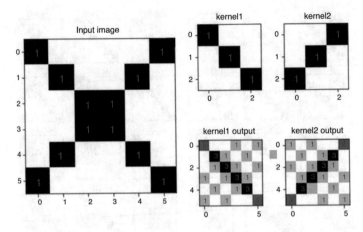

Fig. 4.78 Given two kernels (upper row) and the input image on the left, the output images are shown on the bottom row. Note that each kernel is able to emphasize its feature on the input composite image but other extraneous features appear in the outputs

```
...            for nj,j in enumerate(yi):
...                out[ni,nj]= res[i,j].max()
...        return out
...
```

Programming Tip

The `slice` object provides programmatic array slicing. For example, `x[0,3]=x[slice(0,3)]`. This means you can separate the `slice` from the array, which makes it easier to manage.

Pooling reduces the dimensionality of the output of the convolution and makes stacking convolutions together computationally feasible. Figure 4.79 shows the output of the `max_pool` function on the indicated input images.

Rectified Linear Activation Rectified linear activation units (ReLUs) are neural network units that implement the following activation function,

$$r(x) = \begin{cases} x & \text{if } x > 0 \\ 0 & \text{otherwise} \end{cases}$$

To use this activation properly, the kernels in the convolutional layer must be scaled to the $\{-1, 1\}$ range. We can implement our own rectified linear activation function using the following code,

Fig. 4.79 The `max_pool` function reduces the size of the output images (left column) to the images on the right column. Note that the pool size is 2x2 so that the resulting pooled images are half the size of the original images in each dimension

```
>>> def relu(x):
...     'rectified linear activation function'
...     out = np.zeros(x.shape,dtype=x.dtype)
...     idx = x>=0
...     out[idx]=x[idx]
...     return out
...
```

Now that we understand the basic building blocks, let us investigate how the operations fit together. To create some training image data, we use the following function to create some random backward and forward slashed images as shown in Fig. 4.80. As before, we have the scaled kernels shown in Fig. 4.81. We are going to apply the convolution, max-pooling, and rectified linear activation function sequence step-by-step and observe the outputs at each step.

```
>>> def gen_rand_slash(m=6,n=6,direction='back'):
...     '''generate random forward/backslash images.
...     Must have at least two pixels'''
...     assert direction in ('back','forward')
...     assert n>=2 and m>=2
...     import random
...     out = -np.ones((m,n),dtype=float)
...     i = random.randint(2,min(m,n))
...     j = random.randint(-i,max(m,n)-1)
...     t = np.diag([1,]*i,j)
...     if direction == 'forward':
...         t = np.flipud(t)
...     try:
...         assert t.sum().sum()>=2
...         out[np.where(t)]=1
...         return out
...     except:
...         return gen_rand_slash(m=m,n=n,direction=direction)
...
>>> # create slash-images training data with classification id 1 or 0
>>> training=[(gen_rand_slash(),1) for i in range(10)] + \
...          [(gen_rand_slash(direction='forward'),0) for i in range(10)]
```

Fig. 4.80 The training dataset for our convolutional neural network. The forward slash images are labeled category 0 and the backward slash images are category 1

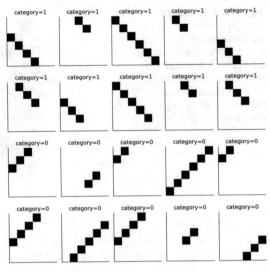

Fig. 4.81 The two scaled feature kernels for the convolutional neural network

Figure 4.82 shows the output of convolving the training data in Fig. 4.80 with `kern1`, as shown on the left panel of Fig. 4.81. Note that the following code defines each of these kernels,

```
>>> kern1 = (np.eye(3)*2-1)/9. # scale
>>> kern2 = np.flipud(kern1)
```

The next operation is the activation function for the rectified linear unit with output shown in Fig. 4.83. Note that all of the negative terms have been replaced by zeros. The next step is the maximum pooling operation as shown in Fig. 4.84. Notice that the number of total pixels in the training data has reduced from 36 per image to 9 per image. With these processed images, we have the inputs we need for the final classification step.

Convolutional Neural Network Using Keras Now that we have experimented with the individual operations using our own Python code, we can construct the convolutional neural network using Keras. In particular, we use the Keras functional interface to define this neural network because that makes it easy to unpack the operations at the individual layers.

Fig. 4.82 The output of convolving the training data in Fig. 4.80 with `kern1`, as shown on the left panel of Fig. 4.81

Training Set Convolution with kern1

Fig. 4.83 The output of the rectified linear unit activation function with the input shown in Fig. 4.82

ReLU of Convolution with kern1

```
>>> from keras import metrics
>>> from keras.models import Model
>>> from keras.layers.core import Dense, Activation, Flatten
>>> from keras.layers import Input
>>> from keras.layers.convolutional import Conv2D
>>> from keras.layers.pooling import MaxPooling2D
>>> from tensorflow.keras.optimizers import SGD
>>> from keras import backend as K
>>> from tensorflow.keras.utils import to_categorical
```

Fig. 4.84 The output of
maximum pooling operation
with the input shown in
Fig. 4.83 for fixed image
kernel `kern1`

Max-pool of ReLU Output for kern1

Note that the names of the modules are consistent with their operations. We also
need to tell `Keras` how to manage the input images,

```
>>> K.set_image_data_format('channels_first') # image data format
>>> inputs = Input(shape=(1,6,6)) # input data shape
```

Now we can build the individual convolutional layers. Note the specification of
the activations at each layer and placement of the `inputs`.‘

```
>>> clayer = Conv2D(2,(3,3),padding='same',
...                  input_shape=(1,6,6),name='conv',
...                  use_bias=False,
...                  trainable=False)(inputs)

>>> relu_layer= Activation('relu')(clayer)
>>> maxpooling = MaxPooling2D(pool_size=(2,2),data_format='channels_first',
...                           name='maxpool')(relu_layer)
>>> flatten = Flatten()(maxpooling)
>>> softmax_layer = Dense(2,
...                       activation='softmax',
...                       name='softmax')(flatten)
>>> model = Model(inputs=inputs, outputs=softmax_layer)
>>> # inject fixed kernels into convolutional layer
>>> fixed_kernels = [np.dstack([kern1,kern2]).reshape(3,3,1,2)]
>>> model.layers[1].set_weights(fixed_kernels)
```

Observe that the functional interface means that each layer is explicitly a function
of the previous one. Note that `trainable=False` for the convolutional layer
because we want to inject our fixed kernels into it at the end. The `flatten` layer
reshapes the data so that the entire processed image at the point is fed into the
`softmax_layer`, whose output is proportional to the probability that the image
belongs to either class. The `set_weights()` function is where we inject our fixed

kernels. These are not going to be updated by the optimization algorithm because of the prior `trainable=False` option. With the topology of the neural network defined, we now have to choose the optimization algorithm and pack all of this configuration into the model with the `compile` step.

```
>>> lr = 0.01 # learning rate
>>> sgd = SGD(learning_rate=lr, decay=1e-6, momentum=0.9, nesterov=True)
>>> model.compile(loss='categorical_crossentropy',
...               optimizer=sgd,
...               metrics=['accuracy',
...                        metrics.categorical_crossentropy])
```

The `metrics` specification means that we want the training process to keep track of those named items. Next, we generate some training data using our `gen_rand_slash` function with the associated class of each image (1 or 0). Most of this code is just shaping the tensors for Keras. The final `model.fit()` step is where the internal weights of the neural network are adjusted according to the given inputs.

```
>>> # generate some training data
>>> ntrain = len(training)
>>> t=np.dstack([training[i][0].T
...                for i in range(ntrain)]).T.reshape(ntrain,1,6,6)
>>> y_binary=to_categorical(np.hstack([np.ones(ntrain//2),
...                                     np.zeros(ntrain//2)]))
>>> # fit the configured model
>>> h=model.fit(t,y_binary,epochs=500,verbose=0)
```

With that completed, we can investigate the functional mapping of each layer with `K.function`. The following creates a mapping between the input layer and the convolutional layer,

```
>>> convFunction = K.function([inputs],[clayer])
```

Now, we can feed the training data into this function as seen in the output of just the convolutional layer, which is shown (Fig. 4.85),

We can do this again for the pooling layer by creating another Keras function,

```
>>> maxPoolingFunction = K.function([inputs],[maxpooling])
```

whose output is shown in Fig. 4.86. We can examine the final output of this network using the `predict` function,

```
>>> fixed_kernels = model.predict(t)
>>> fixed_kernels
array([[0.4104765 , 0.5895235 ],
       [0.50148225, 0.49851784],
       [0.16382763, 0.83617234],
       [0.24335606, 0.75664395],
       [0.34367517, 0.6563248 ],
       [0.12688082, 0.8731192 ],
       [0.43925983, 0.5607402 ],
       [0.20650622, 0.79349375],
```

Keras convolution layer output given kern1

Fig. 4.85 Compare this to Fig. 4.82. This shows our hand-tooled convolution is the same as that implemented by Keras

Keras Pooling layer output given kern1

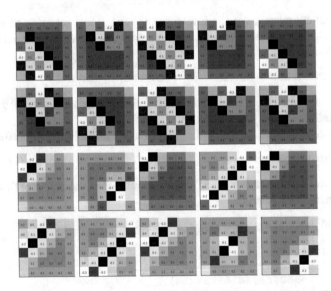

Fig. 4.86 Output of max-pooling layer for fixed kernel `kern1`. Compare this to Fig. 4.84. This shows our hand-tooled implementation is equivalent to that by Keras

```
          [0.50148225,  0.49851784],
          [0.12688082,  0.8731192 ],
          [0.6909458 ,  0.30905417],
          [0.74993294,  0.25006709],
          [0.5368913 ,  0.46310875],
          [0.75481373,  0.24518625],
          [0.5368913 ,  0.46310875],
          [0.7956373 ,  0.20436268],
          [0.8836856 ,  0.11631437],
          [0.7956373 ,  0.20436268],
          [0.39666745,  0.6033326 ],
          [0.741312  ,  0.258688  ]], dtype=float32)
```

and we can see the weights given to each of the classes. Taking the maximum of these across the columns gives the following,

```
>>> np.argmax(fixed_kernels,axis=1)
array([1, 0, 1, 1, 1, 1, 1, 1, 0, 1, 0, 0, 0, 0, 0, 0, 0, 0, 1, 0])
```

which means that our convolutional neural network with the fixed kernels did well predicting the classes of each of our input images. Recall that our model configuration prevented our fixed kernels from updating in the training process. Thus, the main work of model training was changing the weights of the final output layer. We can redo this exercise by removing this constraint and see how the network performs if it is able to adaptively re-weight the kernel terms as part of training by changing the `trainable` keyword argument and then re-build and train the model, as shown next.

```
>>> clayer = Conv2D(2,(3,3),padding='same',
...                 input_shape=(1,6,6),name='conv',
...                 use_bias=False)(inputs)

>>> relu_layer= Activation('relu')(clayer)
>>> maxpooling = MaxPooling2D(pool_size=(2,2),data_format='channels_first',
...                           name='maxpool')(relu_layer)
>>> flatten = Flatten()(maxpooling)
>>> softmax_layer = Dense(2,
...                       activation='softmax',
...                       name='softmax')(flatten)
>>> model = Model(inputs=inputs, outputs=softmax_layer)
>>> model.compile(loss='categorical_crossentropy',
...               optimizer=sgd)
>>> h=model.fit(t,y_binary,epochs=500,verbose=0)
>>> new_kernels = model.predict(t)
>>> new_kernels
array([[5.7992584e-04, 9.9942005e-01],
       [2.0522181e-02, 9.7947776e-01],
       [3.0357076e-04, 9.9969649e-01],
       [3.8909286e-03, 9.9610907e-01],
       [1.5166619e-02, 9.8483342e-01],
       [3.5102875e-04, 9.9964893e-01],
       [1.2327342e-05, 9.9998772e-01],
       [6.3458519e-06, 9.9999368e-01],
       [2.0522181e-02, 9.7947776e-01],
       [3.5102875e-04, 9.9964893e-01],
       [9.9289131e-01, 7.1087363e-03],
       [9.9987829e-01, 1.2172712e-04],
```

Fig. 4.87 Kernels updated during the training process. Compare to Fig. 4.81

Fig. 4.88 Recall that the second half of the training set was classified as category 1. The updated kernels provide a wider margin for classification than our fixed kernels, even though the ultimate performance is very similar between them

```
[9.9026382e-01, 9.7361971e-03],
[9.9617696e-01, 3.8230598e-03],
[9.9026382e-01, 9.7361971e-03],
[9.9812239e-01, 1.8776212e-03],
[9.9886048e-01, 1.1395670e-03],
[9.9812239e-01, 1.8776212e-03],
[9.5428187e-01, 4.5718122e-02],
[9.9982661e-01, 1.7345019e-04]], dtype=float32)
```

with corresponding max output,

```
>>> np.argmax(new_kernels,axis=1)
array([1, 1, 1, 1, 1, 1, 1, 1, 1, 1, 0, 0, 0, 0, 0, 0, 0, 0, 0, 0])
```

The newly updated kernels are shown in Fig. 4.87. Note how different these are from the original fixed kernels. We can see the change in the respective predictions in Fig. 4.88. Thus, the benefit of updating the kernels in the training process is to improve the accuracy overall, but at the cost of interpretability of the kernels themselves. Nonetheless, it is seldom the case that the kernels are known ahead of time, as in our artificial example here, so in practice, there may be nothing to really interpret anyway. Nonetheless, for other problems where there is a target feature in the data for which good a priori exemplars exist that could serve a kernels, then priming these kernels early in training may help to tune into those target features, especially if they are rare in the training data.

4.13 Interpretability

Although machine learning methods are primarily focused on prediction, machine learning for mission-critical applications often require a justifiable understanding of how features were applied to a specific prediction instance. Indeed, in some jurisdictions, the law mandates such interpretations. Beyond consequences, understanding how features interact with predictions allows the machine learning specialist to retain only significant features and discard superfluous ones, thus saving computation, storage, and other maintenance costs.

Partial Dependence Plots Interpretation must start with the results of an opaque optimization process that generates predictions and then somehow convert the results of this process that humans can understand. Humans easily process visual information and partial dependence plots (PDP) attempt to display the relative importance of features visually. This works by showing the marginal effect of a feature on a particular outcome by averaging over the other features used in the prediction.

We denote the particular feature of interest as x_S and the set of all the other complementary features, X_C. The partial dependence function (PD) is defined as

$$PD(x_S) = \mathbb{E}_{X_C} f(x_S, X_C)$$

where f is the prediction based on all the features. As usual, this is usually far too difficult to compute analytically and must be approximated using Monte Carlo integration that averages over the training data,

$$PD(x_S) \approx \frac{1}{N} \sum_{i=1}^{n} f(x_S, x_C^{(i)})$$

where N is the number of samples and $x_C^{(i)}$ denotes the ith instance of the complementary feature set. Because partial dependence integrates over all of the training data, it describes an overall relationship of the given feature x_S and the predicted outcome. Partial dependence has certain advantages: it is straightforward to compute and it yields an intuitive visual. On the other hand, if the given feature x_S is strongly correlated with the other complementary features, then the plot cannot distinguish how the average prediction changes *only* with x_S. Moreover, because the marginal effects are averaged away, the calculation of the average itself may be skewed due to outlier or symmetry effects, which can exaggerate or suppress the value of the given feature of interest, x_S. Fortunately, Scikit-learn provides tools to compute these values.

To get going, let us consider the housing dataset from Scikit-learn,

```
>>> from sklearn.datasets import fetch_california_housing
>>> d = fetch_california_housing(as_frame=True)
>>> d.data.head()
```

Fig. 4.89 Partial dependence plot for a random forest regression on the housing data

	MedInc	HouseAge	AveRooms	AveBedrms	Population	AveOccup	Latitude	Longitude
0	8.325	41.000	6.984	1.024	322.000	2.556	37.880	-122.230
1	8.301	21.000	6.238	0.972	2401.000	2.110	37.860	-122.220
2	7.257	52.000	8.288	1.073	496.000	2.802	37.850	-122.240
3	5.643	52.000	5.817	1.073	558.000	2.548	37.850	-122.250
4	3.846	52.000	6.282	1.081	565.000	2.181	37.850	-122.250

To keep it simple, let us consider a random forest regression,

```
>>> from sklearn.ensemble import RandomForestRegressor
>>> rg = RandomForestRegressor(max_depth=3, random_state=0)
>>> rg.fit(d.data,d.target)
RandomForestRegressor(max_depth=3, random_state=0)
```

Once we have the regressor fitted, we can compute the partial dependence,

```
>>> from sklearn.inspection import partial_dependence
>>> pdep = partial_dependence(rg,d.data,[0],kind='average')
```

The data structure that is returned shows the values of the prediction (house prices) and the average, which are the values of the partial dependence. The [0] argument of the function means that our feature of interest is the 0th feature, which is *Median income*. The partial dependence is evaluated by default on 100 grid points for the given feature. This can be changed using the grid_resolution keyword argument in the function. The resulting partial dependence plot is shown in Fig. 4.89 and indicates that increasing median income is associated with the value of the house prices prediction. Put another way, predicted increasing house prices is dependent upon increasing median income, at least as far as this random forest regression is concerned.

Scikit-learn also provides *feature importances*, which are a way to gauge, at least for those models that provide them, how much a particular future influences the prediction globally,

```
>>> rg.feature_importances_
array([8.49572874e-01, 3.13355677e-04, 1.86084488e-02, 0.00000000e+00,
       7.80734981e-05, 1.30845192e-01, 5.82056545e-04, 0.00000000e+00])
```

which shows that the 0th feature (median income) is the most important overall. Note that these feature importances are computed differently than the partial dependence, so you have to understand the particular calculation for each model. For example, for random forests, they are computed by keeping track of the statistics of the accumulated purity changes (i.e., Gini coefficient) for each of the features. Nonetheless, these can be helpful in determining how to value each feature for a particular model and dataset. For example, the next most important feature according to `feature_importances_` is the *Average occupancy* for the data. We can quickly compute the partial dependence for this features as the following,

```
>>> pdep = partial_dependence(rg,d.data,[5],kind='average')
```

and then plot it on the right side of Fig. 4.89. Keep in mind that this shows the partial dependence of each feature separately and is not accounting for correlation between the two features. This figure shows that the average occupancy does not strongly affect the prediction for larger values of occupancy, at least as much as the prediction is affected by large values of median income. It is possible to compute the joint partial dependence for a pair of features and then plot it as an image density. The `PartialDependenceDisplay` class in Scikit-learn facilitates this kind of graph.

Individual Conditional Expectation Plots Individual conditional expectation (ICE) plots are closely related to partial dependence plots but instead of averaging across values of the complementary features, ICE plots show the dependence of the prediction on a feature for each $x_C^{(i)}$ individually on the same graph. Thus PDP is the average of all of the ICE lines. The distribution of lines is key to dealing with the lack of independence between feature sets, which are obscured by computing the average. These are computed using the same function, but with different arguments,

```
>>> pdice = partial_dependence(rg,d.data.sample(150),[0],
     kind='individual')
```

Figure 4.90 shows the individual lines from the training data. Note that we subsample the data down to only 150 rows out of the 20,640 rows to reduce clutter. The average of these lines is the partial dependence. One problem with this plot is that the predicted values start at different levels, which make it hard to interpret. By getting all the traces to start at zero and then overlaying the partial dependence, it is easier to understand the spread as in Fig. 4.91.

Shapley Values Shapley values emerge from cooperative game theory, distant from the concepts and aims of everyday machine learning. Nonetheless, Shapley values can help discern feature importances on a per-prediction basis. To develop the main theory behind Shapley values would take us deep into cooperative game theory, but we can get a sense for the idea by considering a simple game. Consider a game show where three players must answer a question with multiple choice answers. The first player (A) gets the first attempt and can select one of five possible responses or just pass on the question. If A passes, then the game show host will review one of the *incorrect* answers. This means that the next player B has only to choose out of four possible answers, but can also choose to pass to the final player C. If player B passes,

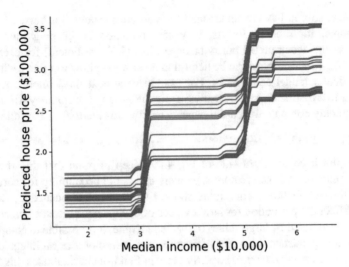

Fig. 4.90 Individual conditional expectation plot for a random forest regression on the housing data

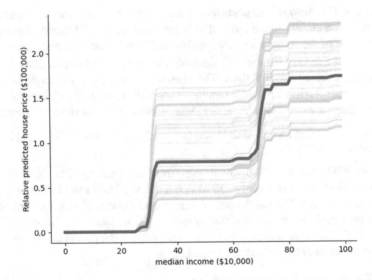

Fig. 4.91 Centered individual conditional expectation plots for a random forest regression on the housing data. The partial dependence is overlaid on the individual traces. Compare this to Fig. 4.90

then C will have only three possible answers to choose from and so on. Whoever can answer the question correctly wins a payoff amount.

It could be that A and C choose to play cooperatively and share equally in the payoff. Likewise, any pair or all or none of them could play cooperatively (i.e., form a *coalition*) and split the payoff. However, instead of splitting the payoff equally, another way is to split based on the contribution of the individual player to the total payoff. Shapley values are a way to determine how to split the payoff averaged over all possible cooperative arrangements. This applies to machine learning by considering each feature as a player where the payoff is the prediction. Then, Shapley values can be used to determine how much of the payoff a specific feature should get. For instance, suppose your prediction value is 100 using three features, x_1, x_2, x_3 for test instance a_1. If the average prediction over all of the instances is 70, then the problem is to determine how much of the individual features contributed to the difference of 30. For example, it could be that each of the three features contributed equally. More likely, the contribution for a particular instance is dependent on the order that the different features interacted jointly. Because there are many possible joint interactions (i.e., *coalitions*) between features, the Shapley value is the average marginal contribution of a feature value across all possible coalitions. Obviously, the more the features, the more expensive it is to compute the Shapley value for all coalitions.

The shap package implements Shapley values and comes with excellent plotting features and can work with machine learning models in Scikit-learn as well as others. The first step is to create an *explainer* for the particular model,

```
>>> import shap
>>> explainer = shap.TreeExplainer(rg)
```

and then compute the Shapley value for a particular test instance,

```
>>> idx = 20 # instance of interest
>>> shap_values = explainer.shap_values(d.data.iloc[idx,:])
>>> shap_values
array([-7.54311256e-01,  3.28829908e-04, -8.21936049e-02,  0.00000000e+00,
        1.65724263e-05, -2.42569670e-02, -3.12794276e-03,  0.00000000e+00])
```

There are as many entries as there are features. Note that some of the values are negative because Shapley values indicate whether, with respect to a particular instance, the feature increased or decreased (negative) the value of the prediction. The shap package comes with truly beautiful visualizations that integrate into Jupyter Notebook.

Permutation Importance Permutation importance comes with Scikit-learn and, although computationally expensive, is straightforward to understand. The first step is to keep a separate set of training data out of the training process. We call this the *validation* set and we use it to compute the permutation importance. We fit the model the usual way with the training data and compute a baseline score. Then, given a particular column (i.e., feature) from the validation set, we randomly shuffle (i.e., permute) the rows for that feature which makes that feature look like noise as far as the prediction is concerned. Then, the permutation importance is the difference between the baseline score and the score from shuffling that column. This process can be repeated multiple times for different shuffles to compute basic statistics for the permutation importance. The basic implementation is shown below but without the separation of the validation set,

Notation

Symbol	Meaning
σ	Standard deviation
μ	Mean
$\mathbb{V}$	Variance
$\mathbb{E}$	Expectation
$f(x)$	Function of x
$x \rightarrow y$	Mapping from x to y
(a, b)	Open interval
$[a, b]$	Closed interval
$(a, b]$	Half-open interval
Δ	Differential of
Π	Product operator
Σ	Summation of
$\lvert x \rvert$	Absolute value of x
$\lVert x \rVert$	Norm of x
$\#A$	Number of elements in A
$A \cap B$	Intersection of sets A, B
$A \cup B$	Union of sets A, B
$A \times B$	Cartesian product of sets A, B
$\in$	Element of
$\wedge$	Logical conjunction
$\neg$	Logical negation
$\{\}$	Set delimiters

$\mathbb{P}(X\|Y)$	Probability of X given Y
$\forall$	For all
$\exists$	There exists
$A \subseteq B$	A is a subset of B
$A \subset B$	A is a proper subset of B
$f_X(x)$	Probability density function of random variable X
$F_X(x)$	Cumulative density function of random variable X
$\sim$	Distributed according to
$\propto$	Proportional to
$\triangleq$	Equal by definition
$:=$	Equal by definition
$\perp$	Perpendicular to
$\therefore$	Therefore
$\Rightarrow$	Implies
$\equiv$	Equivalent to
$\mathbf{X}$	Matrix X
$\mathbf{x}$	Vector x
$\text{sgn}(x)$	Sign of x
$\mathbb{R}$	Real line
$\mathbb{R}^n$	n-dimensional vector space
$\mathbb{R}^{m \times n}$	$m \times n$-dimensional matrix space
$\mathcal{U}_{(a,b)}$	Uniform distribution on the interval (a, b)
$\mathcal{N}(\mu, \sigma^2)$	Normal distribution with mean μ and variance σ^2
$\xrightarrow{as}$	Converges almost surely
$\xrightarrow{d}$	Converges in distribution
$\xrightarrow{P}$	Converges in probability
Tr	Sum of the diagonal of a matrix
diag	Matrix diagonal

References

1. A. Agresti, *Categorical Data Analysis* (Wiley, New York, 2003)
2. A. Agresti, *An Introduction to Categorical Data Analysis*. (Wiley, New York, 2018)
3. E. Alpaydin, *Introduction to Machine Learning* (Wiley Press, New York, 2014)
4. D. Barber, *Bayesian Reasoning and Machine Learning* (Cambridge University Press, Cambridge, 2012)
5. C. Bauckhage, Numpy/scipy recipes for data science: Kernel least squares optimization (1). researchgate.net, March 2015
6. Z. Brzezniak, T. Zastawniak, *Basic Stochastic Processes: a Course Through Exercises*. Springer Undergraduate Mathematics Series (Springer, London, 1999)
7. J. Carpenter, M. Kenward, *Multiple Imputation and its Application* (Wiley, New York, 2012)
8. O. Certik et al., SymPy: Python library for symbolic mathematics (2016). http://sympy.org/
9. A. Chaudhuri, *Modern Survey Sampling* (CRC Press, Boca Raton, 2014)
10. R. Christensen, *Log-Linear Models and Logistic Regression* (Springer Science & Business Media, Berlin, 2006)
11. H. Cuesta, *Practical Data Analysis* (Packt Publishing Ltd, Birmingham, 2013)
12. W.L. Dunn, J.K. Shultis, *Exploring Monte Carlo Methods* (Elsevier Science, Oxford, 2011)
13. W. Feller, *An Introduction to Probability Theory and Its Applications: Volume One* (Wiley, New York, 1950)
14. T. Hastie, R. Tibshirani, J. Friedman, *The Elements of Statistical Learning: Data Mining, Inference, and Prediction*. Springer Series in Statistics (Springer, New York, 2013)
15. S.G. Heeringa, B.T. West, P.A. Berglund, *Applied Survey Data Analysis*. Chapman & Hall/CRC Statistics in the Social and Behavioral Sciences (CRC Press, Boca Raton, 2017)
16. A. Hyvarinen, J. Karhunen, E. Oja, *Independent Component Analysis*, vol. 46 (Wiley, New York, 2004)
17. A.J. Izenman, *Modern Multivariate Statistical Techniques*, vol. 1 (Springer, New York, 2008)
18. N.L. Johnson, S. Kotz, N. Balakrishnan, *Continuous Univariate Distributions*, vol. 2. Wiley Series in Probability and Mathematical Statistics: Applied probability and statistics (Wiley, New York, 1995)
19. F. Jones, *Lebesgue Integration on Euclidean Space*. Jones and Bartlett Books in Mathematics (Jones and Bartlett, Boston, 2001)
20. P. Knottnerus, *Sample Survey Theory* (3Island Press, Washington, 2002)
21. H. Kobayashi, B.L. Mark, W. Turin, *Probability, Random Processes, and Statistical Analysis: Applications to Communications, Signal Processing, Queueing Theory and Mathematical Finance*. EngineeringPro Collection (Cambridge University Press, Cambridge, 2011)
22. H.P. Langtangen, DocOnce markup language (2020). https://github.com/hplgit/doconce

23. H.P. Langtangen, *Python Scripting for Computational Science*, vol. 3, 3rd edn. Texts in Computational Science and Engineering (Springer, New York, 2009)
24. P.S. Levy, S. Lemeshow, *Sampling of Populations: Methods and Applications*. Wiley Series in Survey Methodology (Wiley, New York, 2013)
25. C. Loader, *Local Regression and Likelihood* (Springer, New York, 2006)
26. S.L. Lohr, *Sampling: Design and Analysis* (Duxbury Press, Pacific Grove, 1999)
27. D.G. Luenberger, *Optimization by Vector Space Methods*. Professional Series (Wiley, New York, 1968)
28. D.J.C. MacKay, *Information Theory, Inference and Learning Algorithms* (Cambridge University Press, Cambridge, 2003)
29. R.A. Maronna, D.R. Martin, V.J. Yohai, *Robust Statistics: Theory and Methods*. Wiley Series in Probability and Statistics (Wiley, New York, 2006)
30. W. McKinney, *Python for Data Analysis: Data Wrangling with Pandas, NumPy, and IPython* (O'Reilly, Newton, 2012)
31. T. Mikosch, *Elementary Stochastic Calculus with Finance in View*. Advanced Series on Statistical Science & Applied Probability (World Scientific, Singapore, 1998)
32. A.M.F. Mood, F.A. Graybill, D.C. Boes, *Introduction to the Theory of Statistics*. International Student Edition (McGraw-Hill, New York, 1974)
33. E. Nelson, *Radically Elementary Probability Theory*. Annals of Mathematics Studies (Princeton University Press, Princeton, 1987)
34. T.E. Oliphant, *A Guide to NumPy* (Trelgol Publishing, Austin, 2006)
35. F. Perez, B.E. Granger, et al., IPython software package for interactive scientific computing (2007). http://ipython.org/
36. G.M. Poore, Pythontex: reproducible documents with latex, python, and more. Computat. Sci. Discov. **8**(1), 014010 (2015)
37. W. Richert, *Building Machine Learning Systems With Python* (Packt Publishing Ltd, Birmingham, 2013)
38. C. Robert, *Monte Carlo Statistical Methods* (Springer Science and Business Media, New York, 2013)
39. T. Rudas, *Lectures on Categorical Data Analysis* (Springer, New York, 2018)
40. R.E. Schapire, Y. Freund, *Boosting Foundations and Algorithms*. Adaptive Computation and Machine Learning (MIT Press, Cambridge, 2012)
41. J. Simonoff, *Analyzing Categorical Data* (Springer, New York, 2003)
42. G. Strang, *Linear Algebra and Its Applications* (Thomson, Brooks/Cole, Pacific Grove, 2006)
43. W. Tang, *Applied Categorical and Count Data Analysis* (CRC Press, Boca Raton, 2012)
44. J. Unpingco, *Python Programming for Data Analysis* (Springer, New York, 2021)
45. G.J.G. Upton, *Categorical Data Analysis by Example* (Wiley, New York, 2016)
46. R. Valliant, J.A. Dever, F. Kreuter, *Practical Tools for Designing and Weighting Survey Samples*. Statistics for Social and Behavioral Sciences (Springer International Publishing, New York, 2018)
47. V. Vapnik, *The Nature of Statistical Learning Theory*. Information Science and Statistics (Springer, New York, 2000)
48. L. Wasserman, *All of Statistics: A Concise Course in Statistical Inference* (Springer, New York, 2004)
49. L. Wilkinson, D. Wills, D. Rope, A. Norton, R. Dubbs, *The Grammar of Graphics*. Statistics and Computing (Springer, New York, 2006)
50. C. Wu, M.E. Thompson, *Sampling Theory and Practice*. ICSA Book Series in Statistics (Springer International Publishing, Birkhäuser, 2020)
51. D. Zelterman, *Models for Discrete Data* (Oxford University Press, Oxford, 1999)

Index

Printed in the United States
by Baker & Taylor Publisher Services